Collision Course

Collision Course

Endless Growth on a Finite Planet

Kerryn Higgs

The MIT Press
Cambridge, Massachusetts
London, England

MIT Press books may be purchased at special quantity discounts for business or sales promotional use. For information, please email special_sales@mitpress.mit.edu.

This book was set in Sabon by Toppan Best-set Premedia Limited. Printed and bound in the United States of America.

Library of Congress Cataloging-in-Publication Data

Higgs, Kerryn, 1946–
Collision course : endless growth on a finite planet / Kerryn Higgs.
pages cm
Includes bibliographical references and index.
ISBN 978-0-262-02773-1 (hardcover : alk. paper)
1. Economic history—1971–1990. 2. Economic history—1990– 3. Economic development—Moral and ethical aspects. 4. Economic policy—Moral and ethical aspects. 5. Free enterprise. 6. Sustainable development. I. Title.
HC59.H523 2014
330.9—dc23
2014003794

10 9 8 7 6 5 4 3 2 1

For Meg and Freddy,
in the hope that the world you inherit
will come to its senses soon.

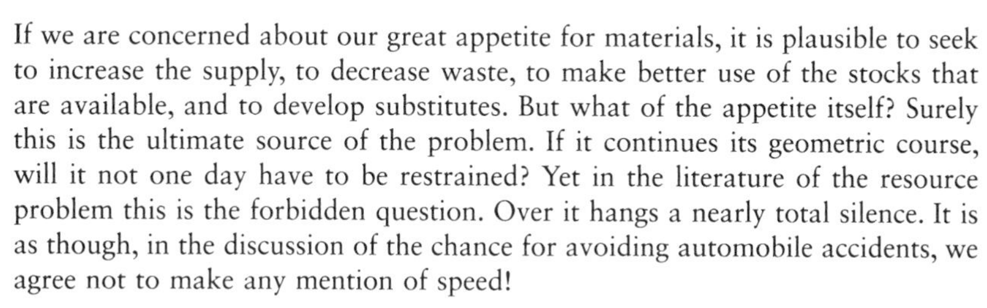

If we are concerned about our great appetite for materials, it is plausible to seek to increase the supply, to decrease waste, to make better use of the stocks that are available, and to develop substitutes. But what of the appetite itself? Surely this is the ultimate source of the problem. If it continues its geometric course, will it not one day have to be restrained? Yet in the literature of the resource problem this is the forbidden question. Over it hangs a nearly total silence. It is as though, in the discussion of the chance for avoiding automobile accidents, we agree not to make any mention of speed!

—John Kenneth Galbraith, 1958

Contents

Acknowledgments

A work of synthesis such as this would have been impossible without the research conducted by many others. To them I owe a great debt, both for their illuminating analysis and for having trawled through segments of the vast primary source material and pointed me in the right directions.

Thanks to all those who read parts of the early drafts—Barbara Bloch, Mary O'Sullivan, Barney Foran, Steve Keen, Laurene Kelly, Graham Wells, and Margot Oliver, as well as to Stan Malinowitz, for running an economist's eye over the final draft of chapter 6. Thanks to Dennis Meadows for providing his historical perspective on the World3 model; to all those who gave permission to include their graphs and tables; and to Riley Dunlap, Ross Buckley, Ramachandra Guha, Rogate Mshana, Bob Ward, Cliff Cobb, Sharon Beder, Pasquale Tridico, Jesse Ausubel, TRK Somaiya, Sharan Burrow and the secretariats of Eurostep and the Club of Rome for prompt and helpful answers to email inquiries. Thanks also to Miranda Martin, Clay Morgan, Deborah Cantor-Adams, and Marjorie Pannell at the MIT Press for their help in seeing the work to publication.

Special thanks to Natasha Topschij for creating the map in figure 9.1, and to the people who read major parts of the original text and offered their thoughts on many aspects—Jen St. Clair, who shares a lifelong enthusiasm for the limits ideas, and Pete Hay, who advised on several versions of the initial draft and gave unfailing encouragement and perceptive critique.

Very special thanks to Harriet Malinowitz, who insisted that my unending research should be written up, commented on drafts at many stages, and gave generously of her time and energy to assist with the revision and re-editing of the entire book over her summer break in 2013.

Thanks, too, to the School of Geography and Environmental Studies at the University of Tasmania for its support, and to the UTAS library staff, who were ever reliable, well beyond the call of duty. Special thanks to the staff at Document Delivery, Launceston; they were terrific, and the research would have been impossible without them.

Abbreviations and Acronyms

AAAS	American Association for the Advancement of Science
ABARE	Australian Bureau of Agricultural and Research Economics
ABC	Australian Broadcasting Corporation
ABC (US)	American Broadcasting Company
ACCI	Australian Chamber of Commerce and Industry
ACF	Australian Conservation Foundation
ACSH	American Council on Science and Health
AEI	American Enterprise Institute
AFL	American Federation of Labor
AiG	Australian Industry Group
AIGN	Australian Industry Greenhouse Network
ALP	Australian Labor Party
APEC	Asia-Pacific Economic Cooperation
ASSC	Advancement of Sound Science Center
AusAID	Australian Agency for International Development
BCA	Business Council of Australia
CASSE	Center for the Advancement of the Steady State Economy
CDFE	Center for the Defense of Free Enterprise
CED	Committee for Economic Development (US)
CEI	Competitive Enterprise Institute
CFCs	chlorofluorocarbons
CIO	Congress of Industrial Organizations
CIS	Centre for Independent Studies (AUS)
CPI	Committee on Public Information (US)
CPS	Centre for Policy Studies (UK)
CRA	Conzinc Rio Tinto Australia
CSIRO	Commonwealth Scientific and Industrial Research Organisation (AUS)
DFAT	Australian Government Department of Foreign Affairs and Trade

EA	Enterprise Australia
EC	European Commission
EIA	Energy Information Administration (US)
EPA	Environmental Protection Agency
EROI	energy return on investment
ETS	emissions trading scheme
EU	European Union
FAIR	Fairness and Accuracy in Reporting (US)
FAO	Food and Agriculture Organization (UN)
FEE	Foundation for Economic Education (US)
FSA	financial services agreement
GATT	General Agreement on Tariffs and Trade
GATS	General Agreement on Trade in Services
GEO	Global Environmental Outlook (UNEP)
I=PAT	Impact = Population × Affluence × Technology
ICC	International Chamber of Commerce
IEA	Institute of Economic Affairs (UK)
IEA	International Energy Agency (OECD)
IIED	International Institute for Environment and Development (FAO)
ILO	International Labour Organization (UN)
IMF	International Monetary Fund
IPA	Institute of Public Affairs (Australia)
IPCC	International Panel on Climate Change
MAI	Multilateral Agreement on Investment
mb/d	million barrels per day
MDG	Millennium Development Goal
MPS	Mont Pèlerin Society
NAFTA	North American Free Trade Agreement
NAM	National Association of Manufacturers (US)
nef	new economics foundation
NRC	National Research Council (US)
NSS	National Sample Survey (India)
OECD	Organisation for Economic Co-operation and Development
OPEC	Organization of the Petroleum Exporting Countries
ORC	Opinion Research Corporation (US)
PBS	Public Broadcasting Service (US)
PPP	Purchasing power parity
SAP	structural adjustment program (IMF, World Bank)
SBS	Special Broadcasting Service (AUS)
SDI	Strategic Defense Initiative ("Star Wars")

SEPA	State Environmental Protection Administration (China)
SEPP	Science and Environmental Policy Project
TASSC	The Advancement of Sound Science Coalition
TISS	Tata Institute of Social Sciences
TNC	Transnational corporation
UN	United Nations
UNCED	United Nations Conference on Environment and Development (Earth Summit/Rio)
UNCTAD	United Nations Conference on Trade and Development
UNCTC	United Nations Centre on Transnational Corporations
UN/DESA	United Nations Department of Economic and Social Affairs
UNDP	United Nations Development Programme
UNEP	United Nations Environment Programme
UN-Habitat	United Nations Human Settlements Programme
UNICEF	United Nations Children's Fund
UNU	United Nations University
USAID	US Agency for International Development
WCD	World Commission on Dams
WCED	World Commission on Environment and Development (Brundtland Commission)
WEF	World Economic Forum
WIDER	World Institute for Development Economics Research (UNU)
WTO	World Trade Organization

Introduction

Since the middle of the twentieth century, the scale of the human enterprise has rapidly escalated, and with it the exploitation of the natural world as a source of raw materials and a sink for the disposal of waste. Though the roots of this explosion lie in the history of the last five hundred years at least (in the rise of capitalism, European colonialism, Enlightenment science, and the Industrial Revolution), the associated disruption of the global biosphere has become evident only over the last half century.

This book is about the story of this growth, its astonishing acceleration since World War II, and its equally astonishing impact on the natural world. Above all, it's about the way the notion of ever-expanding economic growth has gained virtually ubiquitous popularity, both with policymakers and in public discourse, while the idea put forward by physical scientists—that we live on a finite planet that cannot sustain infinite economic expansion—has been treated as an opinion of the lunatic "doom-saying" fringe. Even as concepts such as sustainability and "going green" have, in recent times, paid lip service to the need to act, the commitment to growth without end has not wavered.

From the 1960s on, a succession of books pointed to the perils of pollution, untrammeled population growth, and ignoring ecology in the economic calculus. *The Limits to Growth*[1] was written by MIT researchers in 1972 and commissioned by the Club of Rome, an international think tank promoting "identification and analysis of the crucial problems facing humanity and the communication of such problems to the most important public and private decision makers as well as to the general public." The *Limits* authors, with expertise across many disciplines, including biophysics, system dynamics, and management, found that unmodified economic growth was likely to collide with the realities of a finite planet within a century. They saw grave problems emerging from five major tendencies: accelerating industrialization, rapid population growth, extensive malnutrition, the depletion of nonrenewable resources, and environmental decline.[2] Their modeling of these trends showed that, if we continued along the same growth trajectory, we would be likely to precipitate ecological and social collapse in the second half of the twenty-first century.

I happened upon *Limits* in the year it was published. Its logic was persuasive to me from the outset, and I expected its message to have a significant impact on the subsequent conduct of human affairs. But as the years rolled by, it seemed there was little effect—and then, even less. True, scientists continued to voice alarm, while the evidence began to mount that life on earth was experiencing a sixth extinction pulse and that the planet was warming. United Nations conferences proliferated, drawing attention to a plethora of environmental problems at every imaginable scale and attempting various treaties, protocols, and programs to address them. But outside the scientific community, in governments, bureaucracies, and public debate, an intensifying promotion of economic growth rendered it ever more securely entrenched as the natural objective of collective human effort. Growth became the "commonsense" solution to virtually all social problems—including, paradoxically, the environmental degradation it was causing. This quickening intent was not confined to the developed world but was increasingly emulated by almost all types of state, including communist China.

It was this contradiction between the warnings of scientists and the popularity of growth economics, witnessed over the course of my adult life, that triggered the curiosity that led to this book. How could the advice of the scientific establishment, venerated to a fault during my early life, have been so comprehensively ignored and emphatically discarded a decade or two later by governments and policymakers worldwide? What were the decisive influences that neutralized that counsel of caution? How was such a compelling alert from a hitherto trusted source discounted so successfully? How did the opposite view, that growth was the most essential purpose of human societies, become the accepted wisdom? How had economists eclipsed scientists as preeminent authorities and indispensable voices in the policy sphere?

In the second decade of the twenty-first century, economic growth remains the guiding principle for human endeavor, impeded only by the internal busts endemic to the capitalist system itself and the cascades of environmental degradation that are common in growing economies. The ship of growth sails on virtually unchallenged, as if its consequences had nothing at all to do with it. From international bodies such as the International Monetary Fund to most governments and their oppositions throughout the world, growth is more than ever the prize, the goal, the indispensable foundation, mentioned endlessly in speeches, reports, and press releases. As the idea has continued to flourish, its risks have come to seem less serious, despite the mounting evidence of its dysfunctional outcomes—accelerating species destruction, climate destabilization, toxic pollution of rivers and groundwater, and the depletion of many of the key resources on which the economic edifice depends. In Frederick Buell's words, what was once entertained as a looming apocalypse has been normalized as a "way of life."[3]

The unprecedented economic expansion of the past century and a half coincided with the emergence of modern corporations. These consolidated corporate entities, dependent on a stream of always increasing profits and invested in the continuation of economic growth, were assembled around 1900 in the United States. They went on to band together in various trade lobbies and business organizations designed to influence democratic processes and, later, to propagate an ideology of growth as a universal panacea—even as the solution to the social and environmental problems that followed in the wake of their expansion.

One central aim of the new business organizations of the late nineteenth century and early twentieth century was the limitation of regulation, regarded as an unacceptable brake on growth and profits, whether intended to make the workplace safe or to protect the environment. Neoliberal economics, which has come to dominate political institutions since the 1970s, has been embraced and promoted by the corporations. It shifted the right to make many decisions about the material world away from people and their elected representatives and reassigned this role to unelected bodies such as the World Trade Organization, while suggesting it was merely tapping into the "magic of the market." It also discarded planning and regulation as inefficient, passé remnants of "socialism"—a term often used to describe all forms of social democratic government. The neoliberal "revolution" in public policy set up crucial roadblocks in the path of regulatory strategies and shifted the ideological emphasis from assumptions of pursuit of the common good by a welfare-oriented state to the liberation of the "free market" from state "interference," and the ability of business to dictate government policy. This shift disabled existing democratic methods of addressing environmental questions.

Growth also came to be the preferred solution to global inequality after World War II, initially as "development," later as "globalization." In both variants, economic growth along first world lines was advanced as the commonsense solution to the problems of the postcolonial third world and the guarantee that "millions will be lifted out of poverty," a claim that is frequently made but misleading.

As environmental damage became increasingly apparent and threatened to slow the growth of many industries, scientists who exposed ecological problems found themselves under attack from business organizations and the panoply of think tanks championing the business agenda. Corporations funded attempts to deny the problems, neutralize the science, and delay action.

Powerful propaganda techniques evolved through the twentieth century and were increasingly applied, both to the selling of proliferating consumer goods and to the task of selling private enterprise itself. Economic growth was thus naturalized as the bedrock of prosperity, and prosperity in turn as the meaning of life. From the 1970s, the think tanks amplified this ideological project. Leading corporations funded their expansion, creating a parallel academic universe dedicated to the dissemination of business values and business interests. Scientists and the science they pursued, always struggling for limited funds, had no comparable means of communicating with the public—a situation that has facilitated the denial of numerous health and environmental dangers, from tobacco and thalidomide to DDT and acid rain. In the past decade, scholars have begun to analyze the denial of global warming and the more general attempt to dispute and undermine evidence of environmental decline. These practices have been described as "the cultural production of ignorance,"[4] in which fake experts, assisted by the mainstream press's fixation on "balance," bamboozle the public with spurious doubts.

The book is in four parts, each in roughly chronological order and covering a similar period of history, but with a different focus. Part I sets the stage with an account of the history and science of growth, a history of the dawning awareness of environmental decline, and the emergence of a "limits" literature and debate. Part II returns to the beginning of the twentieth century to explore the pursuit of growth through the construction of consumerism, the neoliberal transformation of policy debate from the 1970s, and the impact of first world growth prescriptions on the rest of the world since World War II. Part III returns again to the early twentieth century to examine the emergence of the large corporation, the propaganda apparatus built over the century, and the consolidation of the idea of economic growth as the overarching goal of human endeavor, especially after World War II. Part IV measures the *Limits* diagnosis against the actual situation forty years later and summarizes my conclusions.

Sources

I have used extensive secondary sources across several disciplines, the chief of which are economics, geography, history, politics, and sociology, though I also touch on the natural sciences, rhetoric, and system dynamics.

I also analyze numerous crucial primary texts written by economists, scientists, other scholars, and public relations advocates, including the work emanating from think tanks, UN agencies, international bodies such as the World Trade Organization, national governments, and nongovernmental organizations; and numerous articles by journalists.

Throughout my research period, I monitored the elite, business, and scientific press of the United States, Australia, Germany, France, and the UK on a regular basis and also reviewed the press of India and China. In this I was assisted by online subscriptions to several premier newspapers and numerous invaluable news digests. I also subscribed to daily and weekly political and environmental journals from the same countries and a number of digests on these subjects. The international scope of this reading provided me with a window onto the major public debates (in key world centers) that have touched on growth, environmental problems, economic beliefs, and the solutions debated in the public sphere.

Key Terms and Emphasis

The work frequently requires me to refer to various nations and to distinguish between the European powers and the countries they colonized over the last five hundred years and more. There are numerous modes for expressing this division: first world and third world (and second world, when the Communist bloc still persisted); industrialized and nonindustrialized; global north and global south; developed and underdeveloped world; developed and developing world; developed, emerging, and less developed world. Immanuel Wallerstein, who theorized the *modern world-system*, used the terms *center*, *semi-periphery,* and *periphery*.[5] There are no doubt others. I use many of these terms, more or less interchangeably, but prefer first and third world in most contexts; old habits die hard and, to me as an Australian, being part of the "north" makes little sense.

The scope of the book is global, and it has been necessary to select and emphasize particular crucial players and nations. My examination of first world countries has focused mainly on the United States and, to some extent, Australia. The focus on Australia results from my being an Australian who has lived through the recent history in question. I also had considerable exposure to the US press through the period of the research; however, the overwhelming reason for the American focus is the preeminent role of the United States in the global economy from early in the twentieth century. *Industry Magazine* celebrated this role in 1921: "among the nations of the earth today America stands for one idea: Business."[6] In line with this reality, Americans also played a leading role in the development of the modern national and transnational corporation, the PR industry, the worldwide selling of "free enterprise," and the proliferation of neoliberal think tanks during the years since 1970. My discussion of capitalism in the twentieth century is often centered on the US version, which produced large integrated national corporations and dominated the spread of transnational business. Transnational corporations headquartered in Europe, Japan, and other OECD countries gradually proliferated after World War II, however, and significant numbers of Chinese corporations now feature in the Fortune Global 500.[7]

For development issues, I have focused largely on China and India, countries that have been regarded as "emerging" or "semiperipheral" in recent decades. They illustrate well the ambiguous benefit of a "growth for prosperity" approach—the air in Beijing, for example, is frequently dangerous to breathe, while Chinese rivers are grossly contaminated and traffic jams last for days or, in one case, well over a week.

I use the term "mainstream economics" to refer to the neoclassical stream of economics, of which neoliberalism became the dominant version beginning in the late 1970s. When I refer to mainstream economists, I allude to this class of economists rather than to those with dissenting views, whether ecological economists or economic historians. Although not strictly mainstream, most Marxists and Keynesians share mainstream growth assumptions.

Gross domestic product also requires a brief explanation. As set out in box 9.4, GDP is not a reliable measure of human well-being and cannot be used as a proxy for it. Neither does it include economic activities, such as housework and subsistence farming, that are conducted outside commodity exchange. At the same time, GDP has several benefits as a metric. It is the available statistic in almost all accounts of economic processes and is a reasonable approximation of the market economy's rate of extraction and depletion; in this respect, it measures gross economic growth without reference to welfare or rationality. I have thus used GDP when discussing the growth in the market economy. It is important to be aware of the usefulness as well as the limitations of the data associated with it.

Dollar values are always expressed in US dollars unless otherwise specified. I have used tons, tonnes, acres, and hectares, depending on the original source.

In the Australian political landscape, governments are usually formed by either the Australian Labor Party (ALP), analogous to the Democrats, or the Coalition, which is an alliance between the Liberals (analogous to the Republican party) and the Nationals (formerly the Country Party and based in rural areas). Conservative governments are described as Coalition governments to reflect this reality. The Australian Greens have participated in some State governments; federally, they exert most influence in the Senate where proportional representation gives them a significant minority bloc and sometimes the balance of power.

I reflect on the problems of poverty in several chapters. Since the World Bank is virtually the only institution generating relevant data, I use its various metrics; the bank's research has, however, been criticized for its bias toward free market economics.[8] For the bank, the "extreme poor" live on less than $1 a day (which was adjusted to $1.25 in 2005 dollars); the "poor" live on less than $2 a day (not adjusted); and the "non-middle-class poor," those without scope for discretionary expenditure, live on less than about $12 a day (or $4,000 a year). All these measures are adjusted for local purchasing power.

Another linguistic choice that may require explanation or defense is that between "skeptic" and "denier" in relation to the vocal dissent regarding the reality of anthropogenic global warming. I subscribe to journalist Chris Mooney's view, mentioned in his address to the Canberra Press Club. He noted that skepticism implies good critical thinking and that the word should not be applied in its absence. He regards the body of climate knowledge that has been investigated and "picked over" numerous times by countless scientists as the best available and unlikely to be overturned. He sees no reason to reject the use of the term "denier" for people who are not qualified in the specific field or who are in alliance with vested interests.[9]

My own assumptions include an intuitive belief in the intrinsic value of the natural world, a science-based belief in the physical grounding of all human economic activities in that same natural world, and a realist attitude toward the laws of physics and chemistry.[10] I also embrace an inclusive democratic paradigm that is suspicious of the buying of influence, representation, and speech, as well as a conviction that any global account of the world must include all seven billion people and explore the relationships between rich and poor.

It is now 2013, more than forty years since *The Limits to Growth* was published. It soon became obvious, as I studied these years and their prologue, that growth was anchored in the profligate burning of fossil fuels and that, unlike the ozone-destroying chlorofluorocarbons, which were more or less easily replaced with substitutes, the fuels that underpin the entire economy are likely to be far more recalcitrant. Not only are many of the most powerful corporations in the world bound up with the extraction and distribution of fossil fuels but virtually the entire productive apparatus depends on them—and this includes agriculture. In addition, first world people and emerging middle classes everywhere have become increasingly habituated to more material objects and an ever-ascending "standard of living." It is difficult to separate us from these proxies for the good life.

In Australia, polls now suggest a decline in willingness to make any sacrifices at all in order to tackle global warming, even though the Gillard Labor government's carbon-pricing scheme was modest and people with lower incomes were well compensated. By September 2013, Australians had elected a conservative Coalition government led by Tony Abbott, which pledged to abolish what it calls "the carbon tax," claiming that this would lower electricity prices. The new prime minister appealed to people's narrow self-interest and a skeptical view of climate change, and won. Our reluctance to acknowledge the needs of the rest of the world or to accept a fair share of the global costs of climate action is emblematic of the influence that has been exerted through multiple channels to block or delay real measures aimed at palliating the ever-increasing consequences of rampant economic growth.

Over the course of the twentieth century, big business on the new US model took charge of the new mass media and succeeded in advancing a profit and expansion agenda while pretending to be merely providing a service. Had the planet been a great deal larger, it might not have fallen to the current generations to choose between recognition of the limits to this project and surrender to terminal decline. As it is, the world expects to support nearly 50 percent more people by 2100 than at present, and to do this we are privatizing and corporatizing all economic endeavor. This includes land, and it consigns families, communities, and other human groupings to subsist through exchange in the marketplace. Good luck to the next few generations; they will need it. I offer my apologies that, though I have tried to limit my own contribution, I did not stop the rot.

I

Growth and Its Challengers

As the absolute load is increased the watermark will reach the Plimsoll line even in a boat whose load is optimally allocated. Optimally loaded boats will still sink under too much weight, even if they sink optimally!

—Herman Daly, 1991

1

Economic Growth: Origins

The discovery of gold and silver in America, the extirpation, enslavement and entombment in mines of the aboriginal population, the beginning of conquest and looting of the East Indies, the turning of Africa into a warren for the commercial hunting of black-skins, signalled the rosy dawn of the era of capitalist production. These idyllic proceedings are the chief momenta of primitive accumulation.

—Karl Marx, 1848

Explosive economic growth is new in human history, and this chapter looks at how it was unleashed in three distinct but related historical developments, with Europe at their center.

First, there was the 500- to 600-year period of Europe's colonial expansion, which enabled Europeans to accumulate great wealth without commensurate cost by appropriating land, resources, and the slave labor of millions, and to solve numerous resource constraints by simply moving on to new frontiers. Second, there were 250 years of coal-based industrialization, which coincided with a massive development of technological capacity (known as the Industrial Revolution), a great wave of urbanization, and the triumph of capitalism as an economic system. Third, and most recently, the past 130 years have yielded oil-based growth—a unique period in the history of civilization and one that is unlikely to be repeated when cheap oil runs low.

These changes involved a wholesale separation of human populations from the land on which they grew their food and constituted a radical shift in the relationship between people and the processes of the natural world. Called *metabolic rift* by Karl Marx, it continues to develop rapidly today, as the rural people of the third world move to newly industrializing cities.

Prelude: The "Age of Discovery"

Well before Columbus's world-altering first voyage to the Americas, sailors had begun to range into the eastern Atlantic; when Columbus set sail, the uninhabited Azores and Madeiras were already in European hands, and Spain had colonized all but one of the Canary Islands. By the time Tenerife was conquered, in 1496, most of the Canaries' indigenous people, the Guanches, had been captured and enslaved, or eliminated by European diseases. The occupation of the Atlantic islands during the fifteenth century provided a foretaste of the patterns of expansion that were to follow. A few rabbits taken ashore on Porto Santo, the smaller of the two Madeiras, multiplied so fast they denuded the place within a few decades, forcing humans to abandon the island, now stripped of their crops as well as of the native flora and fauna. The larger island, Madeira, was named for the "great trees" that covered "every foot" of it. Here the settlers set fire to the forest in order to clear it; over seven years, the entire island was burned. By the middle of the fifteenth century, sugarcane was well established, and the next decades saw an immense explosion of sugar production. By the early sixteenth century, sugar was supporting a population of some 20,000 people, at least 2,000 of them slaves.[1]

Plantation colonies worked by slaves were replicated across the world, accompanied by the destruction of vulnerable native populations, the expropriation of their lands, crops, and natural wealth, and the progressive transformation of native ecologies. Settler colonies were established in more temperate regions, and these too supplanted existing peoples, expropriated their resources, imported slave labor, and replaced forest, woodland, and prairie with crops. Europe exported millions of its own to these new colonies, easing domestic population pressures and importing other people's wealth back into its own metropolitan centers.

Until the so-called Age of Discovery, when the Spanish reached America and the Portuguese rounded the Cape of Good Hope and arrived in India, northwestern Europe had been a backwater of peripheral world importance; before the late fifteenth century it was access to the Mediterranean that gave Europeans global reach. Nonetheless, northwestern Europeans had begun to transform their backwater; by 1300 they had spread out over western and central Europe, cleared the once ubiquitous forest back to some 20 percent of its original cover, and created widespread permanent fields for the first time. Trapping coincided with this great forest clearance and continued eastward after 1300, across Russia, into Siberia, and on to the Pacific coast, driving ever more species of furbearing animals to extinction or its brink. Billions of individual animals were killed. The fur trade reflects as well as any activity the way European expansion relied on ever-new frontiers of exploitation. When Europe's own animal numbers declined, the hunt for fur moved in the same way across North America until the immense populations of the boreal forests had largely collapsed.[2]

Capitalist Accumulation: Engine of Growth

Capitalism is an economic system based on private ownership of the means of production, whether land or technology. The feudal system that preceded capitalism in Europe did not separate peasants from the land that produced both their subsistence and the tribute they owed their masters; they were often tied to the land as serfs, but the land was also tied to them through customary rights to their tenancies and the use of extensive commons for such purposes as grazing, hunting, and collecting firewood. They did not sell their labor for wages; however, they were obliged to pay their masters a significant share of what they produced.

Marx regarded the surge in mercantile wealth associated with colonization as the prelude to capitalism and as one of the primary sources of what he called "primitive accumulation," a translation of Adam Smith's term "previous accumulation." Both Smith and Marx recognized that capitalism required some kind of prelude whereby wealth was first accumulated so that it could be applied to further accumulation. The successive waves of enclosure carried out across Europe over several centuries can be seen as one aspect of primitive accumulation. Landlords, often assisted by the state, curtailed the traditional rights of peasants to their tenant holdings and progressively enclosed the commons and woods on which they had depended. The dispossession of Europe's own peasantry not only turned their land over to sheep farming but created laborers without means of subsistence, the very workforce that industry required. Though originally conceived by Smith as "previous" accumulation, both colonial and internal expropriation went on for centuries, and both continued as industrial capitalism emerged.

Human societies have produced surpluses since the earliest agricultural communities some ten thousand years ago, but precapitalist systems tended to deploy this surplus for consumption, sometimes by elites, priests, the military, relatives, or select allies, sometimes in great monuments or via redistribution among the people. Even though the surplus could and did free certain classes in society from abject toil, it was put to immediate use and, in these circumstances—generalized throughout precapitalist history—economic growth remained slow.

One of capitalism's key innovations was to direct the surplus to reinvestment in production, establishing an ongoing process of accumulation. The influx of plunder from the colonized world underpinned the expanding wealth of the early capitalist nations of Europe, though the wealth was not widely shared among the population in the first few centuries. Once industrialization began, generations of landless laborers—men, women, and children—worked long hours every day of their short, harsh lives. Members of the new industrial working class had a Stone Age standard of living (measured by life expectancy and food intake) and would have been better off in a hunter-gatherer band.[3] It was only much later that capitalist wealth flowed on to considerable majorities of European and settler populations and provided a hitherto unimaginable level of material comfort to the mass of people.

Around the same time as the colonization of the eastern Atlantic in the late fifteenth century, manufacture and technical innovation began to gain ground in England with the introduction of paper and gunpowder mills, cannon foundries, sugar refineries, and various metallurgical works. Commerce, associated with the expansion of long-distance trade, was beginning to apply its profits to investment in agriculture and industry, and to appropriate the artisan's product for trade rather than local use.[4] Technological innovation proceeded in step with the transition from mercantile wealth to industrial capitalism. However, these early industrial processes relied mainly on wind, water, wood, or muscle (both animal and human) for their power, with attendant limitations. Though coal mining began to expand from about 1530,[5] the starting volume was very small, and coal remained a minor power source.

From about 1600 on, a new "mechanics literature" emerged—a torrent of technical works by artisans that proved popular among merchants and businessmen.[6] It is unclear how much the "age of invention" actually depended on the intellectual developments associated with the Scientific Revolution. Both invention and scientific thought were unfolding in the same context—a time of pivotal social transformation, when the feudal economy was in collapse, capital accumulation had started, and knowledge had begun to be equated with utility and technology—with the potential for power over nature.[7] While this trend and the trend away from religious authority to the reason of the individual may have favored inventiveness, the universities were more engaged with deciphering the laws of the universe than with practical invention, which arose in concrete associations of tradesmen and men with capital. The determining factor in the proliferation of the new technologies was "the opportunity for the profitable use of mechanical inventions" by the newly emerging industrial interests.[8]

Industrial capitalism did not become dominant until factory technologies took over production, and by the time they did, in the second quarter of the nineteenth century, coal was the fuel that drove the steam engines.[9] The capitalist system of production, once it was powered by fossil fuels, was set to develop and deploy worldwide an economy capable of accumulating vast amounts of capital and a technology capable of godlike feats such as moving mountains and splitting atoms.

Oil and Exponential Growth

Like coal, oil had been known for centuries, but it was not until the second half of the nineteenth century that oil wells began to deliver petroleum, mainly for use as kerosene in lighting. The United States pioneered the transition to petroleum; the first large American "gusher" was tapped in Texas in 1901, and world production reached approximately 2 billion barrels a year in 1930. This compares, however, with some 25 billion barrels in 1990 and 30 billion in 2006.[10] Thus, significant reliance on oil dates only from the second quarter of the twentieth century and has been with us less than one hundred years, steadily escalating as the decades have passed.

Oil is a miracle fuel. It is compact, liquid, transportable, and cheap to produce; once tapped, large amounts can often be extracted under the oil field's own pressure. Even today, though much of the "easy oil" has already gushed out, the cost of extraction in many of the large Middle Eastern oil fields remains small compared to its market value.

The development of petroleum revolutionized transportation, making viable the individual private car, aviation, and mechanized agriculture. By 1950, oil had replaced coal in the United States as the principal fuel of industry as well as transport and heating. In the following twenty years, similar transitions occurred in the rest of the industrial world.[11] Simultaneously, multiple uses for the range of hydrocarbons found in oil were also developed, until petroleum has become embedded in every aspect of daily life—from fuel to fertilizer and pesticide, from pharmaceuticals and cosmetics to plastics and fabrics, and in an array of industrial chemicals and processes.

The year 1950 represents the approximate moment when the curve of humans' environmental impact may be said to have become dangerously exponential. The world population stood at 2.5 billion, a level at which a doubling period of three or four decades would, for the first time, produce huge annual increments. Estimates of world GDP in 1900 suggest that it had doubled twice since 1500, nearly half of that growth in the thirty years after 1870. In the next fifty years, from 1900 to 1950, it grew almost as much again as in the four hundred previous years, notwithstanding the slowdown of the Great Depression. Then, from 1950, it doubled very rapidly from an ever more immense base—more than twice in the forty-two years between 1950 and 1992.[12] Although world growth slowed after 1973 and reversed in 2009, there has been a further doubling since 1992. The global economy in 2014 is eight to ten times larger than it was at the end of World War II.

The year 1950 also found the mass media on the verge of a rapid expansion into television and, consequently, a singular success as a vehicle for consumerism. The advent of a screen in every home allowed industrial capitalism to maximize its markets in the developed world, as well as to make the consumption of material goods central to everyday life for most people. Over the next fifty years, a progressive democratization of luxury took place in the first world, as aspiration for and access to more and more consumer goods was extended from the rich to the middle class and on to a significant majority of the population. Though exuberant consumption had marked the elites of many cultures throughout history, modern societies of the developed world were the first to extend the option to the mass of people.

Industrial Capitalism and Metabolic Rift

In the *Communist Manifesto*, first published in 1848, Marx eulogized the exponential achievements of industrial capitalism. He celebrated economic growth and the technological advances that it generated and that helped to drive it:

> The bourgeoisie, during its rule of scarce one hundred years, has created more massive and more colossal productive forces than have all preceding generations together. Subjection of nature's forces to man, machinery, application of chemistry to industry and agriculture, steam navigation, railways, electric telegraphs, clearing of whole continents for cultivation, canalization of rivers, whole populations conjured out of the ground—what earlier century had even a presentiment that such productive forces slumbered in the lap of social labor?[13]

Marx regarded the development of such productive forces as the historical predecessor and essential basis for a transition to communism; criticism of this view underpins the widespread characterization of Marx as a technological optimist and cornucopian. The manifest failure of real-world communism to protect its environments lends further support to this critique, though it seems fair to distinguish between the nineteenth-century scholar and the political movements that used his theories. As he was a theorist working before the unprecedented ecological disruption that followed in the wake of twentieth-century growth, it is hardly surprising that Marx did not focus on the environmental impacts of capitalism, though he later addressed its effects on agriculture. Marx concentrated on analyzing capitalism as a system: its history, the nature of its regimes of production, distribution, and exchange, its new class structure, and its tendency toward expansion. He was prescient about such aspects of capitalism's trajectory as the transformation of social relations into market relations, the progressive commodification of all kinds of goods and services, and the emergence of a world market.[14]

Alongside his admiration for capitalism's development of unprecedented productive forces, Marx was aware of its tendency toward exploitation, principally of human beings, but also of the soil.[15] He was particularly affected by his study of the work of the German chemist, Justus von Liebig. Liebig's analysis of soil fertility led Marx to identify the growing split between town and country as "an irreparable break in the coherence of social interchange prescribed by the natural laws of life."[16] This amounted to a kind of ecological disjunction, a rift in the Earth's metabolism:

> Capitalist production, by collecting the population in great centres ... disturbs the circulation of matter between man and the soil, i.e., prevents the return to the soil of its elements consumed by man in the form of food and clothing; it therefore violates the conditions necessary to lasting fertility of the soil.[17]

Liebig pioneered the modern understanding of the sources of soil fertility and laid out the role of the essential nutrients nitrogen, phosphorus, and potassium, which are taken up by crops as they grow. With the advent of widespread urbanization, crops incorporating these nutrients began to be exported across the country and across the world. At the same time, sewage and food wastes, which once were circulated back to the land, where they helped maintain fertility, instead polluted the rivers of the great new towns. Since the last few decades of the twentieth century, an even more massive metabolic rift has occurred as colossal quantities of animal wastes also drain into watercourses. The grain that has fed these animals is usually grown elsewhere, sometimes on the other side of the world, and the fertility stripped from those distant croplands is lost forever.

Until Peruvian guano and nitrate deposits were exploited in the 1840s, European graveyards and battlegrounds were ransacked for bones in an attempt to improve British soils. While Peru's guano and nitrate were mined and shipped back to Europe, numerous Pacific islands were annexed by the North Americans to secure their own sources of fertilizer. The story of the headlong hunt for fertilizer during the nineteenth century reveals a resource scarcity that persisted for a long period without a technological solution being found. In the meantime, colonial possessions again played a crucial role in the accumulation of European-based wealth: ships bent on discovering gold led the way to islands encrusted with millions of years of bird droppings.

This fertility crisis continued for nearly a century. Soluble phosphorus was synthesized from phosphate rock and sulfuric acid in 1842, but Europe relied on imports of guano and nitrates until synthetic nitrogen fertilizer, sourcing nitrogen from the atmosphere, was invented shortly before World War I. While the technical solutions were eventually found and the shortage of fertilizer was banished for the time being, it is clear that the availability of hitherto untapped resources was, for nearly a century, absolutely necessary to avoid severe agricultural decline. The interim solution through the nineteenth century depended on sheer luck in the context of a world largely unexploited. This vast unexplored world no longer exists, and such solutions, though they may emerge, cannot be relied on by living generations today.

The long process of enclosure of peasant land and the gradual transfer of manufacture from artisans' workshops to the urban factories owned by the new industrial bourgeoisie progressively separated European people from the land—a process that is proceeding rapidly in so-called emerging economies today. Industrial solutions to the separation of farming and consumption are themselves in trouble. After a century of intensive inputs in the farming of the developed world and nearly fifty years of similar techniques in the global south, introduced under the Green Revolution, problems such as groundwater contamination and "dead zones"[18] on continental shelves remain unsolved, while many of the raw materials for fertilizer are dwindling.

Phosphate rock is argued to be only twenty-five years from peak or maximum production and likely to reach effective exhaustion before the end of the century.[19] Phosphorus, an element, is one of the indispensable building blocks of DNA, so no substitute is likely to be invented, though it is possible to recover and recycle it from human waste. Nitrogen fertilizers are extracted from the atmosphere or from natural gas feedstock; this requires large energy inputs and uses approximately 1 percent of the world's annual energy budget, still largely derived from fossil fuels.[20] The scale of this industry is such that humans convert more nitrogen into reactive forms than the earth's entire combined terrestrial processes.[21] Difficult or diffuse reserves of both oil and phosphate rock will continue to be found, but the era of each as a cheap resource is almost certainly over. Topsoil is also in trouble. The gross area of degraded land worldwide is reckoned to be in the tens of millions of square kilometers, and about 24 billion tons of topsoil continue to be lost every year.[22] A complex and living medium, soil has not fared well under industrial agricultural regimes—as well as disappearing under urban asphalt.

Marx argued that the rift between town and country could be solved in a socialist society, where the land would be returned to the associated producers, "rationally regulating their interchange with Nature, bringing it under their common control instead of being ruled by it as by the blind forces of Nature." In this world, he saw agriculture and manufacture reintegrated and the distinction between town and country ameliorated by "a more equable distribution of the populace over the country."[23] Whether such a form of socialism might succeed in minimizing destruction of the natural environment is an open question, since it has not been tried—with the partial exception of post-Soviet Cuba. Clearly, neither Soviet Russia nor contemporary China provides a real-world model for such a transformation. Soviet communism replicated metabolic rift, as well as precipitating environmental catastrophe in numerous places, such as the Aral Sea. The Soviet system operated as a form of state-based capital accumulation—partly to compete with the capitalist world during the Cold War period and partly because it embraced industrial production as a bringer of wealth. The establishment of sophisticated technology requires a capital base whatever form of ownership prevails, making full-scale industrialization impossible without capital accumulation. In any case, the Soviet regime was not a system controlled by the associated producers in Marx's sense.

China's situation is similar, though it is less a matter of being pitted against the capitalist world than of imitating capitalism's economic expansion and integrating into the late twentieth-century world market. The industrialization under way in China since 1979 also replicates metabolic rift. Urbanization gallops along just as surely as it does in democratic India. Five or six decades after Chinese peasants were freed from the arbitrary power of their feudal landlords and given access to land, and only forty years after the Cultural Revolution uprooted urban people and drove them into the countryside to be "reeducated," peasants are now routinely separated from their land by party officials bent on taking it over in the interests of industrial expansion, property speculation, and the wealth generated by economic growth. This process proceeds rapidly, especially in eastern China, just as it did in Europe 150 years earlier. Factories cluster in huge urban concentrations along the great rivers, which are just as toxic now as the rivers of the English Midlands were then. Economic growth now, as then, appears to be predicated on the extension of metabolic rift, the separation of most people from the land.

The case of Cuba since the collapse of the Soviet Union is worth noting, though what happened there after the sudden disappearance of Soviet oil and food would be unlikely to have occurred if such imports had remained available. The immediate aftermath was extremely difficult, and many Cubans verged on starvation. Nonetheless, fifteen years later, about half of Havana's vegetables were being grown in the city, a fraction rising close to 100 percent in smaller towns.[24] In response to the crisis, the government restructured over 40 percent of state farmland into 2,007 new cooperatives managed by the workers, who were also allotted a gardening space to grow their own family's food. By 2000, more than 190,000 urban residents had also claimed personal lots on vacant city land. In both arenas, farming methods are organic, easing the need for imported petroleum. The process has reconnected rural workers to the land and helped urban Cubans transcend the town/country divide—producing what sociologist Rebecca Clausen calls a metabolic restoration.[25]

Cuba's struggle after the Soviet collapse brings into sharp focus the consequences of having to do without oil, synthetic fertilizer, or pesticide—as well as imported food. If cheap oil is in decline, as peak oil proponents argue, Cuba will have a head start on the rest of the world—and a big advantage over countries like China and India, which have bet their futures on fossil-fueled agriculture and fossil-fueled economic growth.

The massive economic growth of the twentieth and twenty-first centuries is unprecedented in history. It gathered pace as the mercantile proceeds of the Age of Discovery were applied to domestic production back in Europe. This was in turn boosted to an industrial scale in the early nineteenth century with the mounting use of fossil energy; a further lift was provided from the late nineteenth century when liquid petroleum was tapped. After World War II, a period of explosive economic growth held sway for three decades, a unique event in history. Although it slowed down in the 1970s and was actually arrested in 2009 after the global financial crisis, the immense human economic system continues to double in scale every fifteen to twenty-five years. In the next chapter I look at conflicting views of this incredible phenomenon.

2

Economic Growth: Perceptions

There is no magical New York investment cowboy who's going to suddenly develop a product which overcomes physics and biology. They were designed a long time ago, possibly intelligently—we don't know—but they were certainly designed a long time ago and they are absolutely set in concrete.

—Paul Gilding, 2009

Which Comes First, the Planet or the Economy?

Ecological economists see economic growth very differently from mainstream economists and most policymakers. First, and most fundamental, is the question of which is primary: the economy or the planet's ecological systems? The answer chosen is crucial, since all questions of the limits, boundaries, and scale of the human economic enterprise hinge on whether or not the economic system can be theorized independently of its physical and natural context.

Many standard economics textbooks introduce students to a diagram of "the economy" that includes only the relationship between businesses and households (producers and consumers), depicted as a circular flow and not represented in any wider physical context. The ecological economist Herman Daly contextualized this circular diagram by drawing an outer frame around it to represent the natural world, a world that "contains and sustains the economy" by regenerating renewable inputs and absorbing unavoidable wastes. In *Ecological Economics*, Daly describes his exchange with the World Bank's chief economist:[1]

I asked the Chief Economist if, looking at that diagram, he felt the issue of the physical size of the economic subsystem relative to the total ecosystem was important and if he thought economists should be asking the question, "What is the optimal scale of the macroeconomy relative to the environment that supports it?" His reply was short and definitive: "That's not the right way to look at it."[2]

Daly characterizes this approach, borrowing Joseph Schumpeter's term, as "pre-analytic," meaning that the assumptions involved are implicit and held to be axiomatic—and thus not susceptible to analysis. The key assumption here is that the economy is the overarching system, while nature, if it is considered at all, is a sector of the economy, such as the extractive sector.[3]

This paradigm, which treats nature as a subset of the economy, underlies numerous unexamined verities of mainstream economics: that the planet is functionally infinite, both as a source of materials and as a sink for wastes; that substitutes for depleted resources will be generated automatically by price increases; and that the limitations of the physical world and the laws of thermodynamics are not relevant to economic processes. These ideas inform the conduct of economic activity and the current public debate about it. They have immense consequences for the way people think about economic growth, and they underpin a widely held confidence that economic growth as we know it on earth today can continue indefinitely.

Usually assumed rather than stated explicitly, the idea that nature is a subcategory of the economy gets short shrift from ecological economists. Daly's primary or overarching system is biophysical rather than economic—human endeavor is necessarily proscribed by the laws of physics and the physical constraints of a finite planet. That we live and produce within the boundaries of such an entity means that the human economy has inevitable limitations of scale.

As far back as 1966, Kenneth Boulding, a pioneer of ecological economics, argued that a transition had begun: from the "open" to the "closed" earth, from the "empty" world with ever more frontiers for exploitation to the "full" world where it is no longer possible to go somewhere else when resources fail or pollution destroys. The "illimitable plains" of the endless frontier no longer stretch into the unknown. There are no more "unlimited reservoirs of anything, either for extraction or for pollution."[4]

On the face of it, the economists' basic view—that the human economy is the primary system—seems an astonishing claim. Life on earth, after all, is about 3.8 billion years old, and human life is an infinitesimal fraction of that. The capitalist economy is at most five hundred years old, a small part of the 200,000-year span our species has been on earth. Even what we call "civilization," with settled life in established cities and the cultural complexity that accompanies it, is considerably younger than the 10,000 or so years since we started farming. Understanding the human economy as a primary system independent of the earth it arose upon would appear to defy common sense.

Actual constraints or limitations, of course, were far less obvious before the accelerated growth of the past 250 years—and especially the twentieth century, which by 1999 was delivering an *annual* increase of at least half the entire global economy in 1900. Through most of our history, humans could ignore natural limits. But, as the development sociologist Wolfgang Sachs has put it, the more "the rate of exploitation increases, the faster the finiteness of nature makes itself felt on a global scale."[5]

Self-evident as these boundaries might now seem for ecological economists and allied scholars, they remain invisible or contested in mainstream economics and are of little concern to politicians in most countries and to most of today's citizens. Hardly a news bulletin goes by without reports about growth expected, growth threatened, or growth achieved. Growth is the sine qua non of everyday economic language and expectation and the much-touted solvent for critical problems such as poverty, pollution, and debt. Yet, however necessary ongoing growth is to our current economic arrangements, and however desirable from the point of view of our expectations of material well-being and comfort, it is hardly a practical aim if it is based on a misperception of reality.

The unexamined assumption that a high and increasing level of material consumption is normal stands in the way of a perception of the peculiarity of our times and an ability to engage in the kind of hard scrutiny our path to the future now requires.

The Timing of Economic Change: Which "Growth"?

Some mainstream economists concede that modern economic growth is unprecedented. In his Boyer Lectures of 2006, titled *The Search for Stability*, retired Australian Reserve Bank governor Ian MacFarlane remarked that "viewed against the span of human history, economic growth is a relatively new phenomenon, dating only from the Industrial Revolution in the mid-eighteenth century. In the many centuries prior to that, it had been negligible."

MacFarlane's insistence on the recent emergence of economic growth contrasts radically with that of economists who responded to *The Limits to Growth* when it first appeared in 1972. Many prominent economists characterized growth as a continuum throughout human history. Robert Solow, who later won the 1987 Nobel Prize in Economics, argued that "the world has been exhausting its exhaustible resources since the first cave man chipped a flint."[6] The British economist Wilfred Beckerman thought that problems associated with exponential growth in the use of finite resources have "been true since the beginning of time; it was just as true in Ancient Greece.... This did not prevent economic growth from taking place since the age of Pericles.... There is no reason to suppose that economic growth cannot continue for another 2500 years."[7] This "continuous progress" view allows the economic history of recent centuries to be subsumed under a static notion of transtemporal human culture in which our present system is seen as just a phase of the permanent and normal state of a healthy economy.

Clive Ponting points out in his *Green History of the World* that "for all but the last few thousand years ... humans have obtained their subsistence by a combination of gathering foodstuffs and hunting animals"[8]—a way of life involving virtually no resource extraction, and no changes we would describe as economic growth. Ponting identifies two great transitions in human history—the first, to farming, starting about 10,000 years ago, and the second, in the last few hundred years, to the dominance of fossil fuels and the development of an industrial economy.[9]

The first agricultural revolution took place over millennia as humans began to grow crops and improve pastures and, for the first time, to cause major alterations to the ecosystems around them. People were able to settle in villages and towns and to produce a surplus above their subsistence needs. Though the elites—always a small minority of the population—could be supported without having to produce their own food, for the majority of people these developments led not to plenty but rather to an arduous life of scrabbling in the dirt. It was a system more vulnerable to the vagaries of climate, drought and famine than the hunter-gatherer mode of life had been and, according to Ponting, it was "most definitely not an easier option.... The one advantage agriculture has over other forms of subsistence is that in return for a greater degree of effort it can provide more food from a smaller area of land." Ponting suggests that a world population of approximately four million, reached around 10,000 years ago, was the maximum that could live comfortably by hunting and gathering, and that it was this expanded population that precipitated the shift to farming.[10] Other scholars argue that factors such as climatic variation or intentional risk reduction may have sometimes played a role.[11]

The transition to farming initiated the first episode of severe local ecological damage, mainly as a result of the progressive clearing of the forests.[12] The paleoclimatologist William Ruddiman suggests this might even have affected atmospheric levels of greenhouse gases several thousand years ago, with a discernible impact on climate.[13] Nevertheless, though economic growth quickened, it remained barely perceptible over the subsequent 10,000 years or so. Before 1500, improvements in technique were only occasional, and the world's population took a thousand years or more to double.[14]

In the 1994 edition of Worldwatch's *State of the World*, Alan Durning used the metaphor of a ten-minute satellite-view film of the deforestation of earth over the last 10,000 years to demonstrate the immense change of pace and scale involved in the second great transition to industrialization. There is no obvious sign of change until the eighth minute, with the disappearance of forest around Athens and on the Aegean islands. Forests in Europe, China, India, and Central America can be seen contracting from the beginning of the last millennium (from 1000 CE), but they do not shrink appreciably until the years from 1800 to 1950, coincident with the Industrial Revolution, when about 6 percent of them disappear. From 1950, in just 3 seconds of Durning's film, a further 60 percent of the original forest vanishes. Although trees still grow on about three quarters of the original forest area, less than half of these represent intact ecosystems; the rest are "biologically impoverished stands of commercial timber and fragmented regrowth."[15]

Durning's short film serves as a graphic depiction of the pace of economic expansion through the lens of forest clearing, corroborating MacFarlane's timing of economic growth. Ecological effects of human economies, local in scale for millennia, became global very recently—and very rapidly. The historian John McNeill characterizes modern times, especially the last fifty years, as "bizarre, anomalous and thoroughly unsustainable."[16] Although the earth has changed environmentally for some four billion years and our genus, *Homo*, has altered earthly environments for the last four million years or so, the twentieth century has no parallel. Its peculiarity is largely a matter of scale and intensity; however, a quantitative increment can cross a threshold (or tipping point) and trigger "a grand switch," leading to qualitative change—such as when ice melts as 0°C is exceeded.

This surge in scale and intensity would not have been possible without the exploitation of fossil fuels. By 1700, with the help of domesticated animals, windmills, and water wheels, the most efficient agricultural societies could command four or five times more energy per head than their hunting and gathering predecessors could; even so, the vast majority remained poor and restricted to a life of grinding toil. In the nineteenth century, as the coal-powered steam engine began to transform production, energy availability was again multiplied by five, about the same increment as in the 10,000 years before 1700. But that was still a mere prelude to the boom of the twentieth century, when liquid petroleum fueled a further twelvefold expansion in energy use.[17]

Apart from minor coal-burning, all energy used before the Industrial Revolution was more or less directly solar—from wind, water, animals fed on crops or grasslands photosynthesizing daily solar flows, or human muscle fed the same way. Even wood, as a fuel, concentrates the solar energy trapped by trees over decades, or centuries at most. Coal and oil, on the other hand, are fossilized solar energy, compacted and distilled out of tens of millions of years of sunlight stored by the primitive trees of the vast Carboniferous swamps (coal) or zooplankton and algae thriving in the shallow seas of the Mesozoic (oil). This is immensely concentrated energy, capable of performing colossal amounts of work. It is probably the most basic precondition for the bizarre exuberance of the era of economic growth through which we have been traveling, especially since World War II, often under the false impression that it is the normal and natural state of human affairs.

Human perceptions have always been filtered through prevailing social narratives about how the world is. At least since the emergence of agriculture, the narrative has emanated from the groups who hold social power, and this was just as true after the Enlightenment as it was in the days of Catholic hegemony in Europe. At least in the Western world, the mainstream interpretation of reality moved from being the preserve of God-given authorities to a contest of ideas ideally based on the rational assessment of empirical facts. Notwithstanding the move toward evidence-based beliefs, the tendency to regard the current world as the normal state of human affairs seems not to have been modified. The observed world continues to be understood as "natural." To its inhabitants it appears that it has always been that way.

Until about five hundred years ago, rates of change were infinitesimal and the world one was born into was indeed unlikely to change much, excluding occasional natural catastrophes. The world in 2014 is not such a world. The pace of change has accelerated along the lines described in chapter 1. Economist Greg Clark holds that there was "no economic growth" in preindustrial Europe.[18] Though this may be a slight exaggeration, there is no doubt that the rate since 1750 or so has been unprecedented. At such a rate of transformation, it might seem obvious that there is nothing "natural" about it, and that it represents a sharp departure in human history. Yet little awareness of such a perspective exists. On the contrary, economic growth remains the mantra of global institutions, governments, and the daily media, and ongoing growth is assumed to be not only possible but essential to a desirable future.

Indefinite Growth and the Laws of Physics

Indefinite growth can be seen as viable when human production is assumed to bear no compulsory relationship to the physical world, an assumption that includes an exemption from the second law of thermodynamics. This is the so-called entropy law, developed by the German physicist Rudolph Clausius from Carnot's original observations of engines. In 1850, Clausius formulated the principle that heat always moves in one direction only, from the hotter to the colder body, and will not flow the other way without the application of more energy. Based on this elementary fact, the entropy law holds that, in closed systems, all energy dissipates, becoming useless for further work in the process. In a key departure from classical physics, which dealt with reversible processes such as mechanics, thermodynamics describes one-way qualitative change, which is subject to the "arrow of time."

Nicholas Georgescu-Røegen was the first economist to insist that economics must take the second law into account. He argued that all natural resources are subject to entropic exhaustion, and that human economic production is a physical process that inevitably hastens that dissipation. Irreversible physical transformations are implicit in production, where material resources and available energy are consumed and waste left behind. In Georgescu-Røegen's words, "Coal turns into ashes in the same direction, from past to future, for all humans."[19] Any economy will therefore require a source of materials and a sink where the waste can be dumped, as Daly suggested to the World Bank's chief economist. An economy that produces large quantities of material artifacts will require large sources and large sinks. Yet standard economic formulas include no variables to represent resources or wastes, and so are incapable of reflecting the thermodynamic implications of an expanding economic process.

Mineral ores, for example, are most useful and most accessible for human use in concentrated form. The world's most concentrated ores have already been heavily depleted, while more dilute sources are rendered accessible by the application of new technologies and of more and more energy—and, other than solar power, energy resources are themselves subject to depletion. Gold mining in the twenty-first century illustrates the trend. Much of the gold that remains unmined consists of microscopic particles; in most mines today, something like thirty tons of rock must be excavated, pulverized, and sprayed for years with cyanide drizzle to recover an ounce of gold. Such an undertaking can only be profitable on a huge scale; the energy required to make and drive the machinery is considerable, as are the massive volumes of waste rock and wastewater that must be disposed of, and the acid and heavy metals frequently released into waterways in the process.[20]

Entropy is not necessarily manifested quickly. Georgescu-Røegen notes that it took thousands of years of sheep grazing before "the exhaustion of the soil in the steppes of Eurasia led to the Great Migration."[21] By contrast, indicative of the scale, pace, and intensity of recent economic growth, two decades of intensive cashmere goat grazing in northern China have already compromised the fertility of the Alashan Plateau of

Inner Mongolia. China's rapidly expanded herds of cashmere-producing goats might have slashed the price of sweaters, but they have also grazed these grasslands down to a moonscape, unleashing some of the worst dust storms on record.[22]

In the case of oil—the energy on which the past century of explosive growth has been based—depletion did not take thousands of years, another indicator of the exceptional pace of twentieth-century change. Even if oil supply continues to meet demand for several more decades, as optimists argue, the production of light crude has leveled off, and there is little doubt that depletion will occur fairly soon in any long-run view of the human prospect. In 2010, the International Energy Agency declared that "the age of cheap oil is over."[23] What is also imminent in the second decade of the twenty-first century is the limited capacity of the planet to act as a satisfactory sink to absorb the waste that will be released if we burn the remainder of the fossil fuels.

As far as energy is concerned, the earth is not a closed system, of course. Though matter is not exchanged, the earth receives an immense amount of solar energy every day, making available a generous and indefinite supply of energy, should humans have the means to harness it. This solar flux—not subject to entropic exhaustion for another four or five billion years—is the only rational basis for indefinite human energy use. Industrial capitalism, however, replaced solar-based energy forms such as wind, water, and biomass with the energy stored in finite terrestrial fuels—coal in its initial phases, and petroleum since the unprecedented bonanza generated by the harnessing of oil to the internal combustion engine. These fuels are finite and their availability is governed by the second law, which tells us that energy, once tapped for work and dissipated as heat, cannot be reconcentrated for a second use. In the words of Ehrlich, Ehrlich, and Holdren, those who are waiting for a breakthrough that would overturn the second law might as well "wait for the day when the beer refrigerates itself in hot weather and squashed cats on the freeway reassemble themselves and trot away."[24]

While coal is expected to be available for another century or more, it is widely suspected that oil is near "peak" production—in other words, the maximum level of production we will ever see.[25] Apart from the low-cost oil fields remaining, primarily in the Middle East, oil recovery is now carried on in ever less hospitable contexts—deep seas, Arctic climates, "tight" oil,[26] low-content tar sand and oil shale (kerogen), and low-quality gas liquids. In all these alternative sources, the proportion of energy yield expended in extraction is on the increase. The energy budget is just as relevant as the economic one—a ton of oil from kerogen or tar sands is not worth extracting if it takes equivalent energy from gas or coal to do it; whatever the dollar value, it makes little sense to extract energy once it has negative energy return on investment. This is also a severe qualification on the virtues of corn-based ethanol. Even the most optimistic estimates suggest that, in the United States, it takes seven or eight gallons of petroleum to produce ten gallons of ethanol, while Cornell University's David Pimentel has estimated a significant net energy deficit.[27] Shell's project to extract petroleum from kerogen (or oil shale) deposits in Colorado, for example, is likely to reflect such a deficit.[28] It should be noted, in addition, that oil produced from kerogen not only risks a net energy deficit but yields greatly inflated CO_2 emissions, owing to the large amount of coal burned in the course of its extraction. A similar caveat applies to oil from tar sands where gas is used. The market is not fully sensitive to these drawbacks since the producer does not pay for CO_2 dumped in the atmosphere.

Mainstream economists have ignored or disparaged the appeal to entropy. In his review of Georgescu-Røegen's key text, *The Entropy Law and the Economic Process* (1971), the Michigan economist Robert Solo accused Georgescu-Røegen of peddling an "extreme form of Malthusianism," arguing that his entropic approach rendered him one of "the prophets of doom" who have "followed along behind Malthus" for nearly two centuries. For Solo, the second law is a faith-based dogma in exactly the same way as a belief in the Last Judgment. He repudiates the link between production, on the one hand, and depletion and waste on the other, and denies outright that resource diminution and pollution are related to population size and economic activity. For Solo, advanced economies can "eliminate all noxious wastes," and economic growth is what guarantees the advanced technology that will do it; industry does not necessarily have to produce waste at all and is not limited by the laws of thermodynamics. Solo is silent about the source of its raw material.[29] In short, he does not really engage with Georgescu-Røegen's argument—he simply says it is wrong. This kind of denial, supported by little argument, is the forerunner of the approach taken by many economists throughout the past thirty-five years in which raising questions about the limitations of the physical world is simply "not the right way to look at it."

Scale, Compound Growth, and Herman Daly's "Steady State"

If one regards the human economy as subsidiary to the natural world, rather than vice versa, the scale of the macroeconomy is necessarily circumscribed by the scale of the planet; consequently, endless growth in material extraction and physical waste cannot possibly be sustainable. Herman Daly argues that the physical world is indispensable to economic activity, supplying both the low-entropy inputs and the sinks for discharging high-entropy wastes. The scale of the economic enterprise must therefore be proportionate to the scale of the natural world.[30]

Because of the generous proportions of the original "empty earth," such limitations were not especially obvious while human production was negligible in relation to the global environment. Sources and sinks could reasonably be reckoned to be infinite in this situation[31]—or, if not quite infinite, capable of recovery, given rest. The human ecologist Sing Chew argues that some of the so-called "dark ages" in history represent eras of ecological recovery, redressing periods when overexploitation of natural resources had led to the collapse of populations or economies.[32] Recovery is not guaranteed, even with the passage of time. While the forests of Europe were cleared in pulses of economic expansion and probably grew back again several times,[33] parts of northern Africa, intensively cultivated for Roman consumption, and Mesopotamia, farmed until it was eventually rendered infertile by salt, never fully recovered. Worldwide, however, there has been little sign of a "dark age" since the early centuries of the last millennium.

The mathematics of compound (or exponential) growth indicates that the ongoing growth of any physical process will ultimately collide with physical limits, a key consideration of the *Limits* project. In *The Limits to Growth: The 30-Year Update* (2004), the MIT researchers devoted a chapter to it. The doubling period of compound growth can be calculated by dividing the annual rate of growth into 72. At 10 percent, for example, the volume will double every seven years or so (something like what has taken place in the Chinese economy since 1979). The *Update* recounts a number of folk legends illustrating the counterintuitive surprises built into compound growth. In the Persian legend, in which the king agrees to pay just one grain of rice on square one of the chess board and double the amount on each subsequent square, the effect of protracted doubling is not at all obvious in its early stages, when the base is small. When a trillion rice grains must be supplied at square forty-one, it becomes clear that the debt cannot be paid, confounding the apparently modest undertaking at square one.[34] According to the IMF, world GDP grew by 3.2 percent in 2012, and was predicted to grow by 3.3 percent in 2013.[35] Should growth continue at this rate, world GDP would double in about twenty years.

The discontinuity between a historically modest economy and the novelty of twentieth-century growth helps to explain the shortcomings of the classical economists' preanalytic framework, which underpins current neoclassical economics. Their original understanding of economic realities was established long before the immense bursts of economic growth that have characterized the past 140 years. Daly points out that "our information and control system (prices) assumes nonscarcity … of environmental source and sink functions."[36] Only since about 1870 has the rate of doubling become large enough in a planetary context to call this nonscarcity into question, and only since the end of World War II has the magnitude of the industrial base and of the human population yielded immense increments with each doubling.

According to Daly, economists' attitudes to scale involve a curious inconsistency. While microeconomics recognizes an optimal scale, where marginal costs equal marginal benefits, no such optimal scale is recognized for the macroeconomy: "There are no cost and benefit functions defined for growth in scale of the economy as a whole. It just doesn't matter how many people there are, or how much they each consume." Daly goes on to note the implication: "If the ecosystem can grow indefinitely then so can the aggregate economy. But, until the surface of the earth begins to grow at a rate equal to the rate of interest, one should not take this answer too seriously."[37]

Daly proposed what he called the "steady-state" economy as a solution to these problems of economic scale. The key elements of his model are "a constant stock of *physical* wealth (capital), and a constant stock of people (population) … maintained by a rate of inflow (birth, production) equal to the rate of outflow (death, consumption)"—which amounts to zero growth in both population and material production. Daly's second requirement is for the rate of flow through the system to be minimized, which means minimum replacement of both people and capital plant. Daly argues that human development does not need to cease when the material infrastructure of production is stabilized; he distinguishes between physical or quantitative growth and development, which involves the qualitative evolution of society, including technique, design, and culture. He argues that his system will preserve plenty of scope for development while curtailing the growth of all material flows.[38]

Daly's steady-state economy assumes a free market structure, which he regards as a good system for fulfilling allocation functions. He notes, however, that it fails to solve two critical problems: that of just distribution and that of optimal—or even sustainable—scale.[39] Such an economy, originally mooted as the "stationary state" by John Stuart Mill[40] before the really explosive growth began, may be a desirable plan for the conduct of future economic life—perhaps the only even remotely realistic one from a thermodynamic point of view—but Daly provides virtually no guidance on what, if any, social and political processes could precipitate a transition to such an outcome. Nor does he ask how industrial capitalism could be persuaded to accept such stricture and regulation.

For more than forty years, Daly has explored the inability of mainstream economics to come to grips with the physical dimensions of the real world, to think outside the growth paradigm on which it rests. Yet even though economic orthodoxy underwent significant change during that period, as the postwar synthesis of Keynesianism and neoclassical economics was supplanted by a new neoliberal framework, the expectation of unending economic growth has never subsided.

Modernity and Its Ideologies

The appeal of growth economics is rooted in the ideas surrounding modernity, with its emphasis on material progress and the individual. Modernity may, in fact, be the key driver of ecologically unsustainable practices, as the ecological economist Richard Norgaard argues.[41] Several key assumptions characterize the philosophy of modernity: unbridled optimism about "progress," the validity of "value-free facts," and belief in the preeminence of Western culture, which is expected to sweep cultural differences aside as people discover the effectiveness of a universalized rational approach. Indefinite material progress powered by improved technology is expected to guarantee that future generations will always be better off than their predecessors. Yet there is a circular logic implicit in recommending more industrial growth to solve the problems created by growth in the first place.[42]

The "progress" agenda relies on the epistemological premises and beliefs that developed out of the Scientific Revolution and are taken for granted in Western public discourse and institutions. Policy is developed one issue at a time with the aid of single-field experts, who are thought to be capable of "objective" assessment of "the facts." While this approach has been extremely effective in the design of industrial processes and manufacturing technologies, it is less appropriate to complex systems such as ecologies and cultures. Even though, using the techniques of objectivity and expertise, industrial technologies can yield maximum productivity per worker and smooth operation, their complex interactions with people and communities require an additional level of engagement. A new freeway built using state-of-the-art technology might be a fabulous road from a driver's point of view but its actual value depends on the full range of its environmental consequences over its life span and whether it serves the needs of its community better than alternatives such as mass transit.

A related premise of progress is the idea that a few universal principles govern everything and that Western rationalism has grasped them and continues to refine them. The path of the European-based nations to wealth and prosperity is taken to be the template for a universal human pathway. The development concept, outlined in President Harry Truman's inaugural address, imagined all countries on a single track, where front-runners like the United States would deploy "scientific advances and industrial progress" to help the "underdeveloped" expand their production and "catch up" to the West.[43] Individualistic values, though highly specific to European-derived cultures since the Scientific Revolution, are thought by the people of these cultures to be universal, transcultural truths.

Modernity: A New Symbiosis

Over the past ten millennia or so, from the beginnings of agriculture until the advent of modernity, human history developed a "complex maze of reciprocal causation between environment and culture," a "co-evolution," in Norgaard's words, something like the symbiotic relationship between species such as bees and flowering plants. The modern era progressively transformed this elaborate mosaic of "coevolved traditional farming systems" into a global economy based on large-scale technologies and market links.[44] Instead of growing well-adapted food plants for local consumption, farmers now produce commodities for world markets and must adapt their choices to that market rather than to the strengths and weaknesses of their soil and climate.

For thousands of years, humans in local settings worked at improving the capture of the (solar) energy in their local ecosystems by strategies such as planting strains adapted to specific niches and mixing symbiotic crops to take maximum advantage of the agroecological setting. Industrial farming methods, on the other hand, can succeed with little or no regard for local conditions as long as plentiful energy and inputs of fertilizer and pesticide can be deployed. The exploitation of fossil energy was the key innovation that permitted these great material changes, launching a new "coevolution" between humans and fossil fuels that has supplanted the original symbiosis. This change occurred in both capitalist and communist societies, which have shared the technocratic and cornucopian assumptions associated with progress.[45]

The destruction of local economies appears rational (and often inevitable) from the viewpoint of Western social philosophy, which in Norgaard's analysis "starts with the individual and leaps to the national and on to the global,"[46] ignoring any intermediate levels such as the region or the local community group. This leap carries consequences for all kinds of groups, from families, communities, and others concerned about their local region to indigenous and cultural minorities, and for all types of collective organizations—academic, cultural, environmental, or political.

Choices made by such social units, right up to the level of the province, state and nation, are understood in the philosophy of economic liberalism as hindering individual freedom and erecting barriers to trade.[47] The disputes of the World Trade Organization (WTO) are replete with cases where national preferences were disallowed on the grounds that they constituted unfair barriers to trade. For example, Australia's wish to maintain rigorous quarantine provisions for raw salmon was challenged through the dispute mechanisms of the WTO, as was the right of Europeans to reject and exclude growth-promoting hormones in their meat.[48]

Oddly enough, one major group entity that permeates modern society is never delegitimized under the liberal rubric: the corporation escapes definition as a group entity and is treated as a species of individual, a legal "person." Under US law, corporations have increasingly gained rights once limited to actual human beings. A corporate entity, however, does not represent an individual but the interests of a group—its shareholders, especially those with the controlling interest.

The modern focus on individual "choice" opposes individual interest to the common interest; social and environmental values expressed in intermediate group connections are negated, and the sense of common interest recognized in many traditional societies has withered. Norgaard, however, calls for development based on renewed coevolution, which embraces the rights of local people to determine their destinies in their local contexts and draws on the preindustrial model of communities in dynamic association with their local settings.[49]

Free Market Magic

When the World Bank's chief economist criticized Daly's approach to macroeconomic scale as simply "not the right way to look at it," he was in step with the mainstream economists' assumption that technological advance, alongside price signals that automatically attach to scarce items, will always allow for expansion and substitution. The "right way to look at it" assumes that physical limitations will always be surmounted by combining the fruits of human ingenuity with the magic of the free market. Economists very frequently claim that history "shows this to be true." For example, economist Carl Kaysen argues:

> Resources are properly measured in economic not physical terms. New land can be *created* by new investment, as when arid lands are irrigated, swamps drained, forests cleared. Similarly new mineral resources can be *created* by investment in exploration and discovery. These processes ... have been going on steadily throughout human history [emphasis mine].[50]

While claims of ever-expanding horizons and steady growth may be supported by the evidence of the past few centuries, the longer prospect reveals many dead ends and regressions.[51] Rapid change, whether social or technological, was unknown in pre-capitalist societies. Thus, while it is true that "new land" has been "created" throughout the history of human settlement—for example, as the vast boreal forests of Europe were cleared over several millennia and those of North America were subsequently cleared over a few centuries—and while land is still being "created" with the current clearing of the tropical forests of South America, Southeast Asia, and Africa over much shorter spans, it seems apposite to ask where the next wave of new land is to be created. Only tectonic activities are likely to create new land at this point. Economists like Kaysen stand squarely on a growth platform that recognizes no physical limitations. Investment is said to conjure any productive resource we may need.

Robert Solow argued that "if it is very easy to substitute other factors for natural resources ... the world can, in effect get along without natural resources.... At some finite cost production can be freed of dependence on exhaustible resources altogether."[52] Solow went on to suggest that the prospects for easy substitution are highly favorable, embracing the peculiar possibility that we don't really need natural resources.

The gulf between the assumptions of science and those of economics pervades the assessment of economic growth in the real world. In the next chapter, I turn to the scientists who challenged growth after World War II and the critical response to their work, largely from economists.

3

The Limits to Growth Debate: Precursors and Beginnings

Anybody who believes exponential growth can go on forever in a finite world is either a madman or an economist.

—Kenneth Boulding, quoted by John Steinhart, 1973

Dawning Perceptions of Ecological Crisis

From the end of World War II until the beginning of the seventies, world economic growth exceeded anything ever before seen, with particularly explosive growth in Europe. Earlier growth, even in the nineteenth century, stood still by comparison.[1] In the last decade of the twentieth century, the annual *increase* in world real GDP equaled something like the entire global economy of 1900, $1,000 billion in 1990 US dollars.[2] The world's economic output increased about twentyfold in the twentieth century and, according to the International Monetary Fund (IMF), "exceeded the cumulative total output for the preceding recorded human history."[3] It was the impact of this unprecedented economic expansion on the natural world that began to generate concerns in large sections of the scientific community, launching the twentieth-century debates about environmental quality and limits to growth.

Two biologists, Fairfield Osborn and William Vogt, had already warned of an incipient crisis, in 1948. In 1962, Rachel Carson's *Silent Spring* brought the impact of toxic compounds to public attention. A few years later, in 1966, Kenneth Boulding published his essay, "The Economics of the Coming Spaceship Earth," in which he argued that the "closed earth" of the near future would require an economics different from that of the "open earth" of the past, since the expansionary "frontier" stage of modern history was over. In 1968, Paul Ehrlich's *Population Bomb* warned that the biosphere would not be capable of supporting projected populations of the near future. At about the same time, Herman Daly and Nicholas Georgescu-Røegen began their critiques of an economic theory that was divorced from biophysical constraints. Ecology was becoming established as a branch of the natural sciences during that decade, and scholars from geography and the biological sciences in particular began to put forward the idea that there are obvious physical limits to economic growth.

The sixties saw the start of a legislative transformation in the United States, away from an existing emphasis on resource extraction and the privatization of national resources and toward a stewardship approach to public lands that would limit economic development in selected areas;[4] this shift in values was reflected in innovative laws to conserve land, water, rivers, and wilderness. On January 1, 1970, President Nixon signed into law the National Environmental Policy Act, which declared that "Congress recognize[s] the profound impact of man's activity on the interrelations of all components of the environment, particularly the profound influences of population growth."[5]

During the mid-1960s, the US Congress had also instigated a population policy designed to help finance family planning in the third world. Since population is one of the multipliers contributing to environmental impact (box 3.1), some sort of population policy is necessary. This does not, of course, justify coercive sterilization such as occurred in several countries at this time (including Bangladesh, India, and Indonesia) and which has even been criticized in the World Bank's own history of family planning.[6] The US initiatives of this period contrast with those of the Reagan and George W. Bush administrations, which pursued natalist policies that militated against population control of any kind. Marxist commentators[7] also dispute the need for population policies, arguing that birth control in the global south was merely a diversion intended to avoid solutions involving the redistribution of wealth, and especially land reform.

It was during the late 1960s that, among other disasters, the rubbish and oil on the Cuyahoga River caught fire in Cleveland, Ohio, and Lake Erie seemed in danger of biological death. Public opinion was shifting: in the United States, polls showed a sharp rise in public concern about the environment: membership in such groups as the Sierra Club and the

Audubon Society was on the rise, and media coverage of mounting environmental degradation increased.[8] In the UK, the BBC's 1969 Reith Lectures were given by the celebrated human ecologist Sir Frank Fraser Darling, who identified three key areas of incipient environmental crisis: pollution, damage to the services provided by nature, and the threat of uncontrolled population growth.[9] An avalanche of books appeared at that time dealing with the emerging concept of "the environment" from a variety of angles.[10] At the beginning of 1972, two books outlined an incipient general crisis in the relationship between humans and the natural world.

The editors of the UK journal the *Ecologist* devoted the entire February issue to their essay "A Blueprint for Survival," which was published in book form by Penguin later that year. They argued that governments, however reluctant they might be, had to address the growth of population, consumption, and economic activity and the intensity of the ecological impact involved—and take urgent steps to moderate it. They thought a complete social revolution would be needed, resting on decentralization of population and industry and the adoption of numerous strategies for "the invention, promotion and application of alternative technologies which are energy and materials conservative."[11]

The other book, *The Limits to Growth*, had meteoric popular success at its debut and is cited as the biggest-selling environmental book ever published. It was written by a team of scientists at the Massachusetts Institute of Technology who had been commissioned by the Club of Rome to report on the "predicament of mankind." The MIT team relied on the newly emerging field of systems analysis; their data were fed into and generated out of a computerized model called World3, designed to project into the future the complex relationships among multiple interacting trends. A central concern was exponential economic growth in a finite world, and the researchers identified five key areas of crisis: "accelerating industrialization, rapid population growth, widespread malnutrition, depletion of non-renewable resources, and a deteriorating environment."[12] The team used data observable since 1900 and aimed to model into the future the dynamic interrelationships of trends in that data. They incorporated extensive feedback between factors and the effect of time lags on the impacts delivered. Systems analysis opened an avenue to the modeling of nonlinear processes, a more complex

Box 3.1
Population and the Coalition of the Unwilling

It is hardly surprising that wealthy consumer economies would prefer a raw population emphasis—focusing questions about impacts onto poorer countries where populations were still growing rapidly—to a consumption emphasis that not only challenges the level of affluence enjoyed (and expected) by their people but could challenge the very economic growth that has created it.

Marxist critics such as John Bellamy Foster and Eric Ross show that first world countries preferred population control measures to actual development assistance in numerous parts of the third world. From their point of view, the developed world seized upon population control in preference to alternatives that were far less palatable to the interests of US business. The adoption of what was called "birth control" had two main strengths from a business perspective. The various avenues—from coercive sterilization and the provision of devices of dubious safety to supplying women with methods they were actually requesting—were all far cheaper than wholesale development aid would have been. Perhaps more telling still, a birth control program was both ideologically and practically preferable to land reform. At the end of World War II, many parts of the hitherto colonized world aimed for independence. The regimes of property ownership that existed in these ex-colonies had been influenced by the European colonizers, who had often expropriated land, concentrating ownership in the hands of local elites and foreign corporations. The poverty of millions was intimately linked to their lack of access to land, and the case for redistribution of the land was compelling. Focusing on population no doubt assisted in diverting attention from land reform and toward birth control under the rubric, "Too many people, not enough land."

The growth of the human population is, however, far from irrelevant to questions of poverty and environment. In 1960 the world population reached three billion and was set to double in the next thirty-five to forty years. Though the annual rate of growth eased from a peak of 2.1 percent in the mid-1960s to 1.2 percent in 2006, the gross population grew rapidly. An average of 79.3 million people was still being added annually in the first decade of the new century, a bigger annual gain than that between 1960 and 1975, when the rate of growth was near its peak.[a]

Numerous forces worked against population control remaining the key agenda item it had been until the late seventies, a "coalition of the unwilling" in the words of the UK's then chief scientist, Lord Robert May.[b] The Vatican, which had always opposed contraception, launched a vigorous counterattack against oral contraceptives, exerting substantial influence not only in the United States but also at the UN population conferences that took place in Mexico City (1984) and Cairo (1994). Evangelical Christians, influential in the Reagan and George H. W. Bush administrations, joined forces with Catholics to push US policy at home

Box 3.1 (continued)

and abroad in the same natalist direction, a trend that intensified under George W. Bush. The Saudi Arabians also supported this approach. In Mexico, the United States withdrew its funding for the UN population program, declared that the advance of free market economies was "the natural mechanism for slowing population growth," and announced the sharpening of a "family values" focus.[c] Through the same two decades, the rise of the women's movement brought to the fore issues of women's right to control their own bodies and shifted the focus of population policy away from environmental implications and toward empowerment for women. While an excellent thing in itself, this is not a complete substitute for awareness of the impact of ever-increasing numbers of people on the natural world. The work of Foster and Ross exemplifies how the Left, too, was disinclined to see population as relevant to environmental degradation, arguing that the redistribution of resources would solve problems attributed to overpopulation.

The case of Thailand demonstrates that family planning can be both noncoercive and beneficial. In 1975, Thailand and the Philippines had a similar population size, with a high growth rate, a high fertility rate, and a high proportion of people living below the poverty line. Thailand's GDP was slightly smaller. Thailand's annual rate of population growth fell from 3.2 percent in the early 1970s to 0.06 percent in 2010, largely as a result of its family planning program. The Thai campaign has had the advantages of sustained government support, the work of the inspired campaigner Mechai Viravaidya, and a Buddhist culture that does not forbid contraception. By 2008, Thailand's population was 25 million less than that of the Philippines and its per capita GDP was more than twice that of the Philippines.[d]

Though a tricky area, population is one key driver of environmental impact and cannot be excluded from an examination of the dynamics of unfettered growth. Clearly, the rate of consumption, or "affluence," is an equally significant determinant of impact, and it is often argued that the level of technological sophistication is a third key factor that can reduce impact by minimizing the amount of pollution generated per product made. Ehrlich and Holdren expressed the relationship mathematically in their famous I = PAT formula (Impact = Population × Affluence × Technology).[e] Rather than argue that population itself is not the main issue, it is preferable to acknowledge all aspects of the human impact. Australians and Americans, for example, have a per capita impact many times greater than the world average and as much as seventy or eighty times that of a Bangladeshi;[f] each extra Australian or US birth should be seen in that light.

(continued)

Box 3.1 (continued)

Notes

a. Worldwatch Institute 2006, 75; 2011, 88.
b. May 2007.
c. Kraft 1994, 632.
d. Bello 2011; Bristol 2008.
e. Ehrlich and Holdren 1971.
f. McNeill 2001, 16.

undertaking than the short-term econometric modeling routinely done by economists.

The Limits to Growth was a brief and accessible summary of the team's findings, which they hoped would launch a wide-ranging debate not only among scientists but among governments and people in general. Though their work was ultimately characterized as a "doomsday" scenario by most economists and much of the popular press, their message was more optimistic than that. They warned that unimpeded economic growth would very likely collide with planetary limits within the next one hundred years (by about 2070), but counseled that action could be taken to moderate the impact, and that the earlier this was done, the better the prospects were for avoiding catastrophic decline.

By 1972 the United Nations had also begun to engage with perceived environmental dangers. In 1970 the UN General Assembly voted to convene the first United Nations Conference on the Human Environment and appointed the Canadian geographer Maurice Strong to head it; the gathering was held in Stockholm in June 1972. This event was the forerunner of the Brundtland Commission (1983–1987) and the Rio Earth Summit (1992), as well as a host of regular conferences and specific studies dealing with areas such as water, desertification, and renewable sources of energy. It also gave rise to the United Nations Environment Programme (UNEP) which later became the cosponsor of the International Panel on Climate Change (IPCC) and marshalled scientists from across the world to produce regular reports on the state of the natural world. UNEP is responsible for the *Global Environmental Outlook* (GEO) reports, of which the fourth, GEO-4, was released in October

2007 and the fifth, GEO-5, coincided with the Rio+20 conference in 2012. GEO-4 expressed warnings uncannily similar to those of the Club of Rome, the *Ecologist,* and allied scholars thirty-five years earlier.[13]

During a pre-Stockholm meeting of the Royal Geographical Society in April 1972, the British government's resource expert, Ralph Verney, suggested that it was no longer desirable "to devote our resources to the achievement of the highest possible growth rate."[14] At the same meeting, Strong told British geographers he welcomed the debate brought on by *Blueprint* and *Limits*:

> We still may not know where the limits are, but we know that there are limits to the scale of the present human population and to the scale of its interventions in the natural system.... In the final analysis, any viable solution will be contingent upon bold new steps to bring about vastly improved conditions of life for all people.[15]

Strong's comment reflected his awareness of two crucial aspects of the emerging environmental problem: that natural limits apply to the scale of economic activity and that, notwithstanding such limits, latitude had to be found to ameliorate the grim situations hundreds of millions of people still faced. For Strong, this task would be tantamount to "rational global management of the finite resources of the earth" on behalf of all people.

The broad direction of Strong's approach is worth noting, since the concept of international management was to become a major focus of conflict over what (if anything) needed to be done. Some critics rejected international managers—and even the UN as a whole—because of perceived infringements on national sovereignty and checks on corporate independence; others expressed an opposite fear, that international managers might serve as proxies for unrestrained corporate power.[16] Some also felt the idea of technical experts managing the resources and geophysical processes of the planet was a fantasy arising from runaway hubris. However apt Strong's twin focus on environment and poverty, his solution sounded "technocratic," and it certainly ignored the kind of free market solution that was to gain ground in the 1980s. Strong hoped that the governments assembled for the Stockholm Conference would make decisions about the allocation and distribution needed to conserve and share the resources of the earth and would establish and fund a UN branch to carry them out.[17]

Shortly after his inauguration in early 1977, President Jimmy Carter commissioned the *Global 2000 Report to the President*, which directed the US Council on Environmental Quality and the State Department to liaise with other key government agencies in studying "the probable changes in the world's population, natural resources and environment through the end of the century."[18] Canada's Pierre Trudeau soon sponsored two parallel reports focused on Canada's future.[19]

These initiatives—and those of the UN—demonstrate that the issues raised by scientists and ecological economists were taken seriously among crucial world leaders during the seventies. But by the time President Carter's *Global 2000* was published, in 1980, enthusiasm at this level was waning, and Ronald Reagan was soon to be elected to the US presidency. The 1970s mark a great shift in the ideological framework that buttressed first world national policies and global institutions. From then on, neoliberalism gained ground, bringing with it a crusade against regulation and an exalting of market forces over human agency, both of which were to militate against the ongoing critique of the growth economy (chapter 6). The neoliberals believed that growth would flow from market freedom and that this was the best way to solve pollution and ecological degradation, even though these were themselves the result of prior growth. Growth slowed in this era, compared to the postwar period, but its pursuit remained the dominant paradigm guiding economic policy almost everywhere on earth.

Precursors to the Limits Debate: Discourses of Scarcity

Mainstream and ecological economists diverge sharply in defining scarcity, along the fault lines of the assumptions described in chapter 2. For the neoclassical mainstream, scarcity is always relative, always reparable. Investment, technology, and the operation of the price system are believed to provide substitution for every scarcity. Physical scientists and ecological economists suggest that this account ignores the ultimately finite character of planet Earth, as well as the essential role of energy in all forms of extraction and production; in their framework, *absolute* scarcity encroaches as essential natural resources—especially energy resources—are dismantled.

Until the recent past, almost all human societies suffered food shortages, intermittent in some luckier places and times, frequent in others. Even now this condition has been transcended only by privileged sectors of the global population and still haunts vast numbers of people. It has been argued that hunting and gathering modes of life were less vulnerable than the settled and ultimately stratified and urbanized communities that proliferated over the last 10,000 years.[20] Once communities began to depend on farming, many contingencies could trigger famine—climate shifts, annual weather variations, and the unintended consequences of human efforts to mitigate the problem in the first place. In all stratified societies, including contemporary human society, the consequences of insufficient food production or insufficient food available at affordable prices fall unevenly on people of different rank, class, and gender. The inequalities in the current world economic system reflect this truism just as, though most were affected in the recurrent famines of medieval times, it was the poor who starved.[21] Today, women remain disproportionately disadvantaged. Women make up two-thirds of the world's poor, two-thirds of the world's hungry, and two-thirds of the world's illiterate.[22]

Malthus: Scarcity as a Political Tool

Though Thomas Malthus (1766–1834) was not the first economist to dwell on the likelihood of recurrent food crises, he was certainly the most influential in the Western post-Enlightenment tradition. Malthus lived at the beginning of the industrial miracle, which might have delivered all of humanity from chronic food insecurity had it been differently developed. In the event, after two centuries of an apparently endless upward trend, biologists began warning in the late 1940s that the miracle itself was destroying its natural base and might collapse as a result. "Malthusianism" is an epithet that has been commonly applied to the postwar thinkers who have pointed to the limits of the industrial capitalist expansion of the last 250 years—a description that is not always apt.

Malthus advanced the idea of limits in the late eighteenth century on the basis that population was growing faster than food production. Though his status as a household word flows from this population theory, Malthus's stated agenda was political: to rebut utopian ideas about "the future improvement of society," ideas in the ascendancy since the recent revolution in France. His examination of the mathematics of population growth was, at least initially, incidental to this argument. His broader objective was reflected in the title of the first version of his essay, published in 1798: *An Essay on the Principle of Population as it Effects the Future Improvement of Society*.[23] Malthus's interest in the unfettered growth of human populations arose in relation to another of his key contentions: that social improvement was both impossible and undesirable since more egalitarian social arrangements would only encourage the poor to breed more rapidly and outgrow the food supply—to their own detriment, he asserted, as well as that of the rich.

Malthus did not tackle general limits to economic growth. He lived before the really big surges of industrial expansion that began in the nineteenth century and made no critique of industrial growth, resource depletion, or environmental degradation. The limits he invoked were agricultural limits, and he was unaware of what petroleum would one day do for crop yields. His quasi-scientific theory held that populations increase geometrically while the food supply expands merely arithmetically, and that this imbalance, a putative "law of nature," required curbs on the breeding of the poor. Malthus invoked the population argument to justify a punitive workhouse-centered structure for welfare. The campaign he supported led to the Poor Law of 1834, which placed draconian restrictions on the provision of relief.[24] Malthus's ideas and methods bear little resemblance to the investigations of the *Limits to Growth* researchers, or to the work of most of those who have warned about environmental degradation.

Several notorious passages give insight into the beliefs underlying Malthus's outlook:

> A man who is born into a world already possessed, if he cannot get subsistence from his parents ... and if the society do not want his labour, has no claim of right to the smallest portion of food.... At nature's mighty feast there is no vacant cover for him.[25]

His approach to unwanted children, whether illegitimate or not, was similarly implacable. He would deny all assistance to any child whose parents could not provide support.[26] The costs of assisting the indigent could be avoided by allowing them and their children to starve—for their own and everyone else's good. He also favored separating small farmers from their tenuous but self-supported living so that the land could be properly exploited, advocating that a "great part" of the subsistence potato farmers of Ireland needed to be "swept from the soil into large manufacturing and commercial towns." Not only was he eager to dispossess the Irish peasantry, he was clearly unconcerned about their subsequent fate—as he also reported that the demand for labor was weak.[27] A keen proponent of agricultural enclosure and industrial expansion, Malthus was squarely allied with those who had a seat at the feast. His ruminations on population flowed from these preoccupations.

To be "Malthusian" (i.e., like Malthus) thus involves several key elements over and above a fear of exploding population, and these are more central to his ideas than the population theme. They include resistance to notions of social improvement and social welfare, punitive policies for the poor, a tendency to blame the poor for their own plight, and recourse to speculative theorizing in the service of an essentially political argument. Malthus's population theory was closely associated with his callous and elitist political views, and the label was happily adopted by some American eugenics organizations in the nineteenth and early twentieth centuries; for these reasons, the term "Malthusian" is often used pejoratively. The expression works as a shorthand not simply to denote fear of exponential population growth but also as a label for those who appeal to questionable science, are empirically wrong about limits, favor the rich and powerful over the poor, or are grotesquely inhumane. Some of the scientists who began raising the alarm about ecological damage from 1946 on were vulnerable to parts of that description, but for most it doesn't really fit. Few could be accused of poor science, few were advancing any explicit political agenda, and all had concerns that went well beyond population or its control.

Osborn and Vogt and Their Critics

Fairfield Osborn and William Vogt, both biologists, issued the first in a long line of postwar science-inspired warnings. Both argued that free enterprise and the profit motive were inimical to the natural world and, though they did not much use the word "capitalism," they wanted to see extensive regulation of "free enterprise" in the interests of all citizens.

While concerned about escalating rates of population growth, Osborn's big worry was the degradation of the soil—in cropland as well as forests and water catchments. The soil was a living thing, he argued, not ultimately amenable to cure by chemistry: "the earth is not a gadget." Osborn also pointed to what Marx had earlier called metabolic rift—the "steady movement of organic material to towns and industrial centers—there to be consumed or disposed of as waste but never to go back to the land of origin." Instead, it poisoned rivers and ran to waste in the ocean.[28] The sources of human life were being choked by what Osborn described as "vast industrial systems" proliferating at frightening speed. One can only guess at what he might have said about the situation sixty-five years later.

Long before more recent researchers such as Joseph Tainter and Jared Diamond, Osborn reviewed historical agricultural disasters and declines, and concluded that the main causes of these failures lay in the headlong destruction of forests, the overgrazing of grasslands, and the pressure that was exerted on the land when cash crops were grown for export. He was a forerunner of the emphasis on sustainability and thought that renewable resources were the property of all the people and should be deployed for the benefit of all, through government planning and regulation. In pleading for the protection of the public lands of the western United States, he described as theft the depredations of interest groups such as cattlemen and lumbermen.[29]

Vogt's position was broadly similar, though his use of words like "backward," "ignorant," "illiterate," "hordes," and "Japs" conveyed an explicit belief in white racial superiority. He regarded medical improvements as a tragedy that would only perpetuate the tribulations of the poor by allowing more of them to survive. However, he explicitly called on Europe and North America to curb their populations as well. He described Europe as a long-term parasite on the rest of the world—it had not been able to feed its own for centuries and had survived only because of its extractive relationship to its colonies. He too was deeply worried by soil erosion and the disruption of water supplies by clearing, overgrazing, and overpumping. Like Osborn, Vogt pointed to the irrational "system of sanitation which every year sends millions of tons of mineral wealth and organic matter … to be lost in the sea."[30]

Vogt laid much of the blame for this decline at the door of the "free enterprise system" and its economists, who did not include "the highly vulnerable biotic potential" in their concept of capital. He argued that

America had been living on its resource capital since 1607 and warned that resources are renewable only if they are managed on a sustained-yield basis. The pursuit of profit, he thought, encouraged profligate practices. Vogt also saw the extractive colonial system ruining the land of Africans, who were, he argued, being forced to produce crops for export. Though he did not use the term, Vogt pointed to the immense ecological footprint of European cities.[31]

The responses of economists varied from utter scorn to a plea for economics to embrace and deal with the issues raised, a pattern that would be duplicated when *Limits to Growth* was published twenty-five years later. *Time* magazine panned Vogt's book as "neo-Malthusian propaganda," arguing that soil can easily be "created." The Stanford University agricultural economist Karl Brandt was also contemptuous, calling the book "deliberate propaganda," describing it as "bad," "immoral," "distasteful" and "irresponsible," and arguing that Vogt had no understanding of the "basic issues of the creation of wealth." The resource economist Joseph Fisher, on the other hand, while critical of Vogt's "purple prose" and "sweeping generalizations," accepted the existence of "fundamental dangers" and pointed to the need for economic theory to cover resource conservation more adequately and to look at the economic advantages conservation might bring.[32] Fisher went on to head the think tank Resources for the Future, which explored the policy implications of these issues and was involved in the development of environmental economics as a distinct field. Environmental economists argue that negative externalities (unaccounted costs) can and should be integrated into mainstream economic thinking through various techniques for pricing nature into the economy.

While *Time* magazine and Brandt attacked Osborn and Vogt for calling into question free enterprise economics and the efficacy of scientific progress, equally vehement condemnation has also come from the Left, where Osborn and Vogt have been seen as neo-Malthusians and ideologists for strategies of international domination by Western capital, such as the emphasis on population control instead of land reform and the US-based Green Revolution initiatives that commercialized third world farmers' lands (box 3.2).[33] Marxist anthropologist Eric Ross argued that, as the taint of Nazism drove eugenic ideas underground,

Box 3.2
The Green Revolution: Ambiguities

The Green Revolution of the fifties and sixties commercialized land in many third world countries. Though successful at increasing output per worker (and total output of selected grains) and at connecting third world agriculture to the world market, the Green Revolution did not always provide actual food to the hungry. Soybean production expanded in Brazil at the expense of subsistence crops of black bean and rice, a process paralleled in India, where per capita availability of the coarse grains and pulses eaten by the poor was halved between 1956 and 1987.[a] Where agriculture serves export markets and access to food depends on access to money, overflowing granaries can occur alongside famine. The transition from labor-intensive subsistence farming to energy-intensive crops for sale substituted market risks for the age-old risks associated with the weather.

The Green Revolution's high-yielding varieties of grains required more fertilizer and pesticide and much more water than traditional crops, imperatives that undermined the viability of small peasants, who had no cash for such inputs. As farms were consolidated and cash crops planted, millions of smallholders were severed from subsistence livelihoods. In the Philippines, for example, by the 1970s, 55 percent of the entire farming acreage was devoted to export crops such as sugar, rubber, and coffee, while fertile land in Colombia was turned over to growing carnations for the United States, earning a million pesos a year, compared to only 12,500 for corn. These profitable enterprises were usually owned and controlled not by small farmers but by foreign agribusiness and local landowning elites.[b]

In India, traditional tanks, which had harvested the monsoon rains and provided villages with water for thousands of years, fell into disrepair, supplanted by electric and petrol-driven pumps. The hand pumps of poorer farmers soon failed to reach the falling water table. Aquifers throughout northwest and southern India declined rapidly and continue to fall every year, at rates of a meter or more; by now, many are saline or overexploited, and even electric pumps cannot tap usable water in many districts. The aquifers of China underwent similar depletion, with annual falls of three meters and more reported for Hebei Province on the North China Plain in 2001.[c] The Green Revolution shifted agriculture toward the industrial model, establishing monocultures of water-hungry crops, and mined the groundwater, much of which will not be replenished during the lifetimes of the living.

Box 3.2 (continued)

Notes

a. George 1976, 93; Douthwaite 1999, 250.

b. George 1976, 172. Ownership of land in third world countries is often heavily concentrated in elite hands, partly an effect of colonialism. In 1996, for example, 2 percent of Guatemala's people held 63 percent of the good land, and 0.8 percent of Brazil's owned 43 percent (Athanasiou 1996, 54). The US invasion of Guatemala in 1954 was undertaken to protect unused lands owned by the United Fruit Company which the government wished to redistribute to peasant farmers. This action was taken despite the provision of due compensation (Greer and Singh 2000), an indication of the level of resistance to land reform.

c. Shiva 2002, 14; Sengupta 2006; Brown 2008, 70.

"the principal vehicle for Malthusian fears became, instead, the threat of environmental catastrophe."[34] Osborn's book, he claimed, heralded this shift.

Hardin and Ehrlich—"Neo-Malthusians"?

Among the concerned biologists of the late 1960s were some who put forward extreme policies to combat ecological decline, policies that were destined to alienate nearly everyone across the political spectrum. Though not identical to the prescriptions of Malthus around 1800, there were similarities, perhaps the most striking being the fact that members of the privileged classes of the world were offering solutions exhibiting considerable indifference to the actual real-world fate of the rest.

Garrett Hardin published his enthusiastically reprinted "The Tragedy of the Commons" in 1968 and his notorious "lifeboat" essays in 1974; Paul Ehrlich's *Population Bomb* also came out in 1968. Both Hardin and Ehrlich were biologists, both took more overtly political stances than most of their scientific colleagues, and both had a lot to say about population. Ehrlich's gloom about the immediate future left his predictions (which turned out to be wrong, at least in their timing) open to "doomsday" labeling.

Hardin's so-called tragedy of the commons lies in the notion that, when many individuals compete for their own personal gain in the use of an unmanaged common resource, overuse will inevitably result: "freedom in a commons brings ruin to all."[35] The commons—or common property—of human societies has varied with the nature of the society. Before settled communities evolved, virtually all natural resources, including the land and its fruits, were the common property of a resident tribal group. In these societies, land could not be bought or sold, and people belonged to the land rather than vice versa. Even after the transition to tributary and feudal societies, most peasants had traditional access to large tracts of land. In Europe, it was the successive waves of enclosure that transferred these commons to private hands and deprived Europe's peasants of essential elements of their subsistence, such as pasture, game, and firewood. Indeed, enclosure has been one very prominent aspect of the transformation of property ownership during the history of the past five centuries or so. In 2014 the remaining commons are those that have been difficult or impossible to enclose and privatize. World oceanic fish stocks are an example of this kind of commons, as is the earth's atmosphere—into which carbon dioxide (among other pollutants) has been poured at will, without any check or price paid. Robert F. Kennedy Jr. has argued that the pollution of the atmosphere is an example of de facto privatization, representing the theft of these commons by corporations, especially those operating coal-fired power stations.[36]

Quite erroneously, Hardin used the extensive feudal commons as the template for his tragedy. These were not, in fact, an instance of unmanaged commons subject to overexploitation but a common resource controlled by community mores and traditions and in practice not vulnerable to ruination by overexploitation.[37] Hardin's elision of historical common property with the modern global commons brought into play by postwar capitalism (fish stocks, atmospheric absorption, and so on) served to suggest that human nature itself, assumed to be governed by human greed, was the underlying problem and one consistent across human societies for centuries. The work of historians of the medieval commons and that of Vandana Shiva[38] in relation to traditional Indian irrigation systems—where water was regulated and conserved at the level of the village—shows that humans have been and remain capable of social and political arrangements that protect their shared resources. Elinor Ostrom, who shared the 2009 Nobel Prize in Economics, spent much of her career studying the management of what she termed common pool resources. Ostrom showed that even though cooperative management is not easy and success is not guaranteed, it is still often achieved in practice, through local negotiation and organization.[39] The perception expressed by both Osborn and Vogt—that free enterprise competition in the pursuit of profit has been the main source of extreme pressure on common resources—seems far closer to the actual historical tragedy than Hardin's medieval tale.

Hardin's argument has, however, served as ammunition for those economists who believe that all commons should be privatized to the maximum extent possible. In this view, private individuals should own and control the commons, because private ownership will supply an incentive to use the resource wisely and avoid the "tragedy" of overuse. Under this logic, popular with the emerging neoliberal ideology, private ownership should be extended to water supplies, rivers, forests, and everything for which ownership can possibly be designated and enforced.

In the "lifeboat" essays of 1974, Hardin went further and proposed that rescuing the drowning multitudes in the metaphorical lifeboats of the industrial world might afford "complete justice," but would also lead to a "complete catastrophe" in which everyone would drown (i.e., starve). Hardin thought food aid only encouraged the poor of the global south to breed, similar to Malthus on the subject of aid to the poor in Britain around 1800. While he acknowledged colonial expropriation, he aimed to draw a line in the sand, resisting the notion of redressing past wrongs. Like Malthus, he thought that "we cannot safely divide the wealth equitably between all present peoples, so long as people reproduce at different rates."[40] In saying this, he ignored the question of the differential consumption rates of the "drowning multitudes" of the third world and the lucky people who already occupied the lifeboats.

Ehrlich's *Population Bomb* (1968) examined ecological deterioration, the pollution of air and water, and the depletion of nonrenewable resources, as well as population growth, but emphasized population control as the most pressing requirement to relieve the pressure on the environment. Ehrlich announced that food production could not keep pace with population and that "the battle to feed humanity" had been lost—a claim that focused on the idea of "too many people" and overlooked the relationship between people and their access to land, which might afford them a livelihood. Though his prediction turned out to be somewhat premature, he recently pointed out, in his own defense, that any of the 200 million or so people who have starved to death in the interim would be entitled to argue that the battle had indeed been lost.[41]

At home in the United States, Ehrlich recommended population control—by compulsion if voluntary methods failed. Once successful at home, he proposed "triage" for the rest of the world: the United States would help countries that it judged to be capable of reaching "self-sufficiency" with a program of birth control, agricultural intervention, and industrialization in suitable cases. The rest of the world (which he thought might include India) should be abandoned.[42] In some cases, local insurrection and the secession of more promising parts of the triaged countries would be welcomed and even assisted. Ehrlich's agricultural strategy hinged on training local residents in developed world agricultural methods and introducing these to the poor. Ehrlich did not mention the Green Revolution but advocated its methods—which relied on significant inputs of fertilizer, pesticides, and machinery, largely dependent on petroleum, as well as on far more water than traditional methods had required. Such off-farm inputs had already supplanted human labor and turned first world farming over to corporate control during the twentieth century. While the Green Revolution did increase gross food production, it also had disastrous effects on groundwater, soil fertility, and the viability of small farmers.

Neither Hardin nor Ehrlich shrank from advocating coercion; their politics, which influenced the agenda and values of the early environment movement, has been described as "undemocratic, authoritarian, pessimistic, repressive, illiberal, static and closed."[43] Hardin certainly merited much of this reproachful description—as well as the "Malthusian" tag. The Ehrlich of that era, whose triage recommendations sounded similar, did not, however, believe that the rich should make their getaway in the boats and leave the rest to drown. Subsequently, he revised and reassessed his findings. Moreover, the egregious policies spelled out by Hardin in particular were not broadly generalized throughout the work of other scientists, especially in the work commissioned by the Club of Rome, to which I now turn.

4

The Limits to Growth and Its Critics

The latest wave of environmentalism may turn out to be a fad.... It may also be a result of the first glimmerings of human understanding about total systems and the first human perception of the worldwide negative impact of man's activities on the ecosystem.

—Donella Meadows, Dennis Meadows, Jørgen Randers, and William Behrens III, 1973

Soon after the Club of Rome was founded in 1968, it instigated the *Limits to Growth* project. The Club of Rome was an exclusive think tank, the creation of the Italian industrialist Aurelio Peccei in association with the Scottish scientist Alexander King. They brought together a select group of prominent, mostly wealthy individuals who wanted to address what they called the *problematique*, translated as "the predicament of mankind": how could growing populations, locked into ever-expanding industrialization, avoid immense (if not terminal) environmental degradation, exhaustion of the resources on which everything depended, and the social chaos that would be likely to follow? They were primarily intellectuals, businessmen, and bureaucrats grappling with the problems of a world that had just been through two whirlwind decades of unprecedented economic growth and was apparently on the verge of more. Peccei described himself as "perplexed and worried by the orderless torrential character of this precipitous human progress."[1] They were conscious of the warnings of scientists that all was not well with the natural environment, and skeptical about indefinite expansion of industrial production as the template of economic development for the world's burgeoning populations. Above all, they wanted to stimulate a much longer-term perspective—something little seen in policymaking anywhere.

For this, they aimed to instigate public awareness of these questions and to advance public debate.

In the UK, the editors of the *Ecologist* took up these same questions in their first issue of 1972, with the essay soon republished in book form as *A Blueprint for Survival*.[2] They drew on the work of ecological economists, scientists who were worried about environmental decline, and the *Limits to Growth* team itself.

Criticism of *The Limits to Growth* was multifaceted. It ranged from the accusation that the Club of Rome was a covert agent of international capital, across questions about the nature and validity of the model, to acknowledgment of the need to address very real problems. Much of it, however, especially from economists, was unmitigated abuse and denigration, often mixed with outright denial of the existence of the problem under consideration; much of it also asserted an unwavering faith in economic growth and indefinite material progress.

Rhetorical Attacks

When *Limits to Growth* was published, to considerable fanfare and advance publicity, responses from scholars in the sciences were generally favorable. Scientists also aligned themselves with *A Blueprint for Survival*. The UK edition began with an explicit statement of support from thirty prominent scientists across the natural, physical, and medical sciences from several countries; a further 150 signed a supporting letter to the London *Times* shortly after publication.[3]

Mainstream economists, on the other hand, were unimpressed, so much so that "Club of Rome" became an epithet of scorn—and remains so to the present, especially within economics. Even at the outset, when both the public and many institutions in the developed world were relatively open to these ideas, economists went on the attack. One explanation for their reflexive hostility came from Robert Gillette of *Science* magazine, who covered the star-studded launch of *Limits* at the Smithsonian Institution. A critic of the "messianic impulse" he attributed to the MIT team, Gillette argued that the "assumption of inevitable economic growth" constitutes "the very foundation" of the profession of economics,[4] perhaps explaining some of the intensity of the assault from that quarter. Curiously, in light of its academic context, the criticism from economists was characterized by several lines of attack that fall under the general banner of tirade rather than argument. The most prominent of these were simple ridicule and accusations of both catastrophism and Malthusianism.

Wilfred Beckerman, who had just been appointed to the political economy chair at University College London, adopted a derisory attitude from the start, dubbing *Limits* "brazen impudent nonsense" and a "ludicrous study." He also quoted fellow economists calling it "spurious scholarship" and "computerized mumbo-jumbo" and claimed that "since the natural scientist is not concerned with human beings, the relationships that he studies are not of the kind that are amenable to human policies."[5] His dominant tone was mocking and disparaging, a tone that persisted through his subsequent decades-long defense of the neoclassical orthodoxies. Beckerman published his 1974 celebration of economic growth in the United States under the title *Two Cheers for the Affluent Society: A Spirited Defense of Economic Growth*.[6] He went on to debate Daly and others in academic journals over concepts of sustainability and published a further withering assessment of environmentalists' ideas in 1995 (*Small Is Stupid*). Beckerman was not the only academic who resorted to ad hominem attacks rather than rational critique. For example, John Koehler, then at the Rand Corporation, suggested "fraud," "fantasy," and "fudge" were involved, and the sociologist Robert McGinnis described *Limits* as fostering "that faddish hysteria which passes as concern for The Ecology."[7]

In the same vein, many economists applied the "doomsday" label. Mention of Doom, Doomsters, Doomsayers, Ecodoom, Hysteria, Catastrophism, Disaster, End of the World, Chiliasts, or Apocalyptics (usually capitalized) appeared in a great number of articles on the subject, mainly from economists.[8] The celebrated British physicist John Maddox was one of the few scientists who adopted this terminology; he called his polemic *The Doomsday Syndrome*. The University of Sussex's Science Policy Research Unit published a book of essays by economists and others, including a few scientists; this book too was peppered with "doom" and its cognates, and in the United States was titled *Models of Doom*.[9] So persuasive was this avalanche of "doomster" terminology that even ecologically aware analysts have adopted the doom/doomsday shorthand to refer to the *Limits* warnings of the early 1970s.[10]

"Doomsday" originated as a Christian concept referring to the Last Judgment at the end of the world, an idea based on faith rather than on science. Discourse about doom and apocalypse is inappropriate from the point of view of academic or scientific values and in any case suggests derangement, religious obsession, and irrationality. It implies that the main thrust of the researchers is the circulation of prophecies designed to inculcate fear rather than a serious attempt to explore the likely consequences of continued economic growth. This kind of rhetoric has been effective in blunting the impact of the critique of growth and remains popular with think tanks and economists today.

The epithet "Malthusian" (sometimes "neo-Malthusian") was applied to the MIT team in a manner similar to how it was used against Osborn and Vogt. The accusation that the *Limits* authors were merely repeating the ideas of Malthus acted as another easy shorthand whereby concerns about possible limits could be dismissed without further argument.[11] Again, it was not only the neoclassical economists who dealt out the Malthusian label; many Marxists were equally skeptical about the idea of limitations on human "progress" (box 4.1).

Box 4.1
Marxists and Growth

Despite the exploration of what he called "metabolic rift" in his later work, Marx himself was firmly embedded in the mid-nineteenth-century "material progress" ideology of industrial capitalism and did not foresee any serious limits to economic growth. Following this trajectory, most Marxists embraced the promise of scientific and technological advance fully and saw no constraints on industrialization throughout the world—at least under socialism. Concerns about population growth, environmental damage, and resource depletion were sometimes seen as symptoms of a reactionary collusion with the ruling class and as distractions from the real issues of poverty and imperialism. The "ecology movement" was thought to embody the misguided foolishness of "Malthusian pessimists."[a] Although some Marxists were less dismissive, acknowledging that modern capitalism "threatens all the natural bases of human life," most insisted that a radical redistribution of wealth and resources under socialism was the crucial requirement.[b]

Marxists have been especially dismissive of the case for population policy, an attitude that continues in much ecosocialist thinking today.[c] Some socialists, however, have sought to reconcile their Marxism with concern for ecological decline. The Welsh critic Raymond Williams noted

Box 4.1 (continued)

the "triumphalist arguments about production" embraced by many socialists, especially in North America. Williams accepted what he called the "fact of material limits" and called for negotiated reductions in first world consumption so that redistributive justice could be ecologically accommodated.[d]

Meadows and colleagues never suggested that population was the paramount issue; they saw it as one of several vital problems and were fully aware and deeply disturbed that existing patterns of economic growth were not lowering levels of poverty. They argued that a stationary state economy with a stabilized population would give humanity a better chance of providing for the basic needs of all.[e] Though they did not specify how such redistribution might be achieved, they regarded freedom from hunger and poverty as mandatory and were hardly apologists for international capital.

It is a strength of the ecosocialist analysis, however, that it draws attention to redistribution and makes this a central aspect of realizing environmental objectives. Most ecosocialists have favored global reallocation, and some, like Williams, have advocated a severe contraction of Western consumption as well. These priorities parallel the work of the "ecological footprint" theorists who emerged in the 1990s and of advocates of "contraction and convergence" approaches to carbon emissions.[f]

Ecological commentary from the Left, however, has often argued that redistribution under socialist organization will solve most if not all problems of scarcity and degradation. In this argument, the elimination of production for profit rather than for need will remedy capitalist waste, and concern about natural limits will become redundant. This approach is partial and less helpful than it could be, since both avenues seem vital to any genuinely regenerative outcome.

Notes

a. Fuchs 1970.

b. Enzensberger 1974, 28.

c. Angus and Butler 2011; Monbiot 2009; Pearce 2010.

d. Williams 1982.

e. Meadows et al. 1972, 183–184.

f. Wackernagel and Rees 1996; Athanasiou and Baer 2002.

Early Criticism of the Modeling

The relatively new nonlinear modeling pioneered by Jay Forrester at MIT, and further developed by the MIT team, was one common target of critics. Some were skeptical about the complexity involved in the feedback mechanisms, and one such critic satirized the "mind-bending flowchart" as resembling "a diagram of one of the secret plays that President Nixon sent in to the Washington Redskins."[12] Many argued that the entire model was merely a construct of the Club of Rome's preconceptions—"Malthus with a computer," or "garbage in, garbage out."[13] The MIT researchers conceded that their models were pioneering a new field and were definitely "not perfect," but insisted that an exploration of interconnected processes, including those with delayed effects, was essential to a proper understanding of the dynamic behavior of a complex system such as the real world. Models, they argued, are simplifications by definition, but models offer explicit assumptions that critics can scrutinize far more effectively than the entirely unexamined preconceptions that normally underpin policy decisions affecting the world for decades ahead.[14] The data the project used from the previous seventy years were also panned as inadequate, though no critic indicated where better data could have been obtained—the world was far less forensically measured before 1970 than it is today, and even less again before World War II.

Economists often stated that the model failed to incorporate price mechanisms and technological advance, which are supposed to solve problems of scarcity automatically by creating incentives for innovation and substitution. The MIT team of researchers wrote an extensive reply to their critics in which they acknowledged that price and technology were not specified separately in the long-term aggregated model but were explicitly included in the submodels they had used to explore precisely these factors. They agreed with the economists that price does function as a signal of scarcity and an incentive to solve it. They argued, however, that price is merely a link between scarcity and society's response to it; price was one of many feedback mechanisms explored in the submodels and implicit in the aggregated model. And, they pointed out, it was not a *failure* to specify technological advance that led to collapse in the MIT scenarios but just the reverse: some of the scenarios led to collapse "*because* of the accumulated costs and side-effects of technical successes."[15] Here again we see the gulf between economists and physical scientists. The positive influence of technical innovation is an unchallenged assumption of the growth agenda. Even if it seems to lead to environmental catastrophe, its proponents remain confident of its ability to solve its own problems.

Resource economists were less dismissive than other economists and tended to approach the debate as a conventional argument about facts. Alan Kneese and his colleague Ronald Ridker at Resources for the Future conceded that economics needs to take resource inputs and pollution outputs into account. They were interested in quantifying what they called negative externalities—the uncounted costs of the environmental side effects of production. Like most economists, they believed the MIT model had failed to incorporate price mechanisms and technological advance, but they did not deny that resources are ultimately finite and that collapse is a possible outcome of unfettered growth. They thought that pollution, specifically the increasing concentration of CO_2 in the atmosphere, was more likely than resource scarcity to impose limits on economic growth.[16]

Models: The Clash of Assumptions

The prominent US economist William Nordhaus directed his critique of the modeling at Jay Forrester, who built the initial World1 and World2 models and published *World Dynamics* in 1971.[17] Although he frequently referred to Malthusianism and subtitled his 1973 article "Measurement without Data," Nordhaus did make a more serious attempt to engage with the system dynamics work than most mainstream economists. He and Forrester were, nonetheless, bedeviled by the clash of assumptions discussed in chapter 2.

Nordhaus faulted Forrester for ignoring "standard economic terminology," substituting "vague and often confusing" language, and drawing his conclusions "without the scantest reference to economic theory." For Nordhaus, Forrester's work was a "major retrogression from current research in economic growth theory."[18] In their response, Forrester and his coauthors, Gilbert Low and Nathaniel Mass, were equally unhappy with what they characterized as Nordhaus's "static and geometric frame of reference," which "overlooks the long-term processes of technological change and resource constraints clearly embodied in the *World Dynamics* formulations." They argued that Nordhaus simply "does not recognize the dynamic behavior of multiple-loop feedback systems."[19]

Forrester and his coauthors located resources and prices within a long-term process occurring in a geophysical context; they assumed the ultimately finite character of resources, and focused on the gradual increase in the level of difficulty of extraction: "Greater effort in men, capital and management as stocks are depleted, mines become deeper, oil fields shrink and become harder to find, [and] more waste material must be handled." This is not a concern with relative scarcity, amenable to substitution and reallocation by price adjustment; they are talking about absolute scarcity.[20] As noted previously, neoclassical economists do not acknowledge any absolute scarcity, and even if not all are quite so "cornucopian" as the pro-growth economist Julian Simon, all are convinced that price solves shortage and that the market will manifest solutions. In Forrester's universe, price rises "do not insulate users from shortage ... [but merely] impose the consequences of shortage." Forrester and colleagues cited an eminent economist and reserve bank governor, Henry Wallich, who warned against blind faith in the price system, which, he thought, might not be timely in responding to the threat of actual resource scarcities (see box 6.2).[21]

Forrester and the *Limits* researchers who amplified his model embraced the assumption that it is essential to include the biophysical basis of the human economy if any reasonably accurate account of its longer-term future is to be delivered. They were perfectly open about this. They aimed to integrate economics and natural systems. They assumed that resources are not infinite. Many of the criticisms of economists arose from their assumption that economics is not part of a larger system and ecological factors are not necessary or relevant to economic models (unless environmental issues are under specific investigation).[22]

Criticism by Denial

Beckerman and many other economists denied the existence of any serious problem of any kind at all. These men claimed that technological progress always provides new resources—which must therefore be regarded as infinite—and denied that there is any connection between production and pollution.

Carl Kaysen, then at Harvard, expressed his technological confidence in a nutshell: "it can be shown that the finiteness of the earth does not in itself set limits to what technology might accomplish." Kaysen asserted that, owing to technological advance, limits "are no longer fixed, but grow exponentially." He went on to claim that Robert Socolow had calculated that the earth's "available matter and energy ... could support a population at least 1,000 times the present one at the current US per capita income level." The world's population at that time was just over 3.5 billion, so Kaysen was referring to something like 3.5 trillion people in fairly advanced stages of consumption.[23] Beckerman argued that "even though it may be impossible at present to mine to a depth of one mile at every point in the Earth's crust, by the time we reach 100m AD [the year 100 million] I am sure we will think up something"; he also quoted with approval Neil Jacoby's contention that "there are no foreseeable limits to supplies of basic natural resources including energy at approximately current levels of cost."[24] These claims emerged as articles of faith, often repeated since, with evidence rarely cited. Crucially, as the quotation from Socolow reflects, they treated the earth as "matter and energy" and ignored the self-evident fact that the context that underlies all human economics is a living medium.

Technology was put forward as the solution to both pollution and resource constraints, though some economists simply denied that production involves waste. Solo described the idea that pollution is a function of technology and growth as "facile and ever-so-popular," and claimed that "industrially advanced economies have the capacity to eliminate all noxious wastes ... at a cost that ... is easily bearable because of growth." All that was needed, he said, was "reason in politics and in economic planning."[25] Solow was equally adamant that there is no connection between economic growth and pollution: "that way of looking at the problem is wrong." Another adherent to the notion of the power of prices, Solow thought pollution was not a consequence of production but of an "important flaw in the price system" whereby producers were allowed to dump their waste into the environment without paying the full cost. Solow felt this situation could be very easily fixed by the "simple expedient" of regulating the wastes or charging special taxes.[26] It should be noted that Solow was writing before the neoliberal condemnation of regulation foreclosed this proposed option.

While there was some truth in Solow's claim at that time, Meadows and coworkers noted in their reply to their critics that economists were simply ignoring the scale of the extraction of new natural resources, which was doubling every fifteen years, as was the waste material being dumped into the environment.[27] It is salutary to note that, while some of Solow's "simple expedients" were indeed applied to the cleanup of developed world cities over the next few decades, those improvements were accompanied by the accelerating pollution of industrializing countries such as China and India. Developed environments certainly got cleaner as heavy industry moved elsewhere, but corporations simply shifted the most toxic industries offshore, away from Western eyes. The North Pacific gyre, covering an area the size of Texas (or even of Australia, in one report), with its burden of floating plastic and suspended microplastic fragments, sourced from both North America and Asia, is another example of pollution exported, and thus largely unseen.[28]

Growth Embraced as the Only Solution to Poverty

Numerous critics of *Limits* argued that economic growth is indispensable to relieve the problems of poverty in the third world, suggesting that concern about natural limits tramples on third world interests. The *New York Times* reviewer of *Limits*, Leonard Silk, made explicit a common assumption about the necessity of growth. Writing of the hundreds of millions of people living in desperate poverty, he argued that "their problems cannot be solved by redistribution of existing world income"—an axiom that required no further comment or evidence and that presaged the campaign to "grow the pie, not slice it differently." Thus, "economic growth will be essential to prevent worsening misery, starvation, chaos and war." Silk did acknowledge that some sort of action was needed to conserve the environment and recommended a social revolution in which growth would continue but the goal would somehow shift from "quantity" to "quality."[29]

Most economists, however, placed no such qualifications on their hymn to growth. Most did not even mention redistributive options, and those who did dismissed them out of hand. Solow considered redistribution totally impractical and unrealistic, and, though he acknowledged that equity would not automatically follow from growth, he believed there was no hope at all without it: "The *only* prospect for a decent life for Asia, Africa and Latin America is in more total output."[30] Beckerman concurred: "A failure to maintain economic growth means continued poverty, deprivation, disease, squalor, degradation and slavery to soul-destroying toil for countless millions of the world's population."[31] The so-called "plight of the poor," with accelerated economic growth as its sole solution, would become a core strand of antienvironmental rhetoric throughout the subsequent decades—and remains so today, as commentators endlessly enthuse about growth "lifting millions out of poverty."[32]

These arguments for economic growth drew support in the third world, partly because for centuries, there had been valid grounds to be suspicious of first world plans and prescriptions—limiting growth could be seen as yet another; and partly because wealthy elites remained in control of many developing economies, where their interests coincided more closely with their brethren in the global north than with their own populations, composed largely of peasant farmers. But the rejection of limits thinking and action has not worked automatically in favor of the people of the global south, as it was said to do; rather, the energy- and resource-intensive production techniques of the developed economies have been transplanted almost everywhere; energy-intensive corporate agriculture in particular has dispossessed yet another tranche of the global peasantry and driven it into the megaslums that arose in the late twentieth century (discussed in chapter 9).

The MIT team was explicit in rejecting the idea that economic growth was necessary to assist the poor—or even that it was likely to do so. Quite the reverse: "Historically at least, growth of population and of capital has been correlated with the concentration of wealth and with rising gaps in the absolute income of the rich and the poor."[33]

Several factors influenced the dismissive reaction to the scientists who issued the warnings of trouble ahead. A tendency to overstatement did a disservice to their own cause and facilitated the "doomster" label. Ehrlich's predictions of looming food shortages and concomitant violence in the immediate future turned out to be exaggerated, though the population did grow at rates close to those he predicted, and, though premature, Ehrlich was not really wrong about prices. In the first five years of the new century, key metals such as copper, nickel, and gold doubled or tripled in price, as did oil and natural gas, while grains were not far behind, and continued to rise as the developed world pursued biofuel substitutes for petroleum.[34] These increases reflect the precise situation described by the MIT team: expanding growth in demand unmatched by supply expanding at a similar rate.

Hardin's harsh ideas made ecologists an easy target for Right and Left, north and south—but especially for the Left and representatives of the third world. The "lifeboat" essays were understandably repugnant to people who were seen by Hardin as the irresponsibly breeding poor, much as Malthus had seen the English poor in 1800. His lifeboat ethics—the equivalent of throwing the world's poor to the sharks—were analogous to Malthus's ethics, which had justified incarceration in a workhouse.

The technocrats, such as the Club of Rome founder Aurelio Peccei and the staff of UN agencies concerned with the same problems, were a different case. Despite accusations that they were in league with predatory capitalism,[35] people like Peccei and Strong had no intention of throwing anyone to the sharks. Quite the reverse: like Carter and Trudeau, this group of so-called doomsayers wanted to measure and assess the crisis ahead and build an effective management plan that would preserve hope for the postcolonial world. There is certainly room for a critique of management: it is quite possibly a fantasy that humans can, ultimately, manage the earth, and faceless managers dictating the course of the world economy are not susceptible to any democratic accountability. It is ironic that organizations such as the World Trade Organization, the World Bank, and the International Monetary Fund—as well as the giant corporations themselves—which exert determinate influence on the world economy, are just as undemocratic as the managers envisaged by Peccei, and probably far more so.

The Club of Rome, the Meadows team, and ecological economists in general challenged the wisdom of unfettered growth as the postwar explosion took its toll. In part II, I return to the beginning of the twentieth century to look at the strategies cultivated by business to guarantee that growth: the construction of consumerism early in the century, the neoliberal "revolution" after the crisis of the 1970s, and the pursuit of "development" in the third world from the end of World War II.

II

Chasing Growth

A rising tide that lifts all boats drowns those who have no vessel.

—Modern proverb of uncertain origin

5

Growth and Consumerism

> The development of consumer societies meant the erosion of the traditional values and attitudes of thrift and prudence. Expanding consumption was necessary to create markets for the fruits of rising production.
>
> —Sharon Beder, 2004

Growth and Capitalism

As outlined in chapter 1, one great innovation of capitalism that set it aside from all earlier trading systems was the wholesale application of the economic surplus to the expansion of production, setting in train a regime of accumulation, a dynamic of ongoing growth, and a route to what looked like ever-increasing wealth. It established a system that depends on expansion just to keep going.

The basic financial apparatus for corporate capitalism had already been created during the mercantile era—the limited liability stockholding enterprise. The Australian journalist Murray Sayle, writing about the symbiotic relationship between the capitalist system and hydrocarbon-based industrial technology, locates the genesis of capitalist finance as we know it in the United East Indies Company (Vereenigde Oost-Indische Compagnie, or VOC), founded in the Netherlands in 1602 and chartered by the government.[1] For the first time, investors put money into an entity, not toward a specific venture or voyage; their liabilities were confined to the amount they contributed, and the investment could be traded at any time on the local commodity market. In this way, tradable shares in a corporate body began. The system depended on growth to function:

> The new system showed a well-marked, still powerful feature—to attract more investors, and satisfy the existing ones, the enterprise had to have, or seem to

have, rapid continuous growth, the only inducement to offset the risk being the investors' expectation of getting more out than they had put in. The VOC duly expanded at high speed.[2]

The ventures of the VOC had no pretensions to being beneficial for the broader society; it provided a vehicle for individual gain.

As coal-powered industrialization took hold in the late eighteenth century, however, it began to be argued, famously by Adam Smith, that the individual's pursuit of self-interest is a boon to society as a whole. Though Smith's approach was not quite as clear-cut as this might sound,[3] he did see the stimulation of demand—of what amounts in the long run to unbridled consumption—as an advantage for nations, since increasing the scale of consumption could elicit production on a scale that would invite the efficiencies of specialization and division of labor, and thus drive economic expansion. In this sense, the industrial capitalist era has harnessed desire, with its complex individual peculiarities, to supplant basic survival needs as the driver of economic growth. This switch was central to Smith's new theory of the origins of the wealth of nations. Rather than being a matter of balancing the international books, acquiring territory, or building up reserves of gold and silver (as the mercantilists had thought), wealth could now be understood to flow from unleashing desire and setting it loose in the marketplace of material goods. By the late eighteenth century, it was possible to consider the pursuit of luxury as advantageous for society as a whole.[4]

Even if Smith's approach was ambiguous, neoclassical economists later emphasized self-interest to the point of equating it with greed, and of valorizing greed as a driver of progress and a progenitor of welfare for all.[5] This does not, of course, automatically follow. Capitalist enterprise is run for profit rather than to meet human needs, and even though the two may coincide, there is no intrinsic reason why they should; if there is a conflict, profit will almost always be preferred. Profit can be increased by expanding production to embrace economies of scale, by minimizing costs (including those paid for labor and resources) and maximizing production (often by technological innovation). Efficiency will therefore be pursued to the extent that it achieves these ends. *Efficiency*, a core concept for neoliberalism (see the next chapter), means efficient profit-making. It is not focused on efficient production or the avoidance of waste. There is no motive internal to capitalism that renders the avoidance of environmental degradation a necessary consideration. As the preeminent free marketeer Milton Friedman has argued, "There is one and only one social responsibility of business—to use its resources and engage in activities designed to increase its profits so long as it stays within the rules of the game."[6]

Corporations function on behalf of their shareholders (and other direct stakeholders to some extent); corporate law in Australia, for example, requires company officials to operate "in good faith in the best interests of the corporation," and courts construe this to mean the shareholders. Though other so-called stakeholders, such as employees or customers, might occasionally be regarded as relevant by some directors, corporations are not required to take into account the broader society, the natural environment, or the common good.[7] Indeed, in some circumstances, profit maximization requires that environmental degradation, and even public health, be ignored, as in the notorious case of the Ford Pinto, where hundreds of drivers burned to death because Ford's cost–benefit analysis indicated that settling the resultant lawsuits was cheaper than retooling the assembly lines.[8] When fishermen and their families died of mercury poisoning on the shores of Japan's Minamata Bay in the 1950s, the Chisso Corporation polluting the harbor initially denied any connection, and managed to defer official recognition of the cause of the sickness for twelve years, while people died and fishermen lost their livelihoods.[9] Whether pouring their waste into the rivers of northern England in the nineteenth century or the Ok Tedi (Papua New Guinea) in the twentieth, or into the Yellow River (China) today, businesses operating on the profit system, whether private or state-owned, have saved costs at whatever ecological or health expense has been permitted—or tolerated—by the state.

Rise of the Consumer Economy

Periods of scarcity were endemic in all early human societies, and with the advent of agriculture, settled people were especially vulnerable to annual weather variations, as well as longer-term shifts in climate. One great achievement of the capitalist economy was the amelioration of the problem of periodic scarcity for Europe and its settler offshoots. The travails of getting to this point through the nineteenth century visited grotesque conditions on generations of the workers who built the industrial wealth as they worked the kinds of hours in the kinds of conditions seen today in the mines and sweatshops of the developing world.[10] At the same time, the fantastic expansion of European economies was continuously underpinned by plunder, pillage, and murder throughout overseas empires. For Europe, however, and its settler colonies, such as the United States and Australia, scarcity of daily necessities was greatly reduced by the time of World War I. In these places, the "common man" began to enjoy unprecedented material security.

These material advances were anchored in the first phase of the globalization of industrial capitalism. In his survey of the origins of World War I, Eric Hobsbawm argued that the war had been waged at least in part for the means to continue industrial expansion, to secure access to "world markets and material resources, and ... control of regions such as the Near and Middle East where ... petro-diplomacy was already a crucial factor." The advent of a worldwide industrial capitalist economy "inevitably pushed the world in the direction of state rivalry, imperialist expansion, conflict and war.... Competing national industrial economies now confronted each other." Political power required commensurate economic power. Though Hobsbawm ruled out the notion that individual capitalists or corporations favored war (which usually jeopardized "business as usual"), he argued that the "characteristic feature of capitalist accumulation was precisely that it had no limit. The 'natural frontiers' of Standard Oil, the Deutsche Bank, the De Beers Diamond Corporation were at the end of the universe, or rather at the limits of their capacity to expand."[11]

People Become Consumers: Beginnings

The notion of human beings as consumers first took shape before World War I, but became commonplace in America in the 1920s. Consumption is now frequently seen as our principal role in the world.[12]

People, of course, have always "consumed" the necessities of life—food, shelter, clothing—and have always had to work to get them or have others work for them, but there was little economic motive for increased consumption among the mass of people before the twentieth century. Quite the reverse: frugality and thrift were more appropriate to situations where survival rations were not guaranteed. Attempts to promote new fashions, harness the "propulsive power of envy," and boost sales multiplied in Britain in the late eighteenth century; here began the "slow unleashing of the acquisitive instincts," when the pursuit of opulence and display first extended beyond the very rich.[13] But, while poorer people might have acquired a very few useful household items, a skillet, perhaps, or an iron pot, the sumptuous clothing, furniture, and pottery of the era were still confined to a very small population. In late nineteenth-century Britain a variety of foods became accessible to the average person, who would previously have lived on bread and potatoes—consumption beyond mere subsistence. This improvement in food variety did not extend durable items to the mass of people, however. The proliferating shops and department stores of that period served only a restricted population of urban middle-class people in Europe, but the display of tempting products in shops in daily public view was greatly extended—and display was a key element in the fostering of fashion and envy.[14]

Although the period after World War II is often identified as the beginning of the immense eruption of consumption across the industrialized world, the historian William Leach locates its roots in the United States around the turn of the century. In the United States, existing shops were rapidly extended through the 1890s, mail-order shopping surged, and the new century saw massive multistory department stores "covering millions of acres of selling space." Retailing was already passing decisively from small shopkeepers to corporate giants who had access to investment bankers and drew on assembly-line production of commodities, powered by fossil fuels; the traditional objective of making products for their self-evident usefulness was displaced by the goal of profit and the need for a machinery of enticement. According to Leach, "The cardinal features of this culture were acquisition and consumption as the means of achieving happiness; the cult of the new; the democratization of desire; and money value as the predominant measure of all value in society."[15] Significantly, it was individual desire that was democratized, rather than wealth or political and economic power.

The 1920s: "The New Economic Gospel of Consumption"

Release from the perils of famine and premature starvation was in place for most people in the industrialized world soon after the Great War ended.[16] US production was more than twelve times greater in 1920 than in 1860, while the population over the same period had increased by only a factor of three,[17] suggesting just how much additional wealth was theoretically available. The labor struggles of the nineteenth century had, without jeopardizing the burgeoning productivity, gradually eroded the seven-day week of fourteen- and sixteen-hour days that was worked at the beginning of the Industrial Revolution in England. In the United States in particular, economic growth had succeeded in providing basic security to the great majority of an entire population.

In these circumstances, there was a social choice to be made. A steady-state economy capable of meeting the basic needs of all, foreshadowed by John Stuart Mill as the *stationary state*, seemed well within reach and, in Mill's words, likely to be an improvement on "the trampling, crushing, elbowing and treading on each other's heels … the disagreeable symptoms of one of the phases of industrial progress."[18] It would be feasible to reduce hours of work further and release workers for the spiritual and pleasurable activities of free time with families and communities, and creative or educational pursuits. But business did not support such a trajectory, and it was not until the Great Depression that hours were reduced, in response to overwhelming levels of unemployment.

In 1930 the US cereal manufacturer Kellogg adopted a six-hour shift to help accommodate unemployed workers, and other forms of work-sharing became more widespread. Although the shorter workweek appealed to Kellogg's workers, the company, after reverting to longer hours during World War II, was reluctant to renew the six-hour shift in 1945. Workers voted for it by three-to-one in both 1945 and 1946, suggesting that, at the time, they still found life in their communities more attractive than consumer goods. This was particularly true of women. Kellogg, however, gradually overcame the resistance of its workers and whittled away at the short shifts until the last of them were abolished in 1985.[19]

Even if a shorter working day became an acceptable strategy during the Great Depression, the economic system's orientation toward profit and its bias toward growth made such a trajectory unpalatable to most captains of industry and the economists who theorized their successes. If profit and growth were lagging, the system needed new impetus. The short depression of 1921–1922 led businessmen and economists in the United States to fear that the immense productive powers created over the previous century had grown sufficiently to meet the basic needs of the entire population and had probably triggered a permanent crisis of overproduction; prospects for further economic expansion were thought to look bleak. The historian Benjamin Hunnicutt, who examined the mainstream press of the 1920s, along with the publications of corporations, business organizations, and government inquiries, found extensive evidence that such fears were widespread in business circles during the 1920s.[20] Victor Cutter, president of the United Fruit Company, exemplified the concern when he wrote in 1927 that the greatest economic problem of the day was the lack of "consuming power" in relation to the prodigious powers of production.[21]

Notwithstanding the panic and pessimism, a consumer solution was simultaneously emerging. As the popular historian of the time Frederick Allen wrote, "Business had learned as never before the importance of the ultimate consumer. Unless he could be persuaded to buy and buy lavishly, the whole stream of six-cylinder cars, super heterodynes, cigarettes, rouge compacts and electric ice boxes would be dammed up at its outlets."[22] Edward Bernays, one of the pioneers of the public relations industry, put it this way:

> Mass production is profitable only if its rhythm can be maintained—that is if it can continue to sell its product in steady or increasing quantity.... Today supply must actively seek to create its corresponding demand ... [and] cannot afford to wait until the public asks for its product; it must maintain constant touch, through advertising and propaganda ... to assure itself the continuous demand which alone will make its costly plant profitable.[23]

Edward Cowdrick, an economist who advised corporations on their management and industrial relations policies, called it "the new economic gospel of consumption," in which workers (people for whom durable possessions had rarely been a possibility) could be educated in the new "skills of consumption."[24] It was an idea also put forward by the new "consumption economists" such as Hazel Kyrk and Theresa McMahon, and eagerly embraced by many business leaders. New needs would be created, with advertising brought into play to "augment and accelerate" the process. People would be encouraged to give up thrift and husbandry, to value goods over free time. Kyrk argued for ever-increasing aspirations: "a high standard of living must be dynamic, a progressive standard," where envy of those just above oneself in the social order incited consumption and fueled economic growth.[25] President Herbert Hoover's 1929 Committee on Recent Economic Changes welcomed the demonstration "on a grand scale [of] the expansibility of human wants and desires," hailed an "almost insatiable appetite for goods and services," and envisaged "a boundless field before us ... new wants that make way endlessly for newer wants, as fast as they are satisfied."[26] In this paradigm, people are encouraged to board an escalator of desires (a stairway to heaven, perhaps) and progressively ascend to what were once the luxuries of the affluent.

Charles Kettering, general director of General Motors Research Laboratories, equated such perpetual change with progress. In a 1929 article called "Keep the Consumer Dissatisfied," he stated that "there is no place anyone can sit and rest in an industrial situation. It is a question of change, change all the time—and it is always going to be that way because the world only goes along one road, the road of progress."[27] These views parallel Joseph Schumpeter's later characterization of capitalism as "creative destruction":

> Capitalism, then, is by nature a form or method of economic change and not only never is, but *never can be stationary*.... The fundamental impulse that sets and keeps the capitalist engine in motion comes from the new consumers, goods, the new methods of production or transportation, the new markets, the new forms of industrial organization that capitalist enterprise creates [my emphasis].[28]

The prospect of ever-extendable consumer desire, characterized as "progress," promised a new way forward for modern manufacture, a means to perpetuate economic growth. Progress was about the endless replacement of old needs with new, old products with new. Notions of meeting everyone's needs with an adequate level of production did not feature.

The nonsettler European colonies were not regarded as viable venues for these new markets, since centuries of exploitation and impoverishment meant that few people there were able to pay. In the 1920s, the target consumer market to be nourished lay at home in the industrialized world. There, especially in the United States, consumption continued to expand through the 1920s, though truncated by the Great Depression of 1929. Electrification was crucial for the consumption of the new types of durable items, and the fraction of US households with electricity connected nearly doubled between 1921 and 1929, from 35 percent to 68 percent; a rapid proliferation of radios, vacuum cleaners, and refrigerators followed. Motor car registration rose from eight million in 1920 to more than 28 million by 1929. The introduction of time payment arrangements facilitated the extension of such buying further and further down the economic ladder. In Australia, too, the trend could be observed; there, however, the base was tiny, and even though car ownership multiplied nearly fivefold in the eight years to 1929, few working-class households possessed cars or large appliances before 1945.[29]

This first wave of consumerism was short-lived. Predicated on debt, it took place in an economy mired in speculation and risky borrowing. US consumer credit rose to $7 billion in the 1920s, with banks engaged in reckless lending of all kinds.[30] Indeed, though a lot less in gross terms than the burden of debt in the United States in late 2008, which Sydney economist Steve Keen has described as "the biggest load of unsuccessful gambling in history,"[31] the debt of the 1920s was very large, over 200 percent of the GDP of the time. In both eras, borrowed money bought unprecedented quantities of material goods on time payment and (these days) credit cards. The 1920s bonanza collapsed suddenly and catastrophically. In 2008, a similar unraveling began; its implications still remain unknown. In the case of the Great Depression of the 1930s, a war economy followed, so it was almost twenty years before mass consumption resumed any role in economic life—or in the way the economy was conceived.

The Second Wave

Once World War II was over, consumer culture took off again throughout the developed world, partly fueled by the deprivation of the Great Depression and the rationing of the wartime years and incited with renewed zeal by corporate advertisers using debt facilities and the new medium of television. Stuart Ewen, in his history of the public relations industry, saw the birth of commercial radio in 1921 as a vital tool in the great wave of debt-financed consumption in the 1920s—"a privately owned utility, pumping information and entertainment into people's homes. Requiring no significant degree of literacy on the part of its audience ... radio gave interested corporations ... unprecedented access to the inner sanctums of the public mind."[32] The advent of television greatly magnified the potential impact of advertisers' messages, exploiting image and symbol far more adeptly than print and radio had been able to do. The stage was set for the democratization of luxury on a scale hitherto unimagined.

Though the television sets that carried the advertising into people's homes after World War II were new, and were far more powerful vehicles of persuasion than radio had been, the theory and methods were the same—perfected in the 1920s by PR experts like Bernays. Vance Packard echoes both Bernays and the consumption economists of the 1920s in his description of the role of the advertising men of the 1950s:

> They want to put some sizzle into their messages by stirring up our status consciousness.... Many of the products they are trying to sell have, in the past, been confined to a "quality market." The products have been the luxuries of the upper classes. The game is to *make them the necessities of all classes*. This is done by dangling the products before non-upper-class people as status symbols of a higher class. By striving to buy the product—say, wall-to-wall carpeting on instalment—the consumer is made to feel he is upgrading himself socially [my emphasis].[33]

Though it is status that is being sold, it is endless material objects that are being consumed.

In a little-known 1958 essay reflecting on the conservation implications of the conspicuously wasteful US consumer binge after World War II, John Kenneth Galbraith pointed to the possibility that this "gargantuan and growing appetite" might need to be curtailed. "What of the appetite itself?," he asks. "Surely this is the ultimate source of the problem. If it continues its geometric course, will it not one day have to be restrained? Yet in the literature of the resource problem this is the forbidden question."[34]

Galbraith quotes the President's Materials Policy Commission setting out its premise that economic growth is sacrosanct. "First we share the belief of the American people in the principle of Growth," the report maintains, specifically endorsing "ever more luxurious standards of consumption." To Galbraith, who had just published *The Affluent Society*, the wastefulness he observed seemed foolhardy, but he was pessimistic about curtailment; he identified the beginnings of "a massive conservative reaction to the idea of enlarged social guidance and control of economic activity," a backlash against the state taking responsibility for social direction.[35] At the same time he was well aware of the role of advertising: "Goods are plentiful. Demand for them must be elaborately contrived.

Those who create wants rank amongst our most talented and highly paid citizens. Want creation—advertising—is a ten billion dollar industry."[36]

Or, as retail analyst Victor Lebow remarked in 1955:

> Our enormously productive economy demands that we make consumption our way of life, that we convert the buying and use of goods into rituals, that we seek our spiritual satisfaction, our ego satisfaction, in consumption.... We need things consumed, burned up, replaced and discarded at an ever accelerating rate.[37]

Thus, just as immense effort was being devoted to persuading people to buy things they did not actually need, manufacturers also began the intentional design of inferior items, which came to be known as "planned obsolescence." In his second major critique of the culture of consumption, *The Waste Makers*, Packard identified both functional obsolescence, in which the product wears out quickly and psychological obsolescence, in which products are "designed to become obsolete in the mind of the consumer, even sooner than the components used to make them will fail."[38]

Galbraith was alert to the way that rapidly expanding consumption patterns were multiplied by a rapidly expanding population. But postwar industrial enterprise stoked the expansion nonetheless. The rise of consumer debt, interrupted in 1929, also resumed. In Australia, the 1939 debt of AU$39 million doubled in the first two years after the war and, by 1960, had grown by a factor of 25, to more than AU$1 billion dollars.[39] This new burst in debt-financed consumerism was, again, incited intentionally.

Tapping into the Unconscious: Image and Message

In researching his excellent history of the rise of PR, Ewen interviewed Bernays himself in 1990, not long before he turned ninety-nine. Ewen found Bernays, a key pioneer of the new PR profession, to be just as candid about his underlying motivations as he had been in 1928 when he wrote *Propaganda*:

> Throughout our conversation, Bernays conveyed his hallucination of democracy: A highly educated class of opinion-molding tacticians is continuously at work ... adjusting the mental scenery from which the public mind, with its limited intellect, derives its opinions.... Throughout the interview, he described PR as a response to a transhistoric concern: the requirement, for those people in power, to shape the attitudes of the general population.[40]

Bernays's views, like those of several other analysts of the "crowd" and the "herd instinct," were a product of the panic created among the elite classes by the early twentieth-century transition from the limited franchise of propertied men to universal suffrage.

On every side of American life, whether political, industrial, social, religious or scientific, the increasing pressure of public judgment has made itself felt.... The great corporation which is in danger of having its profits taxed away or its sales fall off or its freedom impeded by legislative action must have recourse to the public to combat successfully these menaces.[41]

The opening page of his 1928 classic, *Propaganda*, discloses his solution:

The conscious and intelligent manipulation of the organized habits and opinions of the masses is an important element in democratic society. Those who manipulate this unseen mechanism of society constitute an invisible government which is the true ruling power of our country.... It is they who pull the wires which control the public mind, who harness old social forces and contrive new ways to bind and guide the world.[42]

The front-line thinkers of the emerging advertising and public relations industries turned to the key insights of Sigmund Freud, Bernays's uncle. As Bernays noted:

Many of man's thoughts and actions are compensatory substitutes for desires which [he] has been obliged to suppress. A thing may be desired, not for its intrinsic worth or usefulness, but because he has unconsciously come to see in it a symbol of something else, the desire for which he is ashamed to admit to himself ... because it is a symbol of social position, an evidence of his success.[43]

Bernays saw himself as a "propaganda specialist," a "public relations counsel," and PR as a more sophisticated craft than advertising as such; it was directed at hidden desires and subconscious urges of which its targets would be unaware. Bernays and his colleagues were anxious to offer their services to corporations and were instrumental in founding an entire industry that has since operated along these lines, selling not only corporate commodities but also opinions on a great range of social, political, economic, and environmental issues. I return to the tactics of these masters of spin in part III.

Though it has become fashionable in recent decades to brand scholars and academics as elites who pour scorn on ordinary people, Bernays and the sociologist Gustave Le Bon were long ago arguing, on behalf of business and political elites, respectively, that the mass of people are incapable of thought. According to Le Bon, "A crowd thinks in images, and the image itself immediately calls up a series of other images, having no logical connection with the first"; crowds "can only comprehend rough-and-ready associations of ideas," leading to "the utter powerlessness of reasoning when it has to fight against sentiment."[44] Bernays and his PR colleagues believed ordinary people to be incapable of logical thought, let alone mastery of "abstruse economic, political and ethical data," and saw the need to "control and regiment the masses according to our will without their knowing about it"; PR could thus ensure the maintenance of order and corporate control in society.[45]

The commodification of reality and the manufacture of demand have had serious implications for the construction of human beings in the late twentieth century, where "people recognise themselves in their commodities"[46] and can be expected to have grave difficulties in reducing the level of the orgy. Herbert Marcuse's critique of needs, made more than forty years ago, was not directed at the issues of scarce resources or ecological waste, although he was aware even at that time that Marx was insufficiently critical of the continuum of progress and that there needed to be "a restoration of nature after the horrors of capitalist industrialisation have been done away with."[47] Marcuse directed his critique at the way people, in the act of satisfying our aspirations, reproduce dependence on the very exploitive apparatus that perpetuates our servitude. Hours of work in the United States have been growing since 1950, along with a doubling of consumption per capita between 1950 and 1990.[48] Marcuse suggested that this "voluntary servitude (voluntary inasmuch as it is introjected into the individual) ... can be broken only through a political practice which reaches the roots of containment and contentment in the infrastructure of man [*sic*], a political practice of methodical disengagement from and refusal of the Establishment, aiming at a radical transvaluation of values."[49]

The difficult challenge posed by such a transvaluation is reflected in current attitudes. The Australian comedian Wendy Harmer in her 2008 ABC TV series called *Stuff* expressed irritation at suggestions that consumption is simply generated out of greed or lack of awareness:

> I am very proud to have made a documentary about consumption that does not contain the usual footage of factory smokestacks, landfill tips and bulging supermarket trolleys. Instead, it features many happy human faces and all their wonderful stuff! It's a study of a love affair as much as anything else.[50]

In the same vein, during the Q&A after a talk given by the Australian economist Clive Hamilton at the 2006 Byron Bay Writers' Festival, one woman spoke up about her partner's priorities: rather than entertain questions about any impact his possessions might be having on the environment, she said, he was determined to "go down with his gadgets."

The capitalist system, dependent on a logic of never-ending growth from its earliest inception, confronted the plenty it created in its home states, especially the United States, as a threat to its very existence. It would not do if people were content because they felt they had enough. Over the course of the twentieth century, capitalism preserved its momentum by molding the ordinary person into a consumer with an unquenchable thirst for its "wonderful stuff."

The postwar growth bonanza, buttressed by consumer incitement, lasted nearly thirty years before faltering amid the oil shocks of the 1970s. It was this crisis that galvanized business to seek new strategies to restore growth and protect its interests. The next chapter introduces *neoliberalism*, or free market fundamentalism, which emerged as the chosen path through that crisis. Under the new doctrine, profit share and wage share became more and more polarized almost everywhere, allowing corporate profits to improve even when the rate of growth remained slower than before. After a postwar average of approximately 5 percent—doubling the economy every fourteen years—world growth averaged approximately 3 percent from 1980 to 2008.[51] Considered barely adequate by the IMF, growth at 3 percent still doubles the scale of the economy every twenty-four years. Even if growth was more modest than hoped, the deregulation and privatization that were presumed to promote that growth were now extensively established, and these changes were to have enormous effects on the political and ideological opportunities for the critique of growth, the protection of nature, and creating just outcomes, especially for peoples outside the first world.

6

The Rise of Free Market Fundamentalism

Like all fundamentalist faiths, Chicago School economics is, for its true believers, a closed loop. The startling premise is that the free market is a perfect scientific system, one in which individuals, acting on their own self-interested desires, create the maximum benefits for all. It follows ineluctably that if something is wrong ... high inflation or soaring unemployment—it has to be because the market is not truly free.... The Chicago solution is always the same: a stricter and more complete application of the fundamentals.

—Naomi Klein, 2007

The emergence of neoliberal economics has radically increased the share of wealth that business is able to capture and drastically altered the ideological climate in favor of business. It has become the philosophical framework that dominates policy thinking almost everywhere, curbing democratic intervention from elected governments and limiting the space for critics of growth—indeed, for critics of any aspect of environmental degradation. The market has increasingly been accepted as the primary institution needed to take care of all aspects of public activity.

Neoliberalism and Its Values

Modern economics is usually seen as originating with Adam Smith's *The Wealth of Nations*, published in 1776, at the very beginning of industrial capitalism. This first phase is known as *classical economics*, though its theorists called it *political economy*.[1] Economics developed as a set of ideas entirely within the context of the great economic expansion of industrial capitalism. Smith's book was prescient: it preceded the wholesale adoption of coal-fueled steam power and the rise of factory production in the nineteenth century, as well as the extraordinary economic growth that accompanied these innovations. The term *economics* was not adopted until the late nineteenth century, when economists introduced the idea that value arises from the rational choices of completely informed individuals, who thus maximize their advantage (or *utility*, as it was termed)—a mechanism thought to match supply with demand and thus yield equilibrium. This suite of economic theories is known as *neoclassical economics* and has formed the basis of the economic mainstream ever since.

The neoliberal thinking of the late twentieth century can be seen as a variant of neoclassical economics; it rests squarely on several central tenets of that theory: individualism; a claim to scientific rigor; the faith that untrammeled markets, by harmonizing the desires of all individuals, will provide an optimal allocation of material things, leading to prosperity for all; and the idea that markets are a "natural," transhistorical arena rather than a human construct.

Freedom, Individualism, and the Right to Property

Individualism lies at the root of all the various versions of laissez-faire economics, beginning with Adam Smith's individual profit-seeker, who benefits everyone without trying:

> He intends only his own gain, and he is in this, as in many other cases, led by an invisible hand to promote an end which was no part of his intention.... By pursuing his own interest he frequently promotes that of the society more effectually than when he really intends to promote it.[2]

Whether or not it is unfair, as some suggest,[3] to characterize Smith as the original free marketeer, his "invisible hand"—mentioned only once in *Wealth of Nations*—became enshrined in neoclassical economics as one of its great underlying truths: all it takes to advance the welfare of everyone is to pursue one's own self-interest. Thus, individual self-interest miraculously produces the best possible world, and individual greed is vindicated as a desirable social behavior. One of the most basic ethical messages of all human religions is relegated to irrelevance. Though obviously a highly contestable proposition, the concept that individual self-interest leads to general welfare supported nineteenth-century laissez-faire economics right through to the Great Depression, and resurfaced with renewed ideological vigor during the 1970s, when Ayn Rand's romanticized version of capitalism enjoyed mounting popularity.[4] Margaret

Thatcher's 1987 contention that "There is no such thing as society,"[5] referring to individuals' obligation to look after themselves rather than to expect help from government, was emblematic of the emphasis.

While the self-interest of atomized individuals is the driver that is thought to underpin the market and produce prosperity, neoliberal economists also contend that the economic freedoms of these individuals (such as secure property rights and freedom of contract) are the underlying prerequisites for all human freedom. According to Milton Friedman, "Economic freedom is a necessary condition for political and civil freedom.... Property rights ... are themselves the most basic of human rights and an essential foundation for other human rights."[6] The Austrian-born economist Friedrich von Hayek argued, in a similar vein, that "the system of private property is the most important guarantee of freedom, not only for those who own property, but scarcely less for those who do not."[7] Indeed, it has been argued that the inversion of the relationship between politics and economics is what distinguishes neoliberalism from its antecedents. The road to political liberty now ran through the free market rather than vice versa.[8]

Freedom is not always understood in this way. When, in the wake of World War II, the UN General Assembly adopted the Declaration of Human Rights, it identified "freedom of speech and belief and freedom from fear and want" as the four fundamental freedoms of human beings. Articles 22 to 25 of the Declaration affirmed the right to work; the right to fair pay; the right to an adequate standard of living, including food, clothing, housing, and medical care; and assistance in the event of lack of livelihood.[9] Hayek, on the other hand, opposed the idea of freedom from want as a core freedom; he saw it as "only another name for the old demand for an equal distribution of wealth," one of the hallmarks of socialism.[10]

Making security of property title the bedrock of ideas about freedom clearly advantages those who hold most of the property, a very small minority of the world's people; its enforcement has ossified many elements dating back to the colonial division of spoils. Optimal allocation, which is thought to be secured by a free market, never addresses prior theft or necessary restitution and simply assumes that payment of the highest price yields the most productive use.[11] Yet what it really reflects is allocation to the person or persons who have the financial edge and the ability to pay. This has no necessary relationship to productive or optimal uses of anything.

Economics Becomes a Science

The neoclassical economists who emerged in the late nineteenth century set out to translate economic ideas into mathematical laws and establish economics as a science just like physics. Léon Walras and William Stanley Jevons, both trained in physics, were responsible for the "thickets of algebra" that have marched across the pages of economics texts since that time. They imagined that human economic systems could be captured in mathematical propositions in parallel with mid-nineteenth-century physics, divorced from the complexity of their history and the agency of the human beings who developed them over centuries. The operation of markets was conceived to be governed by what were virtually analogues of the laws of nature (as understood at the time),[12] existing in a kind of social void, as universal phenomena. Yet the analogy doesn't hold. While mid-nineteenth-century physics deals with reversible processes, economics describes processes that involve irreversible transformations to which the arrow of time applies.

Much of the mystification that has surrounded neoclassical economics is related to this quest for scientific status and the credibility it conferred—supposedly elevating economics above the other social sciences. Underlying it all is the claim that economics is merely descriptive of reality and thus value-free in the same sense as physics is thought to be—a straightforward theory of the workings of the real world. Whether any system of knowledge can legitimately be seen this way is a separate question and one I will not address here. Economics clearly is not akin to physics as it was understood in the mid-nineteenth century, any more than is history or sociology or any discipline that deals with the products of human intervention.

Another flaw in neoclassical economics is its failure to include non-monetary economic contributions, including the services provided in households in all economies, mainly by women, and the subsistence work that dominates rural life throughout the world. Although the rural population is declining as a percentage of all people, about half the world's population still lives outside cities. Any production not reflected in monetary flows is simply left out of neoclassical economics,[13] even though household members, mainly unpaid, produce a third or more of all goods and services in developed world economies and a far larger fraction in less-developed regions.[14] Thus, the wealth quantified by neoclassical economics is little more than half of overall economic activity but is theorized as if it were the whole. Neoclassical economics is also ill-equipped to describe the mixed economies of the postwar system, where governments have funded large fractions of GDP. In the Australian economist Hugh Stretton's analysis, neoclassical economists "selected a few relationships from complicated real life, and modelled those few only. They focussed on sales and exchanges: on markets,"[15] a perfect theoretical basis for the market fundamentalism that dominates current economics.

Optimal Allocation and the Cult of "Efficiency"

Closely related to the endorsement of individual material gratification as the core business of human economic activity is the cult of "efficiency," which revolves around obtaining the biggest profit for the least outlay: getting the most for the least input, maximizing quantities, and minimizing costs. Although mainstream economists do not argue that business *should* attempt to shed the costs of its negative externalities, this is often the outcome in practice. In theory, externalities such as pollution or loss of amenity for citizens can be dealt with via taxes, trading schemes, or compensation, but all affect profit, and corporations prefer to avoid or minimize the bill (box 6.1).

Neoclassical efficiency is defined in "bottom-line" terms; here, considerations based on social, moral, or environmental criteria will only be counted if they can be monetized, an inexact science when applied to phenomena that have no obvious monetary value. The neoliberal era has seen the ascendancy of what is termed cost–benefit analysis as the main instrument for assessing policy outcomes and a concomitant emphasis on monetary values. However "optimal" the allocation of resources (even if optimality could be demonstrated), this is only one aspect of benefit to society.

It might be argued, for example, that the purpose of human economic activity is the care of—or advancement of—all people across a society, a belief commonly found in indigenous societies and somewhat approached during the interregnum between the "gilded age" of the 1920s and that

Box 6.1
Externalities

In economic theory, externalities may be either positive or negative. Here, I deal with negative side-effects such as pollution. Some externalities are unaccounted for, even though they are reasonably quantifiable in money terms; others are unsuitable for monetary quantification. Business resists both categories: when the effect *could* be charged (for example, carbon taxes on CO_2 emissions), business resists to protect profit. When the value is difficult or impossible to reduce to dollars (as when an entire village will cease to exist if open-cut coal mining is allowed), business argues for monetary compensation.

The Indian physicist and environmental campaigner Vandana Shiva described how the quarrying of limestone for cement in northern India in the early 1980s destroyed her native Doon Valley. Originally a region with abundant rainfall, its deep limestone cavities provided a natural reservoir, recharged by rainfall in the Mussoorie Hills. The value of all this was ignored before quarrying began, even though the cost of an artificial structure of comparable depth was later estimated at $500 million. Shiva saw the valley's last perennial stream run dry in 1982.[a] Failure to include the costs of the destruction of a region's water supply in the accounting enables a very large inflation of the profit margin for the miner and cement manufacturer. In this worldview, the destruction of the water resource is a sad but inevitable consequence of "progress," and its real value overlooked.

Such unaccounted costs can include not only the loss of nature and the services it provides to humans but also the costs of cleaning up or coping with pollution, and the lost livelihoods and cultures of displaced people—an estimated 40 million due to big dams in India alone, and tens of millions more in China. "Compensation" has not always been offered and, even when it was, has usually been woefully inadequate (see box 9.1).

Poisoned livelihoods include those of the people who live along the rivers polluted by the Ok Tedi gold mine in Papua New Guinea or the people of Lago Agrio in the Ecuadorian Amazon, where groundwater has been contaminated by toxic waste from Texaco's drilling operations. The returns of cotton irrigators in Central Asia were not reduced by the immense social and ecological costs borne downstream by the Aral Sea and its people, whose fishing ports ended up thirty or forty miles from the shore of a saline, shrinking sea where fish no longer survived. The Canadian government's social welfare bill on behalf of thousands of cod fishermen is another such externality, unreflected in the price of the fish before they disappeared. So is the progressive evacuation of islanders—from Tuvalu, Kiribati, Bougainville's Carteret Islands, India's Sundarbans, and the Bangladeshi delta—as sea levels rise because of greenhouse warming. The immense penalty imposed on islanders who lose their entire islands and everything on them has yet to be included in the costs of the aluminum we use in the first world, or the electricity we squander.

Box 6.1 (continued)

Where externalities are excluded, goods can be sold artificially cheaply, which encourages consumption and suits business. In Australia, the price of bread does not reflect the fact that every tonne of wheat embodies 1,000 tonnes of water and 45 tonnes of topsoil. In addition, the price of a nonrenewable resource, such as oil, does not reflect its status as nonrenewable—its *actual* scarcity—or the real costs of its externalities. Neither does it reflect the fact that the price is also artificially lowered by the massive subsidies showered on the oil and gas industry by governments around the world.

Perhaps the most pressing externality in the second decade of the twenty-first century is the emission of carbon dioxide and methane from a great range of human economic activities. Not only has the market failed to dampen these down—global emissions grow at an increasing rate every year—but the corporations controlling the world economy resist the process (see part III). Global warming was reckoned by the economist Nicholas Stern, who wrote the UK report on the costs of a response, to be the "greatest and widest-ranging market failure ever seen"[b]—where the market reflects neither true environmental nor true economic costs.

Notes

a. Shiva 2002, 2, 5–6, citing a report by India's Ministry of Environment.

b. Stern 2007, executive summary, i.

of the late twentieth century. Similar values were, for example, reflected in President Roosevelt's annual address to Congress in January 1941 (the "Four Freedoms" speech), which stressed equality of opportunity and a "better system" designed to ensure reliable access to employment. Roosevelt's freedoms, like those of the UN's Declaration of Human Rights, included "freedom from want—which, translated into world terms, means economic understandings which will secure to every nation a healthy peacetime life for its inhabitants—everywhere in the world."[16]

Resilience theorists[17] have cautioned that the ability of natural systems to recover from random shock or normal disturbance relies not on efficiency but on redundancy or duplication.[18] The pursuit of efficiency routinely reduces or eliminates anything not immediately essential, in the pursuit of the most profitable way of doing things. The globalized market of the past few decades, for example, has increasingly mediated the food production of the entire world, producing "efficiencies" in such areas as the elimination of national or regional food banks. The steep escalation of commodity prices in 2007–2008 was estimated by the UN to be pushing up to 100 million people "back into poverty."[19] The causes of the price rises are multiple, but the event affords an example of the unintended consequences of swapping a multifaceted system for one with fewer redundancies.

Even if efficiency *were* shown to be the most appropriate criterion for decisions about human affairs and the natural world in which they are grounded, it seems likely that the market's putative efficiency at resource allocation and correct pricing is itself a myth, as suggested not only by the collapse of the financial system in 2007–2008 but also by the financialization of the entire economy over the previous forty years. *Financialization* refers to the dominance of the finance sector in the global economy, the switch from productive to speculative investment, and the ongoing translation of social and environmental processes into tradable entities. Both in the United States and globally, the finance sector's share of corporate profits grew from about 10 percent in the 1950s to between 30 and 40 percent by 2004.[20] What kind of efficiency could possibly be served by displacing capital from the production of food, shelter, and other human necessities and channeling it into speculative derivatives like collateralized debt obligations and credit default swaps?[21]

The economist and veteran investment banker Paul Woolley challenges the efficient market hypothesis, which holds that equity markets "deliver efficient pricing leading to the most productive allocation of resources," even though it was an article of faith throughout the 1980s and continues to be regarded as the bedrock of finance theory up to the present day. Woolley's experience showed him that speculative momentum, rather than actual value, controlled market prices; and that the neoclassical idea that the market provides the only reliable matrix for rational individuals acting on the best information is manifestly misplaced.[22]

The Construction of "Free Markets"

The concept of the "free market economy," a pervasive expression in recent decades (Google yielded nearly 29 million hits in May 2013), implies that the capitalist economic system is a natural phenomenon that is jeopardized by government regulation and will always operate best when liberated from such interference. Yet this idea obscures the fact that the notional free market rests squarely on a multiplicity of institutions created by human agency; the state played an indispensable role over centuries in providing a suitable legislative framework for the operation of capitalism. There was nothing spontaneous or natural, for example, about the early stages of primitive accumulation in Europe, when peasants were progressively separated from their common resources as sheep farming took hold. Enclosure was carried out with the participation of the state, both before and after the transition from absolute monarchy to elected legislatures; further legal curtailment of rights to hunt and gather stripped peasants of residual options for sustenance and drove them into wage labor; governments defined private property and legal contract; the British poor laws of the nineteenth century ensured that landless laborers "freed" by enclosure and artisans displaced by industrial manufacture were confined in Dickensian conditions if they failed to find work in the new factories. In short: none of this developed spontaneously.

The corporation itself is a legal fiction, again created by the state, originally in Amsterdam, and sequentially throughout the world. Limited liability protects the individual shareholder from personal financial responsibility for corporate activities, thus encouraging risk beyond that which most people would take if their personal assets were on the line. The whole edifice of business ownership hinges on privileges bestowed by governments elected, in the main, by propertied males. Governments, especially active in the neoliberal era, have regulated, confined, and neutralized workers' organizations while presenting these measures as essential to freedom. Laws that seek to regulate corporations, on the other hand, even in the interests of the common good, are depicted by free market advocates as pernicious infringements of liberty.

Origins of the Modern Neoliberals

Though rooted in neoclassical economics, neoliberalism is not just a set of theories about how the world works but embraces a very specific program to modify the world, a program that is political in nature and varies accordingly in the settings where it operates.

Modern political neoliberalism is not without precedent. A very similar program was pursued before the Great Depression by corporations threatened by democratic processes that were giving a voice to the interests and values of working people. As Bernays's work suggests, the property-owning elites of Europe and North America were deeply worried about maintaining control over social structures and priorities; contemporary market fundamentalism echoes these concerns. Along with intellectual and academic traditions, neoliberalism has inherited practices of corporate propaganda dating at least as far back as the 1920s,[23] when, for example, the journalist Ernest Hofer ran a business distributing model articles and editorials celebrating free enterprise. He was the largest disseminator of such material at the time. Underwritten by numerous corporations, largely utilities, Hofer sent, free of charge, suitable "news" items and slanted articles to thousands of newspapers across the United States. He explained his purpose (which would sit well with any free market think tank of the current era) as follows:

> First, to reduce the volume of legislation that interferes with business and industry; second, to minimize and counteract political regulation of business that is hurtful; third, to discourage radicalism by labor organizations; ... fourth, a constant fight for reasonable taxation by state, city and county government; fifth, a scientific educational campaign against all socialistic and radical propaganda of whatever nature.[24]

These priorities were displaced by the Great Depression, which demonstrated an urgent need for regulation of business, as well as for adequate taxation and assistance to the poor and unemployed. US citizens voted Franklin Roosevelt into the presidency and embraced his New Deal, which favored a more humane approach to the victims of economic collapse and, it has been argued, "saved capitalism from itself."[25] Yet the policies pursued by FDR are the very ones that have been dismantled by the free market resurgence of the past three or four decades.

Hayek was a prime mover in this demolition project. Though he diverged from orthodox neoclassical economics in some respects,[26] Hayek nonetheless advanced the invisible hand concept of the market as a miraculous mechanism and the bedrock of the society he favored. He was deeply opposed to any system that would "replace the impersonal and anonymous mechanism of the market by collective and 'conscious' direction of all social forces to deliberately chosen goals"; he rejected notions of "common purpose," "common good," or "general interest." For Hayek, economic freedom was the essential foundational freedom that had brought about the increase in European wealth. Though neoliberalism has since masqueraded as a staunch friend of democracy, Hayek was not much interested in electoral democracy; his was a passionate plea for individual economic freedom in a market-based regime. He played a seminal role in the campaign to revive the popularity of the free market after Keynes and to reinstate it as the central influence in economic thinking and social policies. He aimed to construct a liberalism capable of supplanting not only socialism but social democratic forms of capitalism as well.[27]

To that end, he convened a conference in 1947, largely of economists wedded to market economics, plus a few libertarian philosophers and journalists. The attendees included Milton Friedman, who had just begun what would be a very long and influential career in the Economics Department at the University of Chicago. Hayek's meeting marked the founding of the Mont Pèlerin Society (MPS), which aimed, according to long-term member Ralph Harris, to "facilitate an exchange of ideas between like-minded scholars in the hope of strengthening the principles and practice of a free society and to study the workings, virtues, and defects of market-oriented economic systems." Overall, they would "launch an intellectual crusade aimed at reversing the rising tide of postwar collectivism."[28]

The MPS and its growing membership spawned many of the influential free market think tanks that came to dominate policy in the developed world, especially in the Anglophone countries. The first of these offshoots was London's Institute for Economic Affairs (IEA), launched in 1955 by Antony Fisher, one of the business members of MPS and a dedicated popularizer of Hayek's ideas. Fisher assisted in the establishment of further think tanks in the subsequent twenty years, including the Heritage Foundation in Washington, DC, in 1973 and the Manhattan Institute for Public Policy Research in New York, in 1977. An avalanche of such organizations followed after 1981, when Fisher founded his Atlas Economic Research Foundation, aiming "to litter the world with free-market think-tanks."[29]

Although they had a lot to say about what they called freedom, the MPS neoliberals had a much more ambivalent attitude toward the state than orthodox neoclassical economists did. Such a state could be entirely authoritarian, as was Pinochet's Chile, where neoliberals played key roles.

Electoral democracy is desirable when it supports the economic system neoliberals require. The political scientist Dag Thorsen argues, however, that if democracy gets in the way of neoliberal restructuring, "slows down neoliberal reforms, or threatens individual and commercial liberty, which it sometimes does, then democracy should be sidestepped and replaced by the rule of experts or legal instruments designed for that purpose."[30]

America's leading neoliberal theorist and evangelist was Milton Friedman, a key figure in what became known as the Chicago school. He was a founding member of MPS and held views on economic freedom that followed and amplified those of Hayek. He opposed government influence in the economy and government provision of services, which, he thought, the private sector would perform better. In line with his participation in Hayek's club, Friedman was not merely an academic and an adviser to conservative governments but a publicist for "economic freedom." He wrote weekly opinion pieces for newspapers and magazines, gave many public lectures, and, in 1980, with the American Public Broadcasting Service (PBS), he produced a ten-part television series, *Free to Choose,* setting out these views in a slick, persuasive format. A book based on the series was funded by the Scaife and Olin Family Foundations.[31] Friedman's Department of Economics at the University of Chicago teamed up with the US State Department in the late 1950s. Funded by US taxpayers and US foundations, the Chicago School trained hundreds of economists, inserted Chicago-style teachers into Chile's Catholic University, and unleashed a stream of free marketeers into Chilean society. Friedman himself met with Pinochet when he flew to Santiago six months after the military coup to consult with business on the neoliberal makeover of Chile's economy.[32]

The ideas of Hayek and Friedman had little influence on policy before the 1970s growth crisis. When Hayek debated Keynes through the 1930s and 1940s, Keynes tended to win the debate. In the aftermath of the economic calamity of the Great Depression, the alleged miracles of the free market could not command sycophantic approval. Furthermore, it was the policies Keynes championed that facilitated the extraordinary mushrooming of economic growth after the war—an economic triumph that blunted both public and professional memory of the free market disaster of 1929. It was not until the protracted economic crisis of the

1970s, with repeated oil shortages and intractable stagflation, that neo-liberalism found its entry point.

The 1970s: Oil Shocks and Growth Crisis

Just as the Club of Rome and concerned scientists began to focus public attention on doubts about the consequences of indefinite growth, the world economy ran into trouble. Even before the first "oil shock" of 1973, the United States was failing to maintain the value of the dollar in its postwar role according to the Bretton Woods arrangements,[33] as the international reserve currency pegged to gold. The United States had already been expanding the supply of Federal Reserve notes (sometimes known as printing dollars) before August 1971, when President Nixon unilaterally took the dollar off the gold standard; once this was done, the US exchange rate was no longer stable.[34] The world's reserve currency became paper only, backed by confidence alone. As well as providing international liquidity—a role of the reserve currency—the flood of new dollars, before and after 1971, helped pay the mounting debts associated with the war in Vietnam and the domestic antipoverty program instituted by President Johnson. But this fueled inflation at home, which led to inflation in the world economy as a whole.[35]

These trends were well under way before the Organisation of the Petroleum-Exporting Countries (OPEC) began to raise the oil price, causing the first oil shock of 1973, followed by another in 1979. Economic growth stalled. Geopolitical events played a part: the Yom Kippur War in Israel triggered the Arab oil embargo of 1973, and the 1979 Iranian revolution again restricted oil supply; both contributed to steep price rises. Indeed, the price of oil did not again reach an equivalent level in adjusted US dollars until the price spike of 2008.

OPEC had already been moving to nationalize the oil resource in several countries (including Algeria, Libya, Iran, Iraq, and Saudi Arabia) and thus wrest a bigger share of profits and more power over prices from the big seven oil corporations, all based in the developed world. But oil was still priced in US dollars, and the inflation of the dollar, once it was freed from its ties to gold and other currencies, disadvantaged OPEC countries, reducing the purchasing power of the proceeds of their oil sales.

The first price surge, in 1973, tripled the oil price in a few months. Since it had fallen some 20 percent below the inflation-adjusted US price levels of 1955, some increase was clearly warranted, though a threefold increase lifted it far beyond 1955 parity in US dollar terms. Whether this was fair or not depends on the yardstick used. Western analysts have usually viewed the oil price from the buyer's perspective, focusing on the escalating price at the pump; they have rarely mentioned the role played by the declining value of the paper dollar. Relative to gold, the oil price did not rise at all during the 1970s.[36]

Through the 1970s, OPEC's windfall profits added impetus to the inflation pulse, as oil producers poured huge quantities of the "petrodollars" they were reaping into the global investment pool. Capital seeking profit exceeded profitable avenues for investment, especially in contracting economies. As a result, much of this OPEC cash ended up being funneled through US and European banks to be loaned out to countries in the global south, often for major capital projects such as dams, power plants, or ports—projects that would be carried out by the largely US-based global corporations.[37] Many of the immense debts still borne by the borrowing countries originated at this time.

The economic impact of the 1970s oil price blowout involved an unfamiliar combination for the developed world; economists called it "stagflation," where recession (stagnation) coincides with inflation. The stagflation crisis highlighted the pivotal role of oil: as oil prices rose, so did production and transport costs, and as the increased costs depressed economic activity, the price of virtually everything rose. The Keynesian strategy of priming the pump with government spending, which had assisted governments in tackling the Great Depression of the 1930s and rebuilding the world after World War II, did not guarantee renewed growth when inflation was part of the problem.

The price surges of the 1970s reflected a scarcity that was serious but not terminal; by the mid-1980s prices had reverted to rock bottom again. Part of the price collapse followed from exploration and discovery in places like the North Sea, itself enabled by the high prices. With Arab and Iranian oil back on-stream in the eighties, world supply was again more than adequate for immediate demand. Inflation associated with the price of oil subsided. These events seemed to support the general axiom of economists that price, reflecting scarcity, can always conjure new supplies (box 6.2). American oil production had, however, peaked in 1971, exactly as the Shell geologist M. King Hubbert, the original analyst of peak oil, had predicted.

Neoliberals Take Charge: Thatcher and Reagan

By the mid-1980s, improvements in profitability and economic growth owed much to the resurgence of adequate oil-supply at bargain prices. The recovery is often attributed, however, to the economic policies adopted by Prime Minister Thatcher and President Reagan, and then gradually by the entire developed world. These policies adopted the economic approaches proposed by Hayek, which had been dormant since before the Great Depression.

Hayek's central thesis, that government always exercises a detrimental influence on the economy and should not be involved in economic activity, had been put forward in *The Road to Serfdom*, a bestseller that appeared in the United States in a *Reader's Digest* version within a year of its initial publication in 1944[38] and was much admired by Thatcher. Neoliberal rhetoric blamed the 1970s crisis on government "interference" in economic activities, and proposed a new regime of "freedom" for business. According to the neoliberal doctrine, accumulation could be reignited by the triptych of privatization, deregulation, and tax reduction (a program also known as supply-side economics). The practical result was the privatization of the infrastructure of the developed world, the gradual opening of economies worldwide to unrestricted foreign investment, the erosion of progressive taxation, and the celebration of "free trade" as the panacea of prosperity for the globe. This regime remains with us in the twenty-first century. Even after the global financial crisis of 2008 challenged the logic, these prescriptions for economic success still prevail.

Corporate America had never really relinquished its campaign against New Deal policies, or its pro-market, antiregulation, anti-union, anti–social welfare message. However, a modified balance of power between capital and labor did persist in the United States up until the 1970s, exemplified in the greater share of national income held by wage-earners and the far lower share of national income held by the top 1 percent of US households. Wealth follows a similar pattern, but is even more

Box 6.2
When Price Signals Fail

The market occupies a sphere dedicated to short-term profit, divorced from ecological realities, so that when production is robust, prices reflect the immediate glut, not the ultimate scarcity, a situation common both to oil and the Newfoundland cod fishery.

Mainstream economists insist that scarcity is always reflected in price and that price in the marketplace is a reliable mechanism for regulating the flow of resources—indeed, the only efficient one. That alleged efficiency was not demonstrated in relation to oil: though it may have facilitated the recovery of new sources in the 1970s when relatively accessible oil was still available to be exploited, it has been a poor and approximate mechanism as supplies of cheap oil declined (see chapter 14 for the current situation with shale oil and gas). There was no gradual rise in 2007–2008, for example. Instead, prices hit stratospheric levels in a matter of a few months.

The Canadian cod fishery was an early and catastrophic example of the decline of fisheries worldwide. After unprecedented levels of harvesting from the late 1950s, the catch collapsed to nineteenth century levels in the mid-1970s; the fleet renewed its intensity, but cod had virtually disappeared by 1992 and has not recovered (figure 6.1).[a]

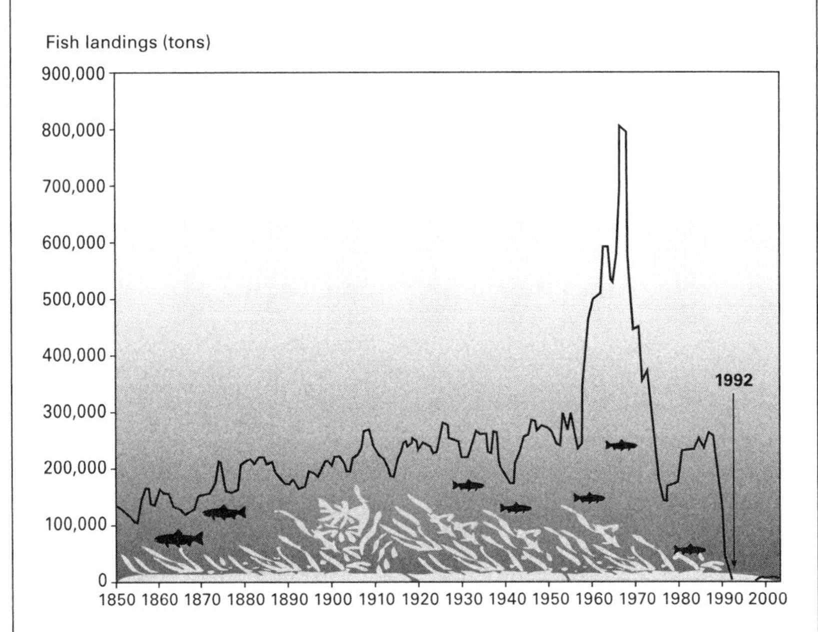

Figure 6.1
Collapse of Atlantic Cod Stocks off Newfoundland in 1992.
Source: Millennium Ecosystem Assessment 2005. Courtesy of World Resources Institute.

Box 6.2 (continued)

Even though the world fishery as a whole is in decline,[b] seafood prices do not properly reflect this fact, and do not much moderate the level of exploitation. Though prices for Patagonian toothfish and bluefin tuna are now extremely high,[c] this has occurred only as the stocks have approached collapse—and there are sufficient ultra-rich consumers to keep paying the prices in any case (see the discussion of Citigroup's Plutonomy Report in chapter 15). Daniel Pauly has pioneered the ecosystem approach to fisheries analysis, exploring the phenomenon of "fishing down the web" to ever-lower trophic levels, a process that may end up offering a harvest of little but jellyfish.[d] Pauly has also stressed that, in estimating the true losses of biomass, original abundance should be taken as the baseline rather than arbitrary benchmarks of a decade or two ago;[e] the destruction involved in ongoing economic expansion is obscured when comparisons are restricted to recent times. In the case of fisheries, price signals, further distorted by widespread government subsidy, have had little or no influence on the conservation of the resource.[f] In the case of exhausted fish stocks, substitutes are unlikely to be generated by high prices.

Notes

a. Millennium Ecosystem Assessment 2005, 58.

b. Pauly 2010.

c. Evans 2012.

d. Gershwin, 2013.

e. Pauly 2010.

f. Pauly 2011, 34–35. Pauly sees the principal obstacles to sustainable fisheries as fleet overcapacity, biomass reductions of at least an order of magnitude for large fish (such as cod, tuna, and large pelagics), wastage of one-third of the global catch on fish-meal, a trade regime that encourages first world importation of third world fish when our own have declined, and the existence of some $30–$35 billion in government subsidies which facilitate overfishing.

polarized. At the time of the 1929 Wall Street Crash, the top 1 percent of US households held close to 50 percent of the nation's wealth, contracting to 35 percent around 1940 and rising again to 40 percent during the war. This share had declined to little more than 20 percent in the mid-1970s before neoliberal measures began to be implemented. By 1995, it was back above 35 percent.[39]

Sections of the business community had never accepted greater relative equality, finance capital in particular. Manufacturers were more likely to settle for the Keynesian approach and the greater share of profits conceded to workers because, on the one hand, there was a grave concern about the spread of communism, and on the other, the devastation of their businesses after the 1929 crash had been triggered by speculative financial dealings during the 1920s.[40] Yet, even if these views generated a degree of compromise in that sector through the postwar boom, there was never much tolerance for unions from US business, and the dissemination of pro-business propaganda hardly missed a beat. The acceptance of union participation in society that characterized some other forms of capitalism (in Scandinavia, for example, and Australia to some extent) came under attack from neoliberals worldwide.

Neither did the finance sector accept for long restrictions such as capital controls and currency rules. By the late 1950s the first steps had already been taken to reestablish capital mobility, so that international financial markets revived through the 1960s, putting pressure on the overvalued US dollar. In Harvard historian Jeffry Frieden's view, everything changed once the shock of the wars, the Great Depression, and the attendant unraveling of the world's formerly integrated economy began to recede.[41] Once the shattered cities of Europe and elsewhere had been rebuilt, the sheer success of the postwar order brought national interests back into conflict with the international economic system. Frieden sees Nixon's move away from gold as a choice for domestic popularity in a pre-election year, in preference to honoring US responsibility as the linchpin of the postwar international system.[42] In any case, free market enthusiasts and their business backers, already organizing themselves into a plethora of think tanks, sought an end to all arrangements extraneous to their priorities. While some sections of some societies saw the erosion of their national ability to control capital flows as a threat, few countries were able to resist successfully.

When the British Conservative Party lost power in 1974, Keith Joseph, a cabinet minister in the defeated government, embarked on a project to make neoliberalism the creed of the party and the nation. Joseph had been familiar with the Hayekian IEA since 1964, and now proceeded to preach the economic policies of Hayek and Friedman. He soon founded his own free market think tank, the Centre for Policy Studies (CPS), with Thatcher as vice president. Joseph went on to elaborate what would ultimately be called *Thatcherism*: breaking trade union influence, fighting inflation with monetary policy, deregulation, privatization, and tax cuts.

Five years later, at the end of the "winter of discontent,"[43] Thatcher won the May 1979 election. According to his biographer, Mark Garnett, Joseph spent those five years on a "crusade to convert the country to his way of thinking," believing "it was his duty to fight back on behalf of the free market."[44]

Thatcher herself was a devotee of Hayek's ideas. At a British Conservative Party policy meeting in the late 1970s, in response to a paper recommending a "middle way" strategy, she slammed one of Hayek's books down on a table and announced, "This is what we believe."[45] Once she was elected, neoliberalism became firmly entrenched in British economic policy. As she remarked in 1993, "The spirit of enterprise had been sat upon for years by socialism, by too-high taxes, by too-high regulation, by too-public expenditure. The philosophy was nationalisation, centralisation, control, regulation.... This had to end."[46]

Across the Atlantic, the Carter administration appointed the monetarist Paul Volcker to head the Federal Reserve in 1979. Closely associated with Wall Street, Volcker immediately tightened money supply and pushed interest rates up to 15 percent—and later close to 20 percent—a measure intended to tackle the inflation problem but also providing a handsome restoration of profitability to financial interests.[47] Throughout the 1970s, business had cut costs in whatever way it could, including moving production overseas, freezing wages for existing workers and lowering wages for new. But the US economy remained mired in stagflation, and people were faced for years on end with price surges, shortages at the petrol pump, and sky-high interest rates.

Reagan was another admirer of Hayek's work and used rhetoric similar to that of Thatcher—curbing welfare, balancing the budget, and cutting taxes. As in the UK, the neoliberal publicists were already well established in proliferating think tanks. Nearly a third of Reagan's economic advisers were members of MPS,[48] and the Heritage Foundation provided the newly elected Reagan administration with a massive 1,000-page volume called *Mandate for Leadership*, analogous to *The Brick*, written by Chicago school economists and adopted by the incoming Pinochet dictatorship.[49] Reagan gave his cabinet members a copy of *Mandate*, a guide to the free market way. It was described by one supportive journalist from the *Washington Post* as "a kind of handbook for the new administration."[50] The Heritage Foundation claims that nearly two-thirds of its policy recommendations were implemented. Some thirty of the authors of this tome were appointed to the administration, including the author of the tax policy.[51] Between them, the Heritage Foundation, the Hoover Institution, and the American Enterprise Institute contributed 150 personnel to Reagan's administration.[52]

The influence and tactics of the neoliberal intelligentsia were similar on both sides of the Atlantic. MPS member Ralph Harris, head of the IEA and mentor to Keith Joseph, saw the role of the think tank enthusiasts this way:

> Americans can best judge the influence of the many MPS members surrounding President Reagan. From Britain I have no quiver of doubt that Margaret Thatcher's central reform of trade unions, state industries, monetary policy, and much else owed a great deal to the advisors and members of Parliament directly instructed in market analysis by IEA publications shaped by Mont Pèlerin principles. But the decisive role was played by our academics and journalists who helped transform public opinion on the market alternative to the failing collective consensus.[53]

"Globalization Round II": Rules Are Relative

Neoliberalism not only drew on ideological precedents and a long history of business propaganda, it also aimed to revive the world market that had emerged before World War I. Thomas Friedman, a *New York Times* columnist and ardent free trade enthusiast, argues that "the first era of globalization and global finance was broken apart by the successive hammer blows of World War I, the Russian Revolution and the Great Depression." He characterizes the "new era" as "Globalization Round II."[54] Friedman, for whom the late twentieth century resumed where the early twentieth century "robber barons" left off, has no trouble being blunt: "The driving idea behind globalization is free-market capitalism," he says.[55] For the citizens of the global south at the World Social Forum at Porto Alegre in 2001, globalization was also understood as the latest manifestation of capitalism, though they referred to a longer history than Friedman's, seeing it as "part of the continuum of colonization, centralization and loss of self-determination that began more than five centuries ago."[56]

Although the globalization narrative, as purveyed by the laudatory media and embraced by most governments worldwide, implies that it is about breaking down barriers and embracing the whole world on one big level playing field, it is also clear that labor rights, economic justice, and the environment are not up for inclusion in the new global rule book (discussed in chapter 13). These values are not to be globalized, as the media critic Norman Solomon has argued:

> The form of "globalization" deemed worthy of the name by media is corporate globalization, which gives massive capital even more momentum to flatten borders and run roughshod over national laws.... Fans of "globalization" routinely contend that protection of labor rights or the environment amount to unfair restraint of trade, retrograde protectionism and antiquated resistance to "reforms."[57]

Thus, globalization reflects the preferences of the global business elite and ignores those of workers, the world's peasant populations, people concerned about protecting the environment, indigenous peoples, or people committed to social justice.

Though Thomas Friedman favors a more liberal application of the "level playing field" where the views of such groups might be heard, he nonetheless celebrates market forces as the driver of Schumpeter's creative destruction:

> The more you let market forces rule and the more you open your economy to free trade and competition, the more efficient and flourishing your economy will be. Globalization means the spread of free-market capitalism to virtually every country in the world.... The essence of capitalism is the process of "creative destruction"—the perpetual cycle of destroying the old and less efficient product or service and replacing it with new, more efficient ones.... Those [countries] which rely on their governments to protect them from such creative destruction will fall behind in this era.[58]

It was instructive to observe, in the course of the October 2008 financial meltdown, that the harsh market prescriptions imposed on the Asian economies in their financial crisis in the late 1990s were not deemed suitable for the global north. In 2008, *USA Today* reported the outrage of Koreans that, back in 1998, the "Americans told them to sell off assets and get the government's hands off the private sector"[59]—no bailout for them. The destruction side of free market creative destruction, though imposed mercilessly on the developing world as bitter but necessary medicine, was assiduously avoided as the first world provided massive government support to its "free market" institutions, apparently quite willing to risk "falling behind," as Friedman puts it. President Bush told

CNN television in mid-December 2008, "I've abandoned free-market principles to save the free-market system."[60] The rules of the free market game are surprisingly fluid; they do not apply to the most powerful players or, at least, are modified when these players' own interests are at stake.

"Global Middle Class" to Save the World Economy

The broad acceptance of deregulation facilitated liquidation of the very natural world that the scientists of the 1960s and 1970s had so urgently argued needed protection. Indeed, unfettered extraction was one strategy in the pursuit of renewed growth. Notions of embracing slower growth, scaling down, or seeking a "steady state" remained outside mainstream concepts of reality—for business and government alike. The neoliberal concept for the world economy advocated extending the consumer template to the whole world.

On the eve of the 2007 meeting of the Asia-Pacific Economic Cooperation (APEC) forum in Sydney, the Economic Analytic Unit of the Australian government's Department of Foreign Affairs and Trade (DFAT) published its report, *APEC and the Rise of the Global Middle Class*, a document that describes APEC as made up of "member economies" (not member countries). Focused squarely on the goal of economic growth, it claims that "international integration and market liberalisation" (popularly known as globalization) have led to the economic growth of recent years, which in turn is driving the emergence of a new global middle class, predicted to expand by some 2.2 billion people by 2030, many of them concentrated in the Asia-Pacific region. The report expresses satisfaction that the "increased purchasing power" of this vast new consumer class "is contributing to the recent strength of global growth and should drive stronger global growth in the future, helping to lift millions more out of poverty and build further wealth ... boosting their living standards as the pool of global consumers grows." The rising global middle class is at one point described as "the dividend" of economic globalization and at another as "a down-payment on ... the fight against extreme poverty."[61] The future of human civilization is viewed through the lens of consumption, which, along with growth itself, is understood to be the solution to poverty.

The strategies for fostering these developments are familiar neoliberal nostrums: free trade, liberalizing service industries, and expanding foreign investment; strengthening the financial sector (there is no hint or suspicion of the financial collapse that was to occur just twelve months later); and finding ways to tackle environmental problems "within a framework of continuing economic growth," since "the key question for policymakers is how to ensure that strong growth continues." The report subordinates all environmental issues, including global warming, to growth, and settles for the hope that more growth will eventually allow attention to be paid to the degradation caused by growth in the first place. The central motive for the report's commitment to environmental sustainability seems to be to "ensure that environmental degradation does not threaten ongoing growth."[62]

Based on the World Bank's modeling for its 2007 *Global Economic Prospects*, DFAT projects that the middle class will double as a percentage of the world population by 2030, while the percentage of poor will decrease by 20 percent. (Middle class is defined here as having some scope for discretionary expenditure over and above basic necessities; in the World Bank's terms, this means their incomes are above about $12 a day.) The figures do indeed project an extra two billion middle-class people in 2030, and this outcome is said to represent "an unprecedented decline in poverty and increase in affluence." Urbanization will increase and an explosion of new consumption is expected, its principal elements being access to meat in the diet, car ownership, tertiary education, mobile telephony, and international travel. Citing Goldman Sachs, the report predicts that China could well have over 500 million cars on its roads by 2050 and India even more—between them far exceeding the entire global passenger fleet in 2008, estimated at 622 million.[63] The report is silent on the implications of such a situation for either greenhouse gas emissions or petroleum consumption; it ignores the danger to climate and assumes resources will be available. According to DFAT, "the emergence of a new consumer class represents the chance for business to tap new markets, creating still further employment opportunities."[64] A brave new world of multiplied consumers will drive economic growth—DFAT's analytic economists go no further.

The APEC report stresses that the poor (non-middle class) will decrease as a percentage of world population. The poor will, however, actually *increase* in gross numbers, though the report does not mention this aspect of the statistics. More than five billion people will still be poor in 2030, according to the World Bank's modeling. More than five billion people who lack the latitude of discretionary spending will still be struggling to keep their families fed, clothed, and housed. This does not seem to be such a "good news" story after all. Although the UN's Millennium Development Goals apply to the "extremely poor" (defined very narrowly as those living on less than $1 a day), it is difficult to see much ground for optimism; if the gross numbers of people living in poverty are not going to decrease, it is hard to share DFAT's satisfaction with the role of APEC or globalization or economic growth in "lifting millions out of poverty." The most that can be claimed about the figures presented here is that most of the 2.2 billion extra people expected on earth by 2030 will be added to the "global middle class" rather than to the mass of the poor. This assumes, however, that roughly three times as many Western-style consumers can be accommodated as occurs today—which seems unrealistic.

The neoliberal—or globalization—agenda so accurately depicted in the DFAT report did not cause the ecological problems that have built up for at least fifty years, but it has compounded them. What we now confront is the moment when the exponential curve has turned the corner and is approaching the planet's physical limits, when, for the first time, we are doubling massive populations and colossal production systems every few decades. The economist Ross Garnaut, in his July 2008 address to Australia's National Press Club, remarked in passing that world economic output will be fifteen times greater in the course of the current century.[65] The financial collapse of late 2008 and the subsequent ongoing recession slowed this trajectory only modestly, with all governments (along with the corporations) frantic to restart the expansion. Since material artefacts will make up the majority of such an expanded output, it is hard to imagine what kind of world would harbor a human economic apparatus fifteen times greater than the present one.

As the sustainability advocate Gianfranco Bologna argued in Wackernagel and colleagues' first *Footprints of Nations* study:

> Western Europe and North America, when entering their period of rapid modernisation after World War II creating a modern consumer economy ... contained "only" 440 million people (280 in Europe and 160 in North America). Today

Asia—the region from Pakistan eastwards till Japan—has 3.1 billion people, more than half of the world's population.[66]

The cornucopian promise of global prosperity needs to be considered in the light of these figures. The US or Australian rate of consumption is not a realistic goal for everyone, suggesting that American and Australian consumption will need to contract. If China alone were to use oil at the per capita rate of the United States, for example, it would require some 82 million barrels a day—only marginally less than the whole world currently uses.[67] In the case of paper, Chinese consumption per person at US levels would take more paper than the world produces.[68] Even consumption at the more moderate European rate, about half the US or Australian rate, is unlikely to be viable for China, let alone for everyone. When large increases in population are factored in, the idea of generating prosperity for all through accelerating economic growth would seem laughable if it were not the apparent intent of governments and businesses worldwide.

The Garrett Hardin approach, where the rest of humanity is abandoned to starve, is obviously unacceptable. But the consumer route to plenty is also fatally flawed. Strategies other than growth are thus clearly required to address the poverty that still prevails for almost half the world's people. After a century that saw twentyfold economic growth, billions of people still lead precarious lives, indicating that growth has not been very successful in addressing their needs. Furthermore, if growth at twentieth-century levels and consumption at first world levels is not universally feasible, growth would be a questionable tool, even if it had a more promising record. Contraction on the part of the developed world appears to be an essential aspect of the only option in this as in almost all other issues of consumption and environment.

In the next chapters, I turn from the neoliberal pursuit of growth in the first world to the global south and the history of the application of the growth template to its people. Two interrelated questions loom. First, who has benefited and how successful has development been in addressing the imperative unmet needs of third world people? Second, what has happened to the natural environments on which billions of these people rely?

7

"Development" and Globalization: Exporting Growth

The house economy is based on livelihood; the corporation's on acquisition.
—Arturo Escobar, 1995

The problem of poverty lies not in poverty but in wealth.
—Wolfgang Sachs, 1999

Colonial Roots

As indicated in the introduction, the world beyond Europe was progressively overtaken by the expansion of the past five hundred years. Unlike many of his disciples of more recent times, Adam Smith was alert to the uneven distribution of benefits from this protracted series of events. A late eighteenth-century witness, aware of the extraordinary commercial success of the colonial enterprise for the metropolitan powers, Smith did not observe any concomitant new-found prosperity among the conquered. Quite the reverse:

> The discovery of America, and that of a passage to the East Indies by the Cape of Good Hope, are the two greatest and most important events recorded in the history of mankind.... To the natives, however, both of the East and West Indies, all the commercial benefits which can have resulted from these events have been sunk and lost in the dreadful misfortunes which they have occasioned.... At the particular time when these discoveries were made, the superiority of force happened to be so great on the side of the Europeans, that they were enabled to commit with impunity every sort of injustice in those remote countries.... In the mean time, one of the principal effects of those discoveries has been, to raise the mercantile system to a degree of splendour and glory which it could never otherwise have attained to.[1]

The "misfortunes" occasioned by Europe's superior force include the horrific fate of the indigenous people of Potosi (now Bolivia), forced to mine silver for Spain in the sixteenth century; the Moluccan Islanders' spice trees, ripped out by the Dutch to preclude competition in the seventeenth; the depredations of the slave trade, visited on West Africa for several centuries; and Congolese rubber tappers, enslaved into the twentieth century by King Leopold of Belgium, who ordered the amputation of their children's limbs when they failed to meet quotas of rubber production for the lucrative new bicycle trade.[2]

The state of the world and its people today is largely the legacy of that history. Whole continents were appropriated, their natural resources repatriated back to Europe, and regimes of property ownership suitable to European goals and interests were frequently imposed. The newcomers carved out entities to suit themselves, imposing boundaries where none had been, boundaries that often proved inappropriate and irrational when the colonies were launched as nations after World War II. As the world outside Europe was progressively reconfigured to harvest this "splendour and glory" for the rich world and its allies, its indigenous peoples were infected, murdered, enslaved, and dispossessed.

The strategy of promoting increasing consumption has been touted as one that will solve the enduring poverty of the erstwhile colonial world. The notion is popular, as *APEC and the Rise of the Global Middle Class* suggests, even though, as the figures demonstrate, the gross numbers of the (non-middle-class) poor are not predicted to decrease under this plan (discussed in chapter 6). Even before the global financial crisis of 2008–2009, poverty was expected to be reduced only in relative terms and, given the risks built into the market economy, might not be ameliorated at all. The financial crisis, when it came, cut across this optimism. The UN Children's Fund[3] warned in June 2009 that an extra 100 million people were going hungry in South Asia alone compared to the situation two years earlier. Later that month, the UN's Food and Agriculture Organization (FAO) stated that more than one billion people were going hungry every day for the first time in human history. Although the FAO noted a reduction during 2010, it warned that renewed increases in food prices would reverse this trend and that economic growth alone would not solve the problem. Food prices did rebound in 2010 and, though they have declined again, have not returned to the levels seen before the price upsurge that began in 2007.[4]

The central strategy offered to the rest of the world by the spokespeople of the rich has been the expansion of the prosperity "pie," though methods of pie inflation have varied. A bigger pie will do the trick, we have been told for sixty years, with planned development as the method in the first thirty years and a version of free trade in the next thirty. In both scenarios, the poor would ultimately get a decent slice and the rich could expand their opulence. In this approach, the rich need not concern themselves with any call for fairness or redistribution; nor need there be any accounting of the historical debt of the European cultures to the people they colonized.

Progress and the Development Discourse

Although the idea of Western-style economic development for poor countries is taken to be natural, inevitable, and the self-evident solution to poverty in 2013, this was by no means obvious as World War II drew to a close. Indeed, during the preceding centuries of colonial rule, the metropolitan powers tended to suppress any competition their colonies might offer, and the European rulers of much of the planet were not particularly distraught about the poverty of the ruled. Australian economist H. W. Arndt's semantic history of the concept "economic development" suggests it made only scattered appearances in public debate through the early twentieth century. "Material progress" and "economic progress" were much commoner terms; these did not imply intentional pursuit but, rather, were expected to flow naturally from profit-making activities.[5] Only after the war did "economic development" gain ground, with the stated agenda of raising the per capita income and the national welfare of entire populations. By 1960, a whole new academic field of development economics had defined people's well-being in terms of economic growth and the exploitation of resources.

President Truman's inauguration speech in January 1949 marked the point where the United States signaled its intention to extend modern industrial production to every corner of the earth:

> More than half the people of the world are living in conditions approaching misery. Their food is inadequate, they are victims of disease.... For the first time in history humanity possesses the knowledge and the skill to relieve the suffering

of these peoples.... I believe we should make available ... the benefits of our store of technical knowledge in order to help them realise their aspirations for a better life.... What we envisage is a program of development based on the concepts of democratic fair dealing ... Greater production is the key to prosperity and peace. And the key to greater production is a wider and more vigorous application of modern scientific and technical knowledge.[6]

The new development economists[7] pursued the Truman vision and regarded technology and capital accumulation as the major elements of human progress. Prominent among them were Yale graduate and high-level government adviser Walt Rostow and Caribbean-born W. Arthur Lewis, who spent much of his academic career at Princeton and shared the Nobel Prize in Economics in 1979.

In *The Stages of Economic Growth: A Non-Communist Manifesto*, published in 1960, Rostow described five stages of economic growth, with apparent confidence in their inevitability. Stage one societies—traditional "pre-Newtonian" societies—lack science and technology, which sets a ceiling on productivity. The second stage prepares societies for "take-off" as modern technique is applied and the centralized nation-state is established. In stage three, when the nation "takes off," savings and investment surge and "growth becomes its normal condition." Take-off is followed by stage four, a longer period of consolidation, as investment and savings increase and the new surge in profits is reinvested. In the fifth stage the economy graduates to "the age of high mass-consumption," considered by Rostow to be the hallmark of economic maturity.[8] Curiously, Rostow never mentioned coal or oil—or any kind of fossil energy—as the driver of the growth he regarded as the ultimate stage of history, preferring human-generated "science and technology" for that role. Nor did he or any of his colleagues question whether the template of the past could be automatically applied to the present.

Lewis's influential ideas were summarized in his 1954 essay, "Economic Development with Unlimited Supply of Labour." He saw the "underdeveloped countries" as composed of two worlds, split between a "primitive" sea of economic darkness and rare islands of urban light where the Westernized men of the future had already embraced suits, European languages, Beethoven, and philosophers like Mill. The gradual displacement of traditional cultures and subsistence livelihoods by the industrial money economy was both necessary and inevitable. "Take-off" demanded increased savings as the first step toward the accumulation of the capital on which all progress depended; these savings could occur only within the elite:

> We are interested not in the people in general, but only say in the 10 percent of them with the largest incomes, who in countries with surplus labour receive up to 40 percent of the national income ... the remaining 90 percent of the people never manage to save.... The central fact of economic development is that the distribution of incomes is altered in favour of the saving class.[9]

This argument, suggesting that income must be concentrated in the elite classes, contradicts Truman's supposed objective of bringing solace to the mass of the people "living in misery"; as for "democratic fair dealing," there was certainly no plan to ask *them* what they thought or wanted. Equally problematic, the home-based savings effort envisaged by Lewis occurred only rarely, and whatever take-off materialized relied on foreign investment and borrowing from Western banks. It was transnational corporations that accumulated most of the capital rather than the local people.

Arturo Escobar, a Colombian anthropologist and critic of these development theories, quotes a 1951 UN committee on which Lewis served:

> There is a sense in which rapid economic progress is impossible without painful adjustments. Ancient philosophies have to be scrapped; old social institutions have to disintegrate; bonds of caste, creed and race have to burst; and large numbers of persons who cannot keep up with progress have to have their expectations of a comfortable life frustrated. Very few communities are willing to pay the full price of economic progress.[10]

The World Bank and various UN development agencies were centrally involved in drawing up plans for numerous countries, often without input from the people. Escobar, who witnessed the bank's first big plan for Colombia, noted how little say Colombians had in the process.

The theories of development on which almost all the interventions were based harbored two key assumptions. In the first place, economic growth was regarded as an inevitable stage of human civilization, a natural and linear progression from more "primitive" social forms to modernity, with European history providing a universal template. Second, economic growth was seen as a process of indefinite duration, with no limits in space or time.

For the descendants of the Europeans (me among them), modernity has been a considerable blessing. Women, certainly, have more rights to independence and fairness than in most other stages of known history. Both men and women in Western countries have been progressively freed from the threat of famine and starvation that recurred through feudal times in Europe, and the majority of first world people have enjoyed some measure of material security during most periods since the early twentieth century.

An immense price was paid for these advances, however. The peasants of England in the first place, and of Europe as time went on, were separated from their livelihoods and cast adrift as landless people seeking work and a wage. One of the main penalties of modernity was exacted from these rural commoners, who were progressively dispossessed, and from their children, who migrated over time into the industrial cities.[11] An even greater penalty was paid by the dispossessed and enslaved populations of Europe's colonies.

Even if the brutal effects could be eliminated from the equation, there are few grounds to believe that the path embraced by early modern Europeans could be replicated in the postcolonial world. From Cortez to Cook, they launched out into a poorly defended world where an apparently infinite supply of natural riches was available for them to expropriate—not a situation in which developing countries found themselves after World War II. The new Indonesia, for example, had access to numerous "empty" islands and West Papua, but none of the new nations had the "superior force" that would allow them to sail off and conquer territories on the other side of the world. The specific historical conditions enjoyed by Europe during its colonial expansion were ignored by the development theorists. These conditions included vast quantities of cheap (often stolen) resources, and captive peoples who could be fashioned into markets and obliged to accept tens of millions of Europe's own surplus population, people who themselves had no livelihood, having been swept from the land into the cities. The development theorists also ignored the role that the entire colonial process had played in the impoverishment of the countries now being designated as underdeveloped in the mid-twentieth century. For Truman, these pitiable people were stragglers who only needed help along the same route. Greater production—a bigger pie—was the key; the scientific and technical knowledge that blossomed out of the European heritage was thought to be the appropriate means to that end.

None of the development economists looked at the primitive accumulation that underpinned European savings and investment, or noticed that most such avenues were not available to third world countries in 1950. Indeed, as noted above, the funds required for the "take-off" and "big push" into the rapid accumulation of capital had to be borrowed from the Europeans themselves, which ultimately led to endemic debt crises and failure to accumulate much at all except debt. Such vastly different circumstances may help to explain why, although significant industrialization occurred in the non-Western world during the development era, prosperity did not follow.[12] With few exceptions, Truman's stragglers did not catch up.

Escobar identifies what he calls the development discourse as a version of Edward Said's "orientalism"—the way Europeans have defined and managed non-European cultures through the lens of their cultural superiority. For Escobar, this attitude was inherent in the idea that it was essential to "modernize" peasants and submit feudal relations to marketplace rationality, to sweep aside "backward" cultures in their own interests.[13]

The development planners of the first thirty years after World War II did indeed seem indifferent to actual outcomes. While the so-called "Brazilian miracle" powered along for several years around 1970, for example, with growth in Brazil's GDP running at 10 percent, poverty and unemployment increased, the distribution of income was further polarized, and low-income groups were worse off in absolute terms than they had been.[14] In this case, the pie (measured as GDP) certainly grew, but only Lewis's saving class benefited.

There is no disagreement among historians and economists of all persuasions that the world experienced a huge surge in economic growth during the 1950s and 1960s. The World Bank measured an annual global rate of GDP growth between 4 and 6 percent in every postwar year until 1974, levels matched again in only a few scattered years between 1975 and 2004;[15] this amounted to a global tripling of industrial production in the twenty years from 1950 to 1970. GDP grew everywhere in those decades, including Africa, but the most rapid expansion occurred in Japan and in Western Europe. Though GDP growth in *percentage* terms was comparable in the third world, the minuscule base dictated a far smaller gross expansion there, and rapid population growth also minimized the per capita increase.[16]

Neither is there any argument that third world countries did see an increase in industrialization in the development era, as the gap between first and third world manufacturing capacity narrowed. What is doubtful is the extent to which these developments actually led to the catch-up in prosperity that Truman advocated and the development economists confidently predicted. Though first and third world levels of industrialization converged between 1960 and 1980, per capita income did not.[17] The hierarchy of wealth between the OECD countries and the rest persisted, with few exceptions. In short, the development project, while fostering industrialization in the third world, had little impact on prosperity there, suggesting that the reliable causality assumed between the two was mistaken. The first development era's rhetoric of saving the world transferred the values of growth economics to the global south, while sharing few of the potential benefits with the people.

Globalization: Over to the Market

The development model fell from favor inside the international institutions of the first world after 1980 and was rapidly replaced with neoliberal nostrums, especially the idea that free trade would solve all the world's problems, including "underdevelopment." During the course of the next decades, the full suite of solutions was offered and in many cases imposed by the World Bank and the IMF. When third world countries were in financial trouble, funds were made dependent on "structural adjustment programs" (SAPs), which required privatization, tax cuts, and the dismantling of whatever meager welfare programs the state might have had in place. The very same institutions that had been the arbiters of the Bretton Woods system from World War II until 1973 were still in charge of the globalization agenda. The changes concerned only the ruling ideology within those organizations, not a realignment of the power of the institutions themselves.

With uncanny echoes of the promises of the late 1940s, purveyors of the new orthodoxy claimed—just as Truman had done—that the means to solve the poor world's problems were finally at hand. In *The End of the Third World* (1986), the economist Nigel Harris argued that the dispersal of the global manufacturing system was bringing the third world to an end and breaking down the old simple distinctions between

"First and Third, haves and have-nots, rich and poor, industrialised and non-industrialised." Instead we have "one world [which] offers the promise of a rationally ordered system, determined by its inhabitants in the interests of need, not profit or war."[18] Eleven years later, in 1997, Renato Ruggiero, the first director-general of the World Trade Organization (WTO), claimed that we now had "the potential for eradicating global poverty in the early part of the next century—a utopian notion even a few decades ago, but a very real possibility today."[19] The claim that millions have been, are being, or will be "lifted out of poverty" by the wonders of globalization still litters the pages of think tank articles, World Bank reports, and the popular press, appears several times in the APEC report discussed in chapter 6 and is endlessly repeated by senior politicians, businessmen, and economists of the first world.

These enthusiastic claims do not, however, tally with empirical research. Political economist Giovanni Arrighi and colleagues found that the free trade era was no more successful in lifting incomes in the third world than the development era had been. A surge in industrialization occurred in third world countries in the period from 1980 to 1998, and though a handful of countries improved their income situation—notably South Korea, Taiwan, and, increasingly, China—the expected improvement was not spread evenly.[20] The prominent World Bank economist William Easterly confirms these findings. From 1980—when SAPs began—to 1999, third world countries stagnated economically "in spite of policy reform." Notwithstanding the inclusion of China's sustained growth in the calculation, median per capita growth in the overall third world was zero, a major reduction from the 2.5 percent figure for the two decades before 1980.[21]

Free Trade: "Kicking Away the Ladder"

Free trade is an axiomatic good in the neoliberal ideological armory. Trade is supposed to add to the wealth of nations by giving each the chance to do a lot of "what it does best"—and to import other things with the proceeds.[22] Nancy Birdsall, a development policy researcher with many decades of experience at the World Bank, the American Development Bank, and development research organizations, challenges the automatic benefits said to be bestowed by free trade. She suggests that the market power of the rich allows them to impose rules and regimes to their own benefit and deploy their immense resources in their own interests. The poorest countries do not stand a chance:

> Highly dependent on primary commodity and natural resource exports in the early 1980s, their markets have been "open" for at least two decades, if openness is measured by their ratio of imports and exports to GDP. But unable to diversify into manufacturing (despite reducing their own import tariffs) they have been victims of the decline in the relative world prices of their commodity exports, and have, literally, been left behind.[23]

Among many difficulties—and despite lowering their tariffs—third world countries have so far been unable to force Europe and the United States to reciprocate by reducing their large agricultural subsidies. Rich developed countries provided more than $250 billion in 2011 in support and protection for their own agriculture;[24] this is a perplexingly large subsidy in what is supposed to be a free trade regime. In the globalization period, earning foreign currency is reckoned to be more important for developing countries than growing food by and for local people, especially in the view of the World Bank and the IMF, who wish to see loans repaid. Free trade has brought another generation of smallholders under terminal pressure, including dairy farmers in Jamaica, tomato farmers in Senegal, and chicken farmers in several West African countries.[25]

The historical evidence from the countries of the developed world does not support the idea that free trade underpinned their rise to economic dominance. As the Cambridge economist Ha-Joon Chang argues in *Kicking Away the Ladder*:

> How did the rich countries *really* become rich? [They] did not get where they are now through the policies and the institutions that they recommend to developing countries today. Most of them actively used "bad" trade and industrial policies, such as infant industry protection and export subsidies—practices that these days are frowned upon, if not actively banned, by the WTO.[26]

Before shifting to free trade in the mid-nineteenth century, Britain had practiced protection for its infant industries for centuries, banning Irish woolen goods and Indian cotton, for example, until British industries could surpass their competitors. Through the nineteenth century and up to the 1920s, the United States was "the fastest growing economy in the world, despite being the most protectionist." Chang concludes that the supposed causal relationship between free trade and economic growth and prosperity is tenuous at best and not apparent from history; in Chang's view, the most plausible explanation for the popularity of free trade in rich world countries is that they are aiming to "kick away the ladder" they have already successfully climbed.[27]

Furthermore, despite prominent claims to the contrary from World Bank researchers, there is little evidence that third world free traders fared better than more protected economies in the last two decades of the twentieth century. As noted above, commodity producers were endangered by open trade. Robert Wade of the London School of Economics notes that the World Bank researchers treat China and India as "globalizers" and attribute their trade growth to openness, even though the definition of globalizer is based on volume of trade rather than on removal of trade barriers. China and India are not, in any case, straightforward examples of liberal trade regimes:

> They began to open their own markets *after* building up industrial capacity and fast growth behind high barriers. In addition, throughout their period of so-called openness they have maintained protection and other market restrictions.... China began its fast growth with a high degree of equality of assets and income, brought about in distinctly non-globalised conditions and unlikely to have been achieved in an open economy and democratic polity.
>
> Their experience—and that of Japan, South Korea and Taiwan earlier—shows that countries do not have to adopt liberal trade policies to reap large benefits from trade.[28]

One further problem with the free trade agenda is its reliance on the consumption of immense quantities of fossil fuels. The fuel for the fleets of ships, trucks, and planes involved in trade is a significant part of the drain on the oil resource and the escalation of greenhouse gas emissions. A UN report in early 2008 estimated that nearly 4.5 percent of all carbon dioxide emissions are attributable to the merchant shipping that moves cheap goods around the world, emitting approximately twice as much carbon dioxide as aviation.[29] Air freight, too, is problematic; perishable and high-value items travel by air (35 percent of the value of all goods traded internationally, according the International Air Transport Association)—an even more intensive expenditure of petroleum-based fuel.[30] All of this movement involves profligate expenditure of energy, contributing 6 or 7 percent of global CO_2 emissions.

Debt

While the first world was assigning the development task to trade and foreign investment, the role of debt in third world poverty was largely ignored. The development phase of the postwar years already involved substantial borrowing by third world countries. Most of this debt was essentially unrepayable from very early on, since new loans were soon being made to pay off the previous loans, and sometimes merely the interest on these, a classic Ponzi scheme.[31] The financial crises of the 1970s drastically deepened that debt. Loans to third world countries increased dramatically through the 1970s, when the surge of petrodollars swelled the global pool of capital seeking profitable investment; when Volcker hiked the US interest rate, these loans soon became onerous or unrepayable.

According to Jubilee Research—a coalition of aid agencies, trade unions, and churches that has fought to cancel third world debt—as much as 20 percent of these loans were spent on arms, which could not be expected to generate any income to finance repayment.[32] In addition, entire loans to many regimes, including to dictators such as Presidents Marcos (Philippines), Galtieri (Argentina), and Mobutu (Zaire, now Democratic Republic of Congo), have been characterized by critics and aid agencies as "odious" or illegitimate lending, in which the money was not spent on the needs of the population that would later be held responsible for paying it back but, instead, was frequently squirrelled away in Swiss bank accounts. The development economist Stephen Mandel has shown in detail that numerous third world countries would actually be owed money if their odious debt were canceled.[33] Significant fractions of the still-mounting third world debt fall into the categories of illegal and illegitimate debt. Notwithstanding certain precedents, most creditors resist cancellation, however extreme the circumstances.[34] Several notorious cases of odious debt involved countries the West wanted to keep on its side during the Cold War, such as loans made to the Philippines when Marcos was in power. In another case, billions were loaned to Zaire by the IMF, even after its own appointee advised the head office that corruption was so serious that there was "no (repeat no) prospect for Zaire's creditors to get their money back."[35]

Even legitimate loans made for infrastructure such as dams and ports benefit a restricted class of people, though serviced by the entire population. And it is the very poor who suffer most when conditions for debt rescheduling include such measures as the abolition of health, education, and farming assistance. Since major infrastructure projects were almost always carried out by global corporations, often US-based, the cash flowed back to the United States or other parts of the developed world and often never left. In 1993, for example, the World Bank's net disbursements to the third world came to just over $7 billion, while the borrowing countries' payments to corporations was $6.8 billion.[36]

When US central bankers began the interest rate hike that aimed to solve the persistent inflationary trend of the 1970s, third world recipients of massive loans suddenly found their interest rates tripled and quadrupled; many could no longer repay even the interest. By the time defaults began, with Mexico in 1982, global financial institutions had adopted the neoliberal paradigm, and SAPs were imposed as the price of rescheduling the debt. State-owned enterprises had to be sold into private, often foreign, hands; agriculture had to be reoriented toward export earnings; taxation had to be reduced; and meager local welfare provisions had to be dropped. These measures masqueraded as rational economic policy for developing nations, but the privileging of export earnings can be better seen as an attempt to protect the interests of the first world bankers whose loans were in jeopardy.

The subsequent bailouts of defaulting countries had similar results. The IMF payments made to countries such as Thailand and South Korea after the East Asian economic crisis of 1997 had to be paid straight out again to their creditors in the first world financial system, while the nations still owed the money to the IMF.[37] Twelve years later, in the wake of the 2008–2009 financial crisis, the G20 provided the IMF with hundreds of billions of dollars, ostensibly to bail out the world's poor. Again, the funds were dispensed as loans to be used for repayment of outstanding debt—described by Ross Buckley, professor of international finance law at the University of New South Wales, as "a stimulus package for the rich countries' banks."[38]

Between 1970 and 2002, the total debts of the poorest countries went from $25 billion to $523 billion, with African debt alone rising from $11 billion to $295 billion. Over this period, African countries fully repaid

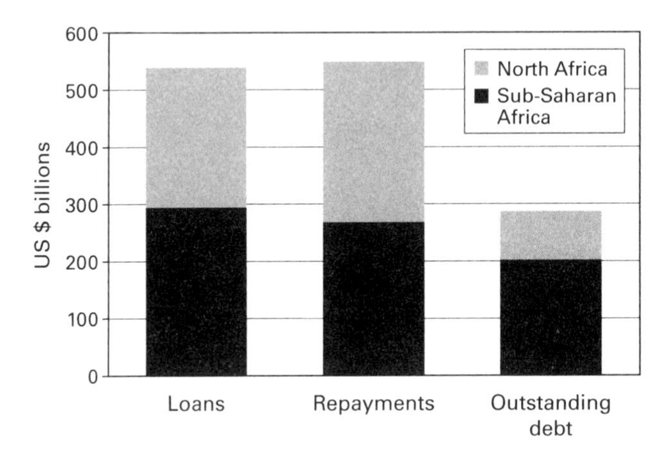

Figure 7.1
Africa's Debts, 1970–2002: Loans, Repayments, and Outstanding Debt.
Source: UN Conference on Trade and Development (2004) based on World Bank data.

$550 billion on loans totaling slightly less; because of interest requirements, however, almost $300 billion remains outstanding (figure 7.1).[39] Latin America also ran on a debt treadmill. It owed $209 billion in 1982; over the next twenty years, despite interest payments of $574 billion (more than it received in extra funding), its long-term debt had mounted to $674 billion.[40]

The net effect of these immense rolling debts, loaded up with compounding interest, endlessly rescheduled, many unrepayable, has been a huge ongoing outflow of funds from the third world to the first, a flow that dwarfs the entire first world's contribution of aid, private investment, and new loans put together. On the basis of OECD figures, the political scientist Susan George has calculated that a net amount of $418 billion flowed back to the first world in *debt service payments alone* during the period from 1982 to 1990 alone, an amount equivalent to six Marshall Plans for the rich at the expense of the global south. As the African case above demonstrates, this avalanche of repayments has done little to defray the debt. As George points out, the total flow of funds to the rich world is in fact far greater if "royalties, dividends, repatriated profits, underpaid raw materials and the like" are added.[41]

Far from enhancing development, debt has trapped most of the developing nations in a vain and marathon attempt to generate sufficient economic growth to repay an ever-bloating debt. This situation encourages enterprise directed not toward the needs of local populations but toward the needs of first world creditors; production is directed to export earnings, generated by extractive industries such as cash crops and mining for rich world consumption.

Sixty Years On: Who Benefits?

To best assess the effects of sixty years of development and the claims of lifting millions out of poverty, this section looks at the evidence of whether Truman's stragglers have indeed been catching up. Has the gap between the wealth of first and third world countries begun to be bridged, and how has any new prosperity been shared between and within countries?

Methodological Difficulties

There are wide disagreements about both the scale of global poverty and inequality and the direction of change.[42] Estimating levels of inequality between countries is beset with methodological difficulties, and quantifying trends in such inequalities with even more. Quantifying poverty is equally challenging. Both exercises involve making choices between methods of measurement and data collection, and matching the chosen data across time. The World Bank has revised its figures on poverty downward in recent years, claiming sharply improved conditions, but whether this claim reflects any real improvement is unclear.

The World Bank does not start out with a clear definition of what extreme poverty means. Its money-based metrics are not based on any agreed-upon minimum requirement that would avoid extreme poverty, such as the cost of adequate nutrition. Instead, it uses an arbitrary international poverty line loosely based on available national poverty lines. Critics of the bank's methodology, such as the economists Sanjay Reddy and Thomas Pogge from Columbia University,[43] warn that the bank's figures most likely understate the extent of global poverty, and that its recent claims of a steep decline in poverty lack adequate justification.

They call for a new definition of poverty based on the actual requirements of real human beings.

Even within the money metric, one must choose between market exchange rates and purchasing power parity (PPP). Purchasing power parity attempts to measure what a set amount will actually buy in different places and is considered to yield a more accurate comparison than market exchange rates would do. World Bank calculations of PPP are, however, cobbled together from disparate national figures and arbitrary base years and distorted by what is included in the basket of goods. Services such as haircuts, which are cheap in poor countries, are included, even though the extremely poor do not buy such services and struggle to afford the food they need. Purchasing power parity also downplays the weight of items governed by market price and exchange rate, which increasingly includes all the staple foods. Reddy and Pogge propose that, even within the bank's money metric, a food-based or bread-and-cereal-based PPP would be more appropriate than the basket approach—and would inevitably raise the poverty line and, thus, the numbers of the poor.[44]

Another problem arises from averaging data over the entire third world. This ignores the fact that the "Chinese miracle" is not generalized across the rest of the world.[45] When China is treated as a separate case and excluded from the calculations, polarization between the first world and the third world has clearly increased, whatever combination of methodologies is used. Thus, much of the self-congratulation noted earlier is based on the changes seen in China over the past thirty years.

The way poverty is defined affects assessments of how many people have actually been "lifted out of poverty." The World Bank's dollar-a-day metric, supposed to equate to "absolute poverty," has been adopted by the UN in its Millennium Development Goal of reducing extreme poverty by half. Robert Wade has pointed out that a dollar a day was about one quarter of the 1999 *world median income* expressed in PPP ($1,690 per annum), an entirely arbitrary benchmark, and a truly pitiful amount in the first place.[46] David Woodward and Andrew Simms of the UK's new economics foundation calculated that someone living on the adult minimum wage in the UK and without access to free services would have to be supporting thirty-six children to experience life as those living on $1 a day in the third world do.[47] The bank has since adjusted the international poverty line figure to $1.25 in 2005 dollars, still "far too low to cover the cost of purchasing basic necessities," according to Reddy.[48] Given that half the world's people, about three billion in 1999, were living on less than $1,690 a year at that time, it is obvious that a minor adjustment of the chosen poverty line will lead to the numbers of the poor fluctuating by hundreds of millions.

Polarization of Rich and Poor Individuals within Countries

At the wealthy end of the spectrum, the economic growth of the past three decades has yielded results beyond the dreams of avarice for a restricted class of super-rich. According to Wade:

> In most countries for which we have data, after-tax income distribution has become much more unequal since about 1980.... The top 1% of income earners has received a rapidly growing and hugely disproportional share of national income. All over the world—from New York and London to Beijing, Mumbai, and Lagos—a small section of the population is gathering vast personal fortunes.[49]

The situation of the top one-tenth of 1 percent in the United States illustrates the point: while workers' pay has atrophied or declined, the top 0.1 percent (a mere 145,000 individuals) garnered an average annual income of $3 million in 2002; they had multiplied their income by two and a half times in the twenty-two years from 1980 and doubled their share of the national income to 7.4 percent of the total. In the period from 1990 to 2002, for every extra single dollar earned by people in the bottom 90 percent of US taxpayers, these ultra-rich individuals brought in an extra $18,000. They were also the only taxpayers whose share of the taxation burden declined in 2002.[50]

The notion of per capita GDP (on which most income studies rely) suffers from the distortions that small numbers of outliers can create—as in the example of George Soros joining a group of wage-earners in a bar and producing an average millionaire. The United Nations Development Programme's (UNDP) Human Development Report for 2005 describes a "divided world":

> The size of the divide poses a fundamental challenge to the global human community.... The world's richest 500 individuals have a combined income greater than that of the poorest 416 million. Beyond these extremes, the 2.5 billion people living on less than $2 a day—40 percent of the world's population—account for 5 percent of global income.[51]

Another factor inflates the scale of this divide. Financial assets hidden away in tax havens by the ultra-rich are estimated to be well over $21 *trillion* as of 2010; these assets are earning their owners invisible and largely tax-free income, and neither asset nor income shows up in the statistics of inequality.[52]

Most studies of inequality use income data or a combination of income and consumption data. The United Nations University's World Institute for Development Economics Research (WIDER), however, has based recent work on household surveys of assets. According to WIDER:

> The richest 2 percent of adult individuals own more than half of all global wealth, with the richest 1 percent alone accounting for 40 percent of global assets. The corresponding figures for the top 5 percent and the top 10 percent are 71 percent and 85 percent, respectively. In contrast, the bottom half of wealth holders together hold barely 1 percent of global wealth.[53]

In the case of China, even though average income has grown in the neoliberal era, inequality within the country has increased.[54] While India is also regarded as a "globalization" success story, there are many places inside India where the reality is very different (discussed in chapter 9). The Indian novelist and essayist Arundhati Roy puts it this way:

> It's as though the people of India have been rounded up and loaded onto two convoys of trucks (a huge big one and a tiny little one) that have set off resolutely in opposite directions. The tiny convoy is on its way to a glittering destination somewhere near the top of the world. The other convoy just melts into the darkness and disappears.[55]

Analyzing Third World "Catch-up"

Perhaps the most telling analysis comes from Peter Edward of the Judge Business School at Cambridge, who used density curves in a 2006 study to examine just which people worldwide have benefited from economic growth under globalization and asked whether growth was shown to be an efficient means to address poverty. This technique allowed him to go beyond averages and identify which deciles of the global population enjoyed expanded consumption. Edward's work for the period 1993–2001 confirms that the changes in China at that time did not apply to the rest of the third world. Generalizations about reduced global poverty masked the real situation—the reduction of extreme poverty in China was offsetting an increase in poverty in the rest of the global south.

Moreover, Edward's analysis shows that the vast majority of the increased income and consumption over this period went to the richer half of the world population. Nearly half of it went to the top 10 percent, almost all of whom live in the first world, while less than 10 percent of it went to the poorer half of the world's population. The emerging Chinese middle class (defined here as those whose incomes range from about $1,000 per year, or $2.75 a day, to $7,000 per year) drew one quarter of the increase, much of this flowing to people close to the $7,000 end of the range. There was a shift toward greater consumption throughout Chinese society, however, and Edward suggests that the extreme poor in China have benefited by moving toward or slightly above $2 a day; at the same time, levels of inequality have increased.[56]

Edward's work demonstrates that it takes an enormous amount of global growth to yield tiny improvements for the very poor. The tactic of pie or cake expansion is still embraced firmly by the global financial entities. Senior IMF economist Anne Krueger is adamant: "The solution is more rapid growth—not a switch of emphasis towards more redistribution. Poverty reduction is best achieved through making the cake bigger, not by trying to cut it up in a different way."[57] For the period from 1990 to 2001, Woodward and Simms calculated that only 60 cents out of every $100 of per capita income growth actually flowed to the poor,[58] making the cake-baking strategy a slow and inefficient means of addressing poverty—and one that is unlikely to be sustainable for long enough to alleviate poverty anyway.

The torrent of avid insistence on poverty-busting success noted above has been based primarily on the rise of the new Chinese middle class rather than on the fate of the poor and very poor in the rest of the third world. This partly explains the triumphalist rhetoric found in the World Bank research, the APEC report, and economics journalism such as that of Martin Wolf in the British *Financial Times* and Rupert Murdoch in his fifth Boyer Lecture, titled "The Global Middle Class Roars."[59] The collective pie is certainly much bigger, eight to ten times what it was in 1950, yet the share of most third world countries, risible at the outset, shows no significant increase, and has in some cases declined.

The development era brought modest economic growth and considerable industrial expansion to the "underdeveloped" countries, and for Korea,

Taiwan, Singapore, and Hong Kong established the basis for their future success as the "Asian tigers." The globalization era that followed brought more industrialization but little economic growth when averaged across the third world as a whole. Extraordinary growth in China and, to a lesser extent, India and Brazil has not necessarily reached the extremely poor of these nations. It is doubtful whether the numbers of people living in the misery noted by Truman in 1949 have been greatly reduced, though it is probable, for what it is worth, that the percentages of desperately poor, poor, and somewhat poor people have decreased.

Growth, touted as the necessary means of catching up, has made little impression on the actual numbers. From the perspective of 2013, the gross number of people in serious trouble (living on less than $2 a day) has fluctuated but not declined decisively in the past thirty years. Even if the burgeoning global middle class is able to extract another decade or two of economic growth from the planet, the World Bank's figures[60] indicate that billions will remain poor (without discretionary expenditure), and a billion or more of these will continue to face the extremes of malnutrition, disease, and early death. At the same time, while growth continues to be the primary tool of improvement, loss of forests, fish stocks, and species cuts away the safety nets of the rural poor.[61]

As Ross Buckley notes, "There are about 195 countries in the world. Fifty years ago, 27 of those countries were developed. Today 32 are. In 50 years, five countries out of about 170 have achieved the goal of development."[62] Buckley's realism throws the growth solution to poverty into stark relief. At this rate, world development, even if it turned out to be ecologically possible, would take a further 1,500 years. To the extent that there *was* any real plan to share wealth or ameliorate poverty, the evidence suggests a verdict of substantial failure for the "bigger pie" approach.

While the business world's free trade campaign was gathering steam, UN agencies attempted to promote the idea of "sustainable development," which professed to reconcile development with ecological protection and somehow ensure both. The next chapter explores the fate of these endeavors in the neoliberal universe that has prevailed since the 1980s.

8

Growth and "Sustainable Development"

Many of the development paths of the industrialised nations are clearly unsustainable. And the development decisions of these countries, because of their great economic and political power, will have a profound effect upon the ability of all peoples to sustain human progress for generations to come.

—World Commission on Environment and Development, 1987

The Brundtland Commission

Sustainable development emerged as an international policy objective from the time of the World Commission on Environment and Development (WCED), also known as the Brundtland Commission, which met in the mid-1980s and filed its report, *Our Common Future*, in 1987. The definition it advanced has been much debated:

The concept of sustainable development provides a framework for the integration of environment policies and development strategies.... Sustainable development seeks to meet the needs and aspirations of the present without compromising the ability to meet those of the future. Far from requiring the cessation of economic growth, it recognises that the problems of poverty and underdevelopment cannot be solved unless we have a new era of growth in which developing countries play a large role and reap large benefits.[1]

From the outset, the WCED accepted a twin focus on environment and development, rejecting an environmentalism that ignored third world problems. To achieve this dual objective, it thought environmental sustainability would have to be made compatible with economic growth. The needs of third world countries were thought to require a new era of growth, but the WCED stressed that "growth by itself is not enough" and that these countries must fully participate and fully benefit. As chapter 7 showed, this kind of benefit is not evident and the rate at which the proceeds of growth have been shared suggests a pace that is unlikely to redress the unmet needs of the world's poorest people in the foreseeable future.

As outlined in chapter 3, the UN took the dawning perception of threats to the environment seriously throughout the 1970s, when the UN Environment Programme was set up and the Stockholm Conference, the first to link environment and the future of humanity, was held. It was as part of this process that the UN appointed former Norwegian prime minister Gro Harlem Brundtland to chair the WCED with the objective of reconciling the environmental problems identified by scientists since the mid-1960s with the intractable "development deficit"—poverty persisted almost everywhere—and to suggest how humanity might best pursue a future that would be more "prosperous, just and secure" for all.

The underlying prescriptions spelled out in *Our Common Future* were that the essential needs of the poor should be given "overriding priority" and that all futures were circumscribed by the capacity of the environment to meet these needs, especially in light of the demands already being made by the affluent. Many people in the first world were living beyond the "world's ecological means," according to the commission, especially regarding energy consumption. It stressed that the energy intensity of economic growth had to be curtailed and per capita consumption in the north had to be reduced.[2] Although some reduction in energy intensity has been achieved (discussed in chapter 14), there has been no sign whatsoever of reduced consumption in first world societies.

The WCED warned at length that the debt regime was forcing third world countries to liquidate their natural resources to pay interest on debts to the first world while forgoing any boost to the welfare of their own people.[3] In examining the role and power of transnational corporations, the commission noted that 80 to 90 percent of the trade in each of the world's key commodities—tea, coffee, cocoa, cotton, timber, tobacco, jute, copper, iron ore, and bauxite—was controlled by fewer than six TNCs. It thought international measures to regulate them were lacking, and recommended the adoption of codes of conduct that would include environmental values; it wanted sustainability addressed by all corporations and relevant international institutions, including the World Bank, the IMF, and the General Agreement on Tariffs and Trade (GATT, which became the WTO). It stressed the need for third world countries to retain sovereignty over their resources in all cases.[4] Yet in the new world of the neoliberal economic orthodoxies, such measures were regarded as unacceptable barriers to trade or as unwelcome regulation; in this world, prosperity could only be guaranteed by liberating the "free market" to work its wonders.

The commission was sharply aware of environmental degradation of many kinds and of necessary limits to expansion in the use of fossil energy. It pointed to the immense scale of the growth already experienced (a fiftyfold increase in industrial production in one century, 80 percent of it since 1950) and the unimpressive level of improvement that had resulted in third world countries. It described the situation as one of "interlocking crises" and "a threatened future." It acknowledged that the first world had "already used much of the planet's ecological capital" and that population was growing faster than the capacity to provide for all. Questions of distribution, it concluded, would need to be tackled, since growth alone was insufficient. Part of the increases in the income of the rich should be diverted to the very poor, it declared.[5] Although the idea was to redirect only part of the *increment*, not current wealth and not the entire increase, nothing of the kind has occurred. As explained in chapter 7, the world's rich monopolized most of the increased income and consumption during the 1990s.

Growth was accepted as essential, if not sufficient, to address deepening worldwide poverty.[6] The WCED's new era of economic growth was to be "based on policies that sustain and expand the environmental resource base." The commission believed that the concept of sustainable development would allow environmental policies to be integrated into development strategies.

Though the Brundtland Commission's definition of sustainable development has been criticized by environmentalists for its emphasis on growth and its optimism about the sustainability of economic growth,[7] it should be conceded that the commission recognized the failure of the development era up till then, and was far more attentive to issues of equity and ecological limits than such institutions as the World Bank, the IMF, and GATT. It was these bodies, however, along with the TNCs themselves and the US Treasury, that would continue to shape the world's fate over subsequent decades. Most of the considerations canvassed by the commission, apart from the broad call for economic growth, were ignored.

The Earth Summit

The Conference on Environment and Development (UNCED)—known as the Earth Summit—took place in Rio de Janeiro in 1992. Intended as a conference to move the conclusions of the Brundtland Commission into the practical sphere, it planned to address such pressing environmental concerns as species loss, depletion of energy resources, global warming, and the looming water crisis; its various statements, such as *Agenda 21*, spoke of sustainable development as if it were the most crucial priority. But even as UN agencies moved to incorporate concepts of ecological sustainability and equity into their approach to development, the neoliberal program was already sweeping the world.

The Earth Summit has been characterized by the development analysts Neil Middleton, Phil O'Keefe, and Sam Moyo as leaving the poor behind while addressing the first world's obsession with environmental decline.[8] But a close reading of its two general texts, *Agenda 21* and *The Rio Declaration*,[9] suggests that the diplomats who signed them had by this time accepted the market as the answer to decades of failure on poverty and were handing *both* equity *and* environment over to a "rising tide" solution. Clearly, after the poor performance of the development era, anything that promised that the first world could hold on to its affluence and still solve the twin problems of equity and ecology was an offer too good to refuse. Consequently, although the introduction to *Agenda 21* spoke of "equity" numerous times and acknowledged that debt, poor commodity prices, and poor terms of trade had made developing countries even poorer, it went on to install global free trade as the central strategy for the key task of "international cooperation to accelerate sustainable development." The neoliberal code words "optimal" and "efficiency" occurred throughout the document.

The Rio preparations were similar to the IPCC process, where major meetings are preceded by a series of prior negotiations allowing for reports and texts to be worded in such a way as to enable agreement to be reached at the final meeting. According to the Cambridge economist Michael Grubb and colleagues, who wrote an analysis of the Earth

Summit, the drafting of *Agenda 21* was a process of "Byzantine complexity," beset with the "difficulties raised by trying to address the concerns and problems of very disparate countries." The result was "an ungainly compromise, with specific *caveats* for special concerns and interests."[10] Several sources indicate that the US negotiators managed to water down key elements, in particular chapter 4, which was intended to acknowledge that the consumption patterns of the rich world were unsustainable, and chapter 16 which dealt with sustainable agriculture and biotechnology.[11] President George H. W. Bush is widely quoted as commenting before the conference that "the American way of life is not up for negotiation."[12] First world governments also refused to modify their protection of intellectual property rights: despite much talk of "technology transfer," nothing in the documents provided for waiver of or discounts on such commercial rights.

Middleton, O'Keefe and Moyo, who argue that Rio shifted "away from development and towards the environment," are unduly optimistic about the Rio conference actually tackling environmental problems. Their critique of Rio's failure on equity is, however, quite apt. "We are operating in a political world from which morality has been banished," they wrote. "In its place … we find simple greed masked by the euphemisms of 'management' and 'efficiency.'"[13] Just as land reform was discarded as an option in the 1950s, here again questions of ownership of land and the distribution of wealth were not permitted to surface. Third world countries had wanted the Rio Declaration to acknowledge "sovereign rights of nations to exploit their own resources … and the right of individuals to have freedom from hunger, poverty and disease"; they also called on the rich to accept the main burden of repairing the environmental damage they had caused. Only hints of these priorities survived the preparatory phases.

Even so, the United States still issued formal objections to several articles in the final draft which it felt gave poor countries too much latitude, including the one that referred to "common but differentiated responsibilities"; this was thought to imply acceptance of obligations and perhaps liabilities, and was rejected.[14] President Bush refused to sign the framework Convention on Biodiversity, despite the repeated protections for intellectual property written into all the Rio documents. US economic interests, narrowly perceived, would not be subordinated, whether to development or equity, sustainability or the environment. Bush, it should be remembered, was not in conflict with the US Senate, which never ratified the Convention on Biodiversity, even after the newly elected President Clinton signed it. The non-negotiable "American way of life" was and remains popular among Americans. The second President Bush declined to sign the Kyoto Protocol on the same grounds: "We will not do anything that harms our economy, because first things first are the people who live in America."[15]

No reform of the international economic system was envisaged in the Rio documents, though the Brundtland Commission had thought it essential. No changes were recommended to the debt regime or the terms of trade, and the IMF was given the nod to continue imposing "liberalization" in the guise of structural adjustment plans for poor countries in default. Funding for all the programs set out in *Agenda 21* was estimated in the final paragraphs of each chapter, but no mechanisms to raise actual amounts were agreed upon. What was offered was a pledge to try to meet the aid commitment of 0.7 percent of GDP that had been promised since 1970 but never delivered, except occasionally by some Scandinavian countries. The other assistance offered was "technology transfer," but this can hardly be regarded as generous since it made no commitment beyond commercial exchange at market prices. Aid has never approached the target suggested at Rio (box 8.1), and technology is transferred only to

Box 8.1
Aid

Popular concepts of aid are beset with misconceptions. Leaving aside both nongovernment aid and emergency aid such as is dispensed for earthquakes and tsunamis (less than 20 percent of aid), the focus here is on official development assistance, which began to emerge in the late 1950s. In line with President Truman's approach, official aid was initially conceived as technical rather than financial, with military aid available to some countries aligning themselves unambiguously with the United States against the Communist bloc.[a] By the end of the 1950s, however, the usefulness of development aid to the donor countries began to be recognized. The development authority Walt Rostow, in collaboration with ex-CIA man Max Millikan, argued that the needs of the first world for markets and raw materials made development aid a good investment.[b]

Box 8.1 (continued)

This sort of aim was described baldly by economist Hollis Chenery: "The main objective of foreign assistance, as of many other tools of foreign policy, is to produce the kind of political and economic environment in the world in which the United States can pursue its own social goals."[c] Though Chenery's comments refer to the United States, there is little reason to suppose that other nations' objectives have been any more altruistic. Such forthright descriptions of the purposes of aid are not common these days, and the purpose is often misunderstood by the populations of the first world, who see aid as giving money away. It is sometimes argued, especially on the political right, that aid should be abolished and the money spent at home.[d]

The majority of government-financed aid is not, in any case, dispensed to the poor countries to spend according to their own priorities but is usually paid directly to the rich countries' own corporations to carry out projects often nominated by these companies. In the case of the United States, for example, about one-third of the 1992 foreign aid budget (some $5 billion) was devoted to grants or loans for purchasing US military equipment.[e] The *Sydney Morning Herald*, reporting on the Australian aid budget of approximately $3 billion in the 2007 tax year, found that "much of the money has never left Australia, and that 10 private companies held almost $1.8 billion in contracts let by the Government's official aid delivery agency—AusAID."[f]

While Rio recommended that aid be boosted to 0.7 percent of GDP, as originally promised in 1970 and repeated many times since, the volume of aid from the rich to the poor world, as a proportion of their GDP, has actually declined in the past forty years, from 0.51 percent in the late 1960s to approximately 0.3 percent in 2009.[g]

Notes

a. Payer 1991, 42.

b. As noted in Payer 1991, 43.

c. Chenery 1964, 81.

d. DeLay cited in Kristof 2006; Joyce cited in Grattan 2010.

e. Hartung 1992.

f. Jopson 2007.

g. Riddell 2009.

those who can pay. Corporations were not burdened with regulation or even oversight.

Soon after Rio, in 1993, the UN Centre on Transnational Corporations was dissolved and its work on a corporate Code of Conduct was abandoned. Its residual website inside the UN's Trade portfolio notes that "its work reflected the changing times and became more focused on the positive, rather than the negative, effects of FDI [Foreign Direct Investment] and TNCs."[16] In 1994, GATT concluded what was known as the Uruguay Round of negotiations, strengthening intellectual property rights, and the WTO came into being with far stricter means of enforcing its trade rules than GATT had ever had (discussed in chapter 13). The liquidation of life on earth has continued unimpeded by Rio and its successor conferences in Johannesburg in 2002 and Rio in 2012.

Sustainable Development: A Dubious Proposition

The kind of development that has transpired since Rio has not reflected principles of sustainability, in the sense of being able to continue a course of action indefinitely without jeopardizing the ecological ground of the enterprise. It is more likely, as some have argued, that the notion is an oxymoron. Herman Daly makes a sharp distinction between sustainable development (qualitative) and sustainable growth (quantitative), which is, in his view, the oxymoron. If development is to be sustainable, Daly believes, it must be "development without growth."[17] Certainly, the idea that the habits of the affluent can be extended to all of the earth's seven billion people and rising is fantasy (as exemplified in the discussion of cars and paper in chapter 6), even though the celebration of the swelling "global middle class" apparently envisages such a dream. Wolfgang Sachs asks, aptly:

> Is sustainable development supposed to meet the needs for water, land and economic security or the needs for air travel and bank deposits? Is it concerned with survival needs or luxury needs? Are the needs in question those of the global consumer class or those of the enormous numbers of "have-nots"?[18]

These are questions that the Brundtland Commission touched on but did not press, and questions that Rio shelved despite its best intentions. It appears that these questions, too, are forbidden under the dominant economic regime. Whether sustainability could have been useful remains an unanswerable question, since the idea has been disabled. As British playwright Jeremy Seabrook put it, "Sustainable now means what the market, not the earth, can bear.... Sustainable is what the rich and powerful can get away with.[19]

In the next chapter, I turn to what are regarded as great development successes in China and India, where economic growth has risen by more than 7 percent in most years over the past two decades (three in China's case). I ask whether these transformations have met any criteria of sustainability, and how they have affected the poor and the very poor.

9

Growth and Its Outcomes for the Poor

We all have a responsibility to create the conditions for the poor to be less poor and then to be middle class and beyond.

—Rupert Murdoch, 2008

God forbid that India should ever take to industrialism after the manner of the west. The economic imperialism of a single tiny island kingdom is today keeping the world in chains. If the entire nation of three hundred million people took to similar economic exploitation it would strip the world bare like locusts.

—Mahatma Gandhi, 1928

China and India are regarded as success stories for economic growth in the developing world over the past twenty to thirty years, but there are disturbing indications that growth has not relieved extreme poverty in India and that the improvements in China are not being shared equitably. In both countries, the pursuit of growth is accompanied by serious environmental damage and the dispossession or forced removal of tens of millions of the poorest people.

Twenty-first Century Enclosure: Tribals, Farmers, and Slums

Dispossession has been part of the transformation of subsistence farming into profit-making agriculture since the earliest British enclosures turned farmland over to sheep grazing. Sir Thomas More in 1516 condemned the "unreasonable covetousness" with which the rich evicted the peasants from their houses and fields and replaced husbandry and tillage with sheep that "eat up and swallow down the very men themselves."[1]

Enclosure today is little different. Already imposed in the development era and intensified in the "free market" phase, it has taken two main forms: a transformation of small-scale subsistence farmland into profit- and export-oriented agriculture (analogous to the European enclosures of past centuries) and direct expropriation of the land of indigenous peoples and peasant farmers (this time carried out by the ruling elites of the postcolonial societies, often in collaboration with Western corporations). The same process occurs on land that is not inhabited by humans, where ecosystems sheltering other species have been relatively unthreatened; these places too are being "enclosed," logged, or cleared for palm oil or similar plantations. Motives for these dispossessions range from establishing corporate mining operations and agribusiness to setting up industrial complexes and special economic zones (SEZs) or clearing valleys for dam building. The poor and the powerless are routinely losing what land they have (or have access to) across the world, both in countries considered to be in transition to modernity and wealth (such as China and India) and in those still regarded as underdeveloped (such as parts of Latin America and most of Africa).

Though still ruled by the Communist Party, China's development path involves an urbanization parallel to that which characterized industrialization in Britain. Much of the recent confiscation of land has been along the eastern coast, where the great new industrial cities are swallowing adjacent villages and fields. By late 2004 the state-run Xinhua News Agency was reporting that 20 million of China's 900 million farmers had been displaced from their land by commercial projects. Developers allied with government officials were "gobbling up" the land of powerless farmers.[2] More recent reports suggest far greater numbers. In 2011 the Landesa Rural Development Institute conducted a survey in seventeen provinces where some three quarters of China's rural population live. The study found that 43 percent of villages surveyed had lost land to compulsory acquisition for nonagricultural purposes since the late 1990s. The number of "takings" rose steadily through the current century, and the median compensation for the farmer was a tiny fraction (less than 2.5 percent) of the proceeds for the officials.[3]

Though unevenly reported—the government is reluctant to have unrest publicized—millions of Chinese peasants have resisted the appropriation of their land, especially during the twenty-first century, when the scale of Chinese economic growth has ballooned. The Communist Party magazine *Outlook* reported 58,000 major incidents of social unrest in 2003, an average of 160 per day and an increase of 15 percent over the year before. About half the unrest at that time was related to land disputes, but demonstrators also opposed chemical plants, power plants, and the pollution of their rivers and fields.[4] According to the US online magazine *Grist*, China's public security minister Zhou Yongkang "told a closed meeting that 3.76 million Chinese took part in 74,000 mass protests last year [2004] alone."[5] In 2011 the official paper, *China Daily*, reported on research into forced demolitions and relocations; conducted by the Research Center for Social Contradiction in Beijing, the study showed that nearly 70 percent of the respondents were dissatisfied with the outcome.[6] The *Global Post* correspondent Kathleen McLaughlin reported in 2012 that over half of the tens of thousands of protests each year are directly related to land loss and forced resettlement. People particularly objected to lack of notification and poor compensation, with nearly one quarter receiving no compensation at all.[7]

Though one might imagine that a democratic system such as India's would allow farmers a greater say in the disposition of their lands, the path of development has been no happier for India's peasant farmers, Dalits (also known as untouchables) and *Adivasis* (indigenous or tribal people). As anthropologist Felix Padel and political ecologist Samarendra Das have reported,

> In India, industrialisation has already displaced an estimated 60 million villagers in the past 60 years.... A shocking 75 percent comprised *Adivasis* and Dalits. Very few of them have been adequately compensated; most report no improvement in their standard of living though such displacement is unabashedly presented as a precursor to development. The poverty that they have been reduced to is just as painful as the erosion of their cultural values and traditions, which invariably accompanies the forced separation from the land that they and their forefathers cultivated.[8]

Displacement on behalf of miners is prevalent in the mineral-rich states of Jharkhand, Chhattisgarh, Karnataka, Andhra Pradesh, and Orissa. In the northwestern Niyamgiri hills of Orissa state, Vedanta Resources built an aluminum smelter and planned open-cut mining of bauxite in 600 hectares of mountain forest where springs and the headwaters of two rivers sustain the local Dongria Kondh people and the forest they live in. Local people claim that when Vedanta built its smelter at Lanjigarh, some one hundred indigenous families were evicted and their villages razed. Caustic runoff from the refinery has contaminated their crops and livestock and made them sick.[9] In April 2013, after years of litigation, the Indian Supreme Court ruled that Vedanta must consult the Dongria before mining. In an August 2013 referendum, all twelve village communities voted unanimously against the mine and in January 2014 the Indian federal government ruled that mining would not proceed. This is an isolated victory, but mining has been halted and, for once, an indigenous people's land and way of life has been protected.

Indigenous people have always resisted being removed from the lands to which they are culturally and historically connected, so compensation is rarely satisfactory. When Indian tribals *are* forced off their land, suitable resettlement is rare and compensation is minimal. Common property, analogous to the commons of medieval Europe and making up at least half the commandeered land, is simply excluded from compensation calculations, while payouts for *patta* land, acknowledged as tribal property, are minimized.[10]

Growth and the "Greater Good"

At Kalinga Nagar, also in Orissa, twelve rice farmers of the Ho tribe were killed while demonstrating against being deprived of their lands for Tata Steel's new plant in 2006. Tata is the Indian TNC famous for the Nano, India's very small car. Tata spokesman Sanjay Chowdry told SBS *Dateline* that "there's a lot to say about a person wanting to till his land and not move.... But the greater good is what is important."[11]

This interpretation of the greater good—that it flows from rapid industrial development and justifies the government in sacrificing individuals or groups who are in the way—is the ruling concept. It echoes Indira Ghandi's 1984 reply to a social worker concerned about displacement and the drowning of 200,000 acres of dense forest by two new dams. Ghandi wrote that although she was "most unhappy that development projects displace tribal people from their habitat ... sometimes there is no alternative and we have to go ahead in the larger interest."[12]

Growth through industrial expansion is an objective frequently repeated by India's top officials. In 2006, Orissa's governor, Rameswar Thakur, told India's *Business Standard* that "Orissa is committed to create an industry-enabling and investor-friendly climate in the state with a view to accelerating industrial development, employment opportunity and economic growth."[13] In an extensive interview in 2008, India's then finance minister, P. Chidambaram, also stressed "the imperative need of growth over a long period of time.... We must develop those iron ore mines, we must mine that coal, we must build industries." When asked for his ideas on eliminating poverty, he outlined a vision in which the vast majority of Indians (85 percent) would live in cities.[14] Chidambaram did not explain how this move from country to city would address India's widespread poverty.

There are serious doubts as to whether the pursuit of the "greater good" along these lines has led to actual improvements in the lives of the Indian population. Sociologist Michael Goldman, who studied the inner workings of the World Bank, found that the poorest people did not necessarily benefit from the mines, industries, dams, and irrigation projects that have characterized much development since World War II. When, for example, the World Bank planned and financed irrigation canals in the Thar Desert in 1958, the project was hailed as a great development success in official bank reports, but the reality was more ambiguous. Wealthy landowners did indeed produce high yields of export crops, but smallholders got neither water nor government help, and fell into debt. Many of the herders, weavers, traders, and rain-fed farmers who had managed their communal village lands for centuries became landless laborers working for the rich. Some, no doubt, were bound for the slums of the cities.[15] The history of big dams also shows how the large-scale development approach has favored corporate interests and agribusiness ventures and directed much of the water and electricity the dams produced to industry, while neglecting and displacing the poorest, least powerful people and rarely, if ever, compensating them properly (box 9.1).

In 2010, Binayak Sen, a pediatrician who serves the poor of the state of Chhattisgarh, pointed to the findings of India's own National Nutrition Monitoring Bureau: 33 percent of Indians have a body mass index of less than 18.5, meaning they are underweight, below the normal range; more than half of all Dalits and *Adivasis* were in this malnourished category.[16] In her recent research, the economist Utsa Patnaik looked into what food is actually eaten by the average family and what caloric intake is being achieved. While official government poverty statistics, based on a monetary poverty line similar to the World Bank's benchmark, claim a substantial reduction in poverty since India adopted neoliberal economic

Box 9.1
Big Dams

After fifty years of frantic dam building—during the 1970s two or three large dams were being commissioned somewhere in the world *every day*—the World Commission on Dams (WCD) delivered a report that echoed the criticisms of the opposition to big dams that had been growing throughout the period.[a] Among their case studies, the WCD examined the Tucurui Dam in Brazil, where the aluminum industry draws more than half the electricity generated, and the Kariba Dam in Zimbabwe, which was built to serve foreign copper miners' requirements for water and power.[b] Job creation, often claimed as one of the positive effects, was found to be largely confined to the construction phase, and thus ephemeral.[c] While suggesting that dams have sometimes offered significant benefits, the report also stressed endemic failure to consult the affected, endemic failure to compensate them, and endemic failure to share much of the benefit with those who had given their land, their homes, and their livelihoods for the cause. At the same time, 60 percent of the world's rivers have been greatly modified, many of the world's wetlands have been destroyed, and numerous species and ecosystems have been irreversibly lost.[d]

The commission was composed of a cross section of diverse interests, from dam proponents representing business and government to an Oxfam representative and the Indian activist Medha Patkar, who led the decades-long resistance to the Sardar Sarovar Dam on the Narmarda River in Gujarat. They were nonetheless able to produce a consensus report that made several recommendations for future dam projects. Perhaps the most radical of these were the policies of "free, prior and informed consent" from affected peoples and, at the outset, a "comprehensive and participatory assessment of the full range of policy, institutional, and technical options [in which] ... social and environmental aspects have the same significance as economic and financial factors."[e] Under such planning constraints, many of the 42,000 dams the commission scrutinized would never have been built.

In 2014, notwithstanding these findings, dams continued to be built, and many more are planned. For example, India and China have plans to dam most of the rivers flowing out of the Himalayas and Tibet. India plans 292 Himalayan dams, affecting twenty-eight of the thirty-two river basins in India's control; 80 percent of these are in dense, undisturbed forests. The government has not reviewed future water and energy needs systematically or strengthened public participation in decision-making processes, and resettlement compensation is still not guaranteed. Other countries, including China, have at least 129 additional projects planned, many affecting downstream nations such as those dependent on the Mekong. There has been little consultation between nations as each hurries for "prior appropriation." The melting of glaciers and snowfields, accelerating

Box 9.1 (continued)

with global warming, is likely to jeopardize the future viability of Himalayan hydropower, but its effect has not been assessed.[f]

Notes

a. World Commission on Dams (WCD) 2000, xxix.
b. WCD 2000, 170, 173.
c. WCD 2000, 133.
d. WCD 2000, xxx–xxi.
e. WCD 2000, 112, 221.
f. Grumbine and Panjit 2013; Vidal 2013.

policies in the early 1990s, Patnaik found that actual nutritional intake had declined drastically. Indeed, the level of food grains available per capita of total population had fallen by 2005 to levels not seen since the early 1950s: "Forty years of successful effort to raise availability has been wiped out in a mere dozen years of economic reforms."[17] Furthermore: "In actuality, the average Indian family of five in 2005 was consuming a staggering 110 kg less grain per year compared to 1991.... Not only has calorie intake per capita fallen, there is also a steep decline in protein intake for four-fifths of the rural population over the period 1993–94 to 2004–05."[18] These figures are all the more shocking, Patnaik points out, when it is recognized that the average encompasses "a sharp rise in intake for the wealthy minority," indicating a catastrophic decline for the poor majority.

A survey conducted by the Indian Health Ministry and UNICEF in 2006 confirmed that malnutrition was widespread among India's small children, with some 43 percent of them undernourished (figure 9.1).[19] Commenting on that survey, the London *Times Online* noted that the average rate of malnutrition in sub-Saharan Africa was about 35 percent, significantly less than the figure for India, even though India's economic growth had exceeded 8 percent in the previous three years, "a shocking illustration of how India's recent economic gains, while enriching the social elite and middle classes, have failed to benefit almost half of its 1.1 billion people."[20] So widespread is malnourishment in India that the Indian government legislated in September 2013 to deliver a feeding

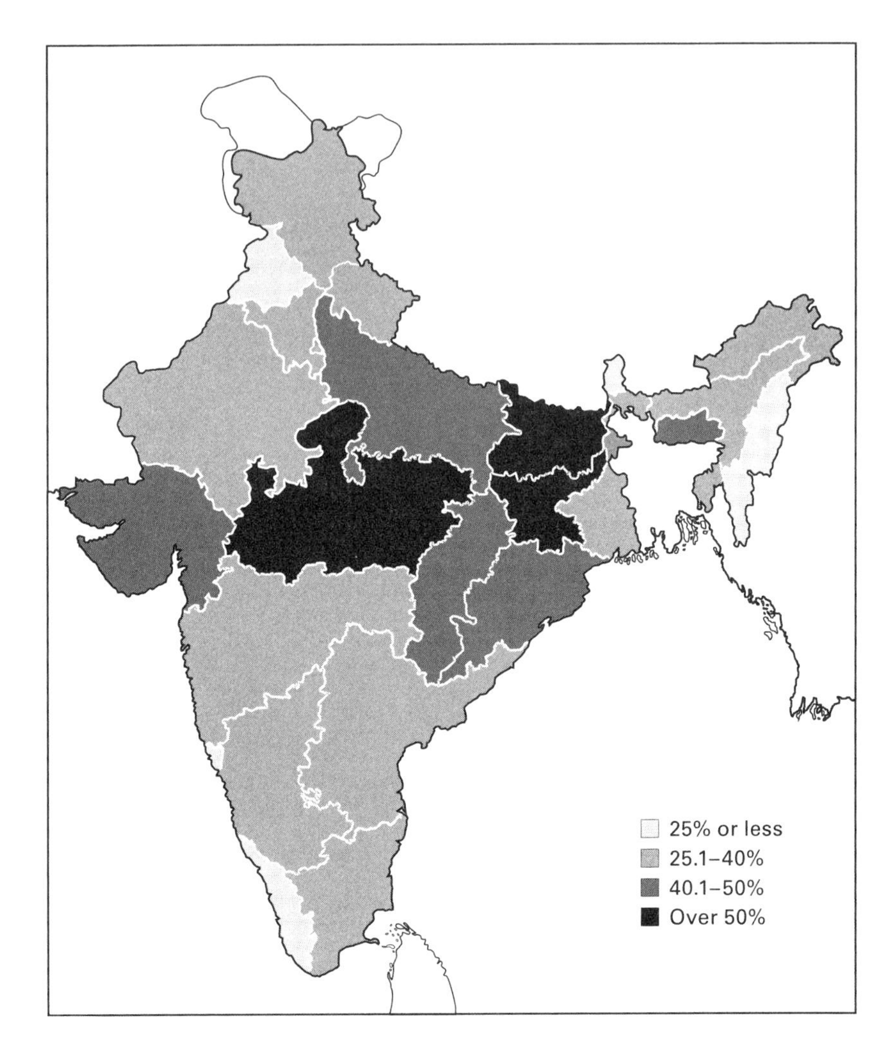

Figure 9.1
Malnourished Children Younger Than Age 5, India.

program to 800 million Indians—two-thirds of the population—in an attempt to improve their situation. The initiative may be politically motivated, coming soon before an election, but the severity of the hunger it purports to address is uncontested.[21]

The growth path to the "greater good" has left immense numbers of India's people behind, forfeiting the poor to the ambitions of India's elites. A belief in harsh remedies and the idea that "progress" can involve the sacrifice of some for the benefit of others, or of one generation for the benefit of the next, are part of the myth of inevitable ascent and serve to justify brutal measures. Milton Friedman in his television series, *Free to Choose*, talks about his eastern European forebears working long hours in the late nineteenth-century sweatshops of New York City:

> They are not going to stay here very long or forever. On the contrary, they and their children will make a better life for themselves as they take advantage of the opportunities that a free market provides to them.
>
> The irony is that this place violates many of the standards that we now regard as every worker's right. It is poorly ventilated, it is overcrowded, the workers accept less than union rate—it breaks every rule in the book. But if it were closed down, who would benefit? Certainly not the people here. Their life may seem pretty tough compared to our own, but that is only because our parents or grandparents went through that stage for us.[22]

Friedman did not mention people who *actually died* "for the sake of the next generation," as was the case with five-year-old chimney sweeps in nineteenth century Britain; or the 1,127 garment workers caught in the 2013 Bangladesh factory collapse (they were ordered to work that day even though an engineer had ruled the place unsafe); or, indeed, the 146 garment workers, most of them eastern European immigrant girls, who were burned to death in New York City in the Triangle Shirtwaist Factory fire of 1911 where they had been locked in—Friedman's mother presumably not among them.

The Urban Future

Urbanization has been a hallmark of the economic growth of an industrializing economy. In the early twenty-first century, half the world's population lives in cities for the first time in history. Between 1950 and 2001, the number of cities with more than one million people grew from 86 to 400 and is expected to be 550 or more by 2015.[23] Megacities of more than 10 million people have multiplied by ten between 1950 and 2005, from two to twenty,[24] and, alongside the swelling numbers of smaller cities, "hypercities" of 20 million have begun to emerge, with some commentators predicting "a continuous urban corridor stretching from Japan/North Korea to West Java."[25]

Mainstream economists regard poverty as primarily a rural phenomenon, typical of premodern societies. The UN Department of Economic and Social Affairs (UN/DESA), for example, believes that the urban poor of the current era usually do better than their rural counterparts and counsels against viewing urbanization as a negative phenomenon: "Cities are ... engines of economic, social, political and cultural change. Urbanization can thus be viewed as an indicator of development, with higher urban levels generally associated with more industrialized and technologically advanced economies."[26] This judgment is in line with the postwar development agenda already outlined, which gave industry, technology, and economic growth the task of human social improvement; mainstream economists remain confident that a largely urbanized world population of nine or ten billion can be supported.

Mainstream economics, however, has rarely considered land reform as a possible path to rural well-being. The UN/DESA report quoted above does not examine the relationship between rural poverty and landlessness and consequently ignores the influence of the distribution of land on rural poverty and the possibility that people given access to land may be able to support themselves with at least as much and probably more security and dignity than is possible in the slums and *favelas*. The theory of "greater good," however, sees the ongoing dispossession of peasants and indigenous peoples as a necessary price to be paid, along the same lines as those put forward by Rostow and Lewis in the 1950s.

If humans are to be as heavily urbanized as P. Chidambaram desires and most demographers predict, then very few of us will be involved in growing food; an adequate supply will have to be mass-produced along the lines of industrial agriculture, which has shown itself able to produce food (or foodlike substances) with very few workers. However, despite the cheery optimism of Chowdry and Chidambaram, there is little sign of what livelihoods will absorb the billions of people in the families of peasants still farming today, nor is there any persuasive rationale for ignoring the most abundant resource of countries such as India or Ethiopia—their people—in order to make a tiny regiment of mechanized farmers more "productive." While agricultural productivity measured as output per person has steadily expanded in the capitalist era (box 9.2), actual people have frequently been rendered as destitute as the husbandmen observed by Sir Thomas More in 1516.

Calls from the corporate mainstream, in response to the food crisis of 2008, emphasized technical and investment assistance to facilitate productivity growth and "market participation in agriculture."[27] Whether industrial agriculture is a feasible solution to feeding the projected urban masses of the future depends heavily on the continuing availability of sufficient topsoil and adequate inputs of water, petroleum, and phosphate, as well as on the capacity of the surrounding countryside and watercourses to absorb the wastes. All these elements are under challenge, as noted in chapter 1. Just as important, if commodity production of food is to feed the world, the burgeoning urban masses will need to have a capacity to pay, something sorely missing in many places during the 2008 food crisis; although there was often enough food available—in Haiti, for example—the urban poor were unable to pay the escalating market price.[28]

In the absence of large social expenditures on facilities in the overcrowded cities, optimism about the urban future seems misplaced. Much of modern urban life is lived in slums. The UN Human Settlements Program (UN-Habitat) made a study of slums worldwide in 2003; it found that 924 million people were already living in slums in 2001 and predicted that the figure would reach two billion by 2035. The report was cautious about the prospects for urban improvement and found that, while rural poverty is generally assumed to be far greater in both extent and intensity than urban need, "the locus of poverty is moving to cities. ... Depending on the individual countries and cities, between 40 and 80 percent of urban dwellers in the world are living in poverty, with very little or absolutely no access to shelter, basic urban services and social amenities."[29]

UN-Habitat also found increasing urban unemployment and increasing inequality in third world cities in the last two or three decades. It argues that the great range of improvements needed by the urban poor requires not only "robust growth" in the national economies but also equitable policies, both national and international: "Instead of being a focus for growth and prosperity, the cities have become a dumping ground for a surplus population working in unskilled, unprotected and low-wage informal service industries."[30] These "dumping grounds" are already swelling rapidly in cities that will need to accommodate *billions* more if demographic predictions are correct and the industrialization of agriculture continues. Few of these third world cities have water or sanitation, and governments are reluctant to provide these, especially where

Box 9.2
Agricultural Productivity

Agricultural productivity, measured as output per worker, has been the benchmark of progress in capitalist farming, and it is commonly argued that the productivity of industrial agriculture is the only alternative to feed a world of seven billion people. This argument is advanced against proponents of a more organically based agriculture that would minimize fossil fuel inputs and restore the natural fertility of living soil. It is also pressed against those who advocate supporting small farmers in their traditional settings rather than taking their land for the mass production of food and consigning the farmers to slums. However, output per hectare of land is far more crucial to farmers with access to a piece of land than is output per worker.

The capitalist definition takes for granted the multiple inputs of energy, fertilizer, pesticide, and irrigation water and ignores the losses of topsoil and fertility and the pollution of streams and groundwater that are involved in its practices; many of the social and ecological losses of industrial agriculture are unpriced, so market price is often unrealistically low. Capitalist enterprise has shown itself to be reluctant to accept the inclusion of such hidden costs in its budget. This reluctance is illustrated by the resistance of Australia's fossil-fuel-dependent industries to government attempts to put a price on carbon, however gradualist and partial.[a]

For peasant cultivators, by contrast, the essential input is human labor, and where labor is plentiful, the crucial criterion of productivity is the yield per hectare of land and unit of water, and its adequacy to feed the family or community working it. Success cannot be measured with reference to output per worker. Labor is not a limiting factor when populations are large. The availability of water or options for fertilizer are more important in many contexts. Peasant farmers use local resources such as manures to fertilize their crops and the intercropping of grains with legumes to provide nitrogen. Moreover, as fossil fuel becomes more expensive and more polluting, and given the significant contribution of industrial agriculture to global warming, alternative forms of agriculture are well worth exploring.

Western measures of productivity not only stress yield per worker but are reckoned using only one crop at a time, and overlook the complex mosaics of edibles produced in traditional agriculture, such as the rice paddy where fish and leafy greens coexist with the rice harvest, or the many different intercropping systems described by Shiva in India. It is argued by agroecosystem farmers that the yield per hectare is greater than the tonnage of the principal crop and rivals or betters monocultural outputs per hectare, if not per person. Shiva provides detailed statistics suggesting yields from multiple cropping systems perform significantly better per hectare than monocultural maize or millet,[b] a claim supported by studies of the "home garden" of many small farming cultures, most

Box 9.2 (continued)

highly developed and productive in Java.[c] While industrial farming produces commodities that are only available to those with the money to pay, the "Java garden" grows food for direct local use rather than for a global market, so that distributional problems are minimized. Small-scale organic farming in Cuba after Soviet support collapsed has had similar benefits.

Average "productivity" in the Western system of vast, highly mechanized single-crop farms is 1,000 to 2,000 tonnes of grain per farmer per year. Some 40 to 50 million such farmers worldwide produce colossal quantities of grain. This "productivity" is now orders of magnitude greater than a yield per person of about one tonne of grain for traditional peasants and up to 50 tonnes for those using Green Revolution techniques such as high yielding varieties and some associated extra inputs of water and fertilizer.[d] Western techniques rely, however, on massive inputs rather than labor and, if generalized to world agriculture, can be expected to make some three billion people from peasant farming families redundant. In the twenty-first century, few will have the option of destinations in "empty lands" far away, such as the Irish or eastern Europeans enjoyed in previous centuries. Survivors of the urbanization of the third world have little or no choice but to aggregate in the already massive conurbations of their own countries.

Notes

a. Gittins 2008; Kohler 2009; Symons-Brown 2009.

b. Shiva 2008, 115–118.

c. Soemarwoto and Conway 1992.

d. Amin 2003, 32.

the tenure is informal. Kinshasa, with a population approaching ten million, has no sewage system at all; only 10 percent of Manila has sanitation; and "flying toilets" are prevalent in such cities as Nairobi—these are plastic bags thrown on roofs, which burst or decay in the sun and rain.[31] Women, burdened with expectations of modesty, are especially oppressed by such circumstances.

It has been argued that slums that are allowed to mature, such as Dharavi in Mumbai, where the film *Slumdog Millionaire* is set, do produce a self-sustaining economy and gradually become less poor[32]—but this is not a guaranteed outcome. Plans to redevelop Dharavi will raze the shanties and build high-rise apartment buildings. While these new structures will have sanitation, and there are plans to provide town facilities, the residents will lack the community structure that supports the economic resilience of the so-called mature slum, where business is conducted from home, close to the street.[33] All informal settlements are vulnerable to this fate as real estate in the cities becomes hot property for developers: the Bassac River shantytown in Phnom Penh, for example, was demolished in 2006 to make way for luxury apartments and shopping centers; its residents were evicted.[34] Several other Phnom Penh slums burned down five years earlier, in fires probably deliberately lit, leaving thousands homeless.[35]

The Consumption Route to Prosperity

The extension of current trajectories of modernization, industrialization, and urbanization assumes that the Western template of progress can be applied indefinitely until the stage that development economist Rostow called mature has been attained, the consumption-driven wealth of the past century or so,[36] which all people can then enjoy.

"The Global Middle Class Roars," Rupert Murdoch's lecture on his vision of human progress, supports the impression that everyone can and will catch up and eventually become middle class—or even "beyond." Everyone is running in the same race, a race without material constraints. In Murdoch's worldview, the guarantee of this advance is an ever-expanding productivity (defined as output per worker; see box 9.2); if the third world could emulate the West in this respect, "the world would grow fantastically richer and everyone would be better off."[37]

There is a kind of precedent for this; it is, after all, very close to what occurred over the past century or so for most of the working class of the first world, despite horrific conditions in the previous century. Ninety-five percent of the first world's population now fall within the top quintile of world income, making first world middle classes and employed working classes part of the global rich.[38] Thus it can be argued that a precedent exists for a gargantuan extension of middle-class consumption to 95 percent of the world's huge population during the current century: if Europe and its offshoots could do it, why not everyone else? This confidence about catching up, however, rests on the rickety assumption that what has happened before can always happen again. In this scenario, the entire world population of seven billion joins the stairway to imitative affluence described in chapter 5. I have already examined one of the fallacies involved here, noting that the empty world into which Europe expanded is long gone, transformed by the European capitalism that depended on its fabulous riches.

The vision advanced by Murdoch also skips over the gulf between the poor of the third world and the "battlers" of the first, offering an effortless transition, automatically provided by "market forces." According to Murdoch, "When the poor are given access to the global economy they build a better life for their families and a brighter future for their countries." These views mirror the APEC report examined in chapter 6. Mobile phones, airborne holidays, and a diet rich in meat are the hallmarks of the ascending future, and this is what "lifting millions out of poverty" is suggesting. The meaning of the word "poor" in Murdoch's lectures conflates the colloquial first world meaning (you can't afford a dentist) with the World Bank's quantified definition of income below $1 a day. Though this benchmark was updated to $1.25 a day in 2005,[39] it remains immiseration at a very different level from poverty defined as "below middle-class consumer status." The emphasis on the roaring of the new middle class implicitly suggests that poverty is about lacking disposable income for consumption. In reality, any transition to the flexibility of discretionary expenditure is far beyond the people the World Bank calls poor, who need adequate daily food, clean drinking water, and sanitation long before they will fly off to Bali for a break.

It is important to remember Peter Edward's careful analysis of just who has received the proceeds of the economic growth of the recent past: half went to the world's top decile, almost all in the first world, another quarter went to the already established Chinese middle class, and most of the rest edged many of the poor of China toward or just over the $2 a day mark.[40] China may continue to be the exception (box 9.3), but Edward's data throw grave doubt on the assurances of the Indian finance minister that industrial development is providing a direct and indispensable route to the end of poverty. Clearly, the neoliberal market regime is not providing prosperity for all. Rather, it has ameliorated extreme poverty in China and provided Western levels of consumption to a small sector of the Chinese population—and to a smaller segment of the Indian and other third world populations.

Box 9.3
China Takes the Consumption Route

In China, it is no longer simply a matter of expecting urbanization to follow from economic growth (a trend already in progress) but of fostering urbanization on a massive scale in an attempt to perpetuate the growth of the past three decades. Just prior to becoming China's premier in 2013, Li Keqiang explained that moving people into cities would spur investment and consumption, since urban residents spend 3.6 times more than rural dwellers. Though he also spoke of protection for arable land, fair treatment for the farmers being moved, and provision of urban services for the new arrivals, success in these respects is not guaranteed.[a]

According to the *New York Times*, Chinese authorities plan to relocate a staggering 250 million rural people over the next twelve or so years, into high-rise apartments in the hundreds of small and medium-sized cities currently under construction. This is partly intended to move people out of the path of the massive series of dams, canals, and tunnels that are being built to divert water from the south to the arid north, where groundwater has been drastically depleted. These removals will help to meet the urbanization ambitions of the Chinese premier and, it is hoped, fuel ongoing economic growth by rapid expansion of the consumer middle class. The new wealth of these new city dwellers is expected to finance the immense infrastructure and services they will require. In a little more than a decade, China hopes to expand its consumer class by a number similar to the entire current US urban population.[b]

Notes

a. Li 2012.

b. Johnson 2013a, 2013b.

Growth and Poverty: Smoke and Mirrors

Equitable, or even reasonably fair, distribution is not relevant to market solutions, since ongoing growth is supposed to fix extreme poverty eventually without bothering about questions of equity. It continues to be the preferred route to achieving the UN's Millennium Development Goals (MDGs), minimal as they are, including the halving of extreme poverty and hunger by 2015. Like the World Bank's 2007 predictions of reductions in the non-middle-class poor, the MDGs use percentages as their benchmark. While the original undertaking in 1996 was to cut the *numbers* of undernourished people in half by 2015, the MDGs adopted in 2000 promised to cut the *prevalence* of undernourished people in half, a change that reduced the target by hundreds of millions.[41]

The FAO's 2011 revision of its criteria for defining hunger has further minimised the task. In 2010 the FAO estimated the number of hungry (undernourished) had fallen to just under 800 million in the mid-1990s, had risen again to nearly 850 million by 2007, and had jumped to over a billion after the financial crisis of 2008.[42] The 2009 MDG report warned of "a long-lasting food crisis" and noted that though international food prices had decreased during 2008, this did not flow through to local markets: "Consumer access to food in many developing countries, such as Brazil, India and Nigeria, and to a lesser extent China, did not improve as expected."[43] Using the new 2011 definition, the FAO identified far more hungry people back in 1990 than in previous estimates, so it was able to report a steadily falling number ever since.[44] Thus, by 2013, the annual MDG report was claiming that Goal 1, the halving of extreme poverty and hunger, had already been achieved—though it noted on the same page that 1.2 billion people were still living in extreme poverty.[45]

The success claimed here relies in part on the fact that it is the percentage of extreme poor that is being halved; the actual number is not being halved and remains large. A second influence on the success claimed is the FAO's redefinition of hunger, using narrow measures that seem certain to underestimate its extent. The 2011definition measures caloric needs for a *sedentary* lifestyle; if the threshold were set to requirements for *normal* activity, the number of undernourished people would rise from 868 million to 1.33 billion. The new definition, moreover, includes only periods of undernourishment lasting more than one year, and so excludes any shorter food crisis or price spike.[46] Averaging extreme poverty and hunger across the world also allows the distortion discussed in chapter 7, whereby significant improvements, especially in China, mask static or declining situations in most other countries.

The FAO's *State of Food Insecurity in the World 2012* (SOFI12), on which the MDG hunger estimate is based, depicts the halving of hunger as "within reach" and argues that restored economic growth is the number one requirement to get there.[47] Like the Brundtland Commission twenty-five years earlier, the FAO points out that growth alone is not sufficient and that appropriate public policy is needed to make sure the poor can share in it, but this aspect is left entirely in the background. Frances Moore Lappé, founder of the Small Planet Institute, and her research team observe that the word *growth* appears over 250 times in SOFI12, whereas there are only nine references to equity,[48] and *sharing* appears once. As their succinct analysis shows, the FAO seems to be blind-sided by the idea of the efficacy of growth, a view not unexpected in the neoliberal universe. It is stressed again and again, even though SOFI12 explicitly concedes that the "linkage between economic growth and nutrition has been weak."[49] Lappé and colleagues point out that the idea of growth has been wedded to ideas about "removing government and privileging the private sector," but making this aspect so prominent could lead a reader of SOFI12 to overlook completely the more crucial need for equity and government action.

The Lappé team's research into where hunger has actually been reduced contradicts the supposed efficacy of private enterprise and the growth it is assumed to generate, and indicates that government policies have probably been more crucial. The most dramatic success occurred in Ghana, where growth has been low over the last twenty years; there, the government has had programs to pay cash to mothers living in extreme poverty and supports both export and food-growing agriculture. Vietnam's success is connected to land reform, irrigation subsidies, support for smallholders, and control of price fluctuations. Brazil's progress, also significant, again relies on cash payments to mothers living below the poverty line, land grants, and support for smallholders.[50] None of these programs would be favored under free market rules.

The MDG target of decent work for all also remains elusive. By the end of 2005, well before the world recession, global unemployment was already at its highest level ever and had grown by 25 percent in the previous ten years, suggesting that growth in the free market era has never generated sufficient jobs. Growth is advanced as the primary provider of employment, but its performance in this respect was not impressive, even before the global economic crisis. The trend toward higher unemployment continues, with the 2013 MDG report recording an additional 67 million people without jobs since 2007.[51]

Overall, then, economic growth is not closely related to improvements in the welfare and well-being of the poor, including the extreme poor and the hungry. After thirty years of galloping economic growth worldwide and another thirty years of growth at a more modest rate (though very rapid in China), progress for the least advantaged is still lagging. The global economy is eight to ten times the size it was in 1950, but its fruits have not reached the poor at anything like the same rate and, in many places, not at all.

Wealth: Straggler's Destination or the Other Side of Poverty?

There are two closely interrelated problems with the cornucopian notion embedded in the straggler metaphor, where everyone will ultimately catch up and live the affluent life. In the first place, except for China, the past sixty-five years do not provide much confidence that such a catch-up will be achieved in the foreseeable future, even if the percentage of extremely poor people (under a relaxed definition) *is* halved by 2015. Second, it relies on infinite economic growth and does not take into account any depletion of resources or loss of sinks for waste. In chapter 6 I stressed the enormous volumes of paper and oil needed for China alone to attain a US per capita level of consumption (or even a European level, about half that of the US and Australia). The economist Paul Ekins looked at the overall arithmetic and concluded there was no prospect of finding sufficient resources and sinks for the *universal opulence* outcome. Even to ensure one-fifth of current Western affluence for the people of the third world would require, at an absolute minimum, an immediate freeze on per capita consumption in the first world.[52]

The same dilemma dogged the 2009 Copenhagen climate conference and affects the entire climate crisis. As Bolivia's chief negotiator, Angelica Navarro, explained to Amy Goodman of US independent radio's *Democracy Now!*, the first world has only 20 percent of the world's population, yet it has produced more than 90 percent of the existing greenhouse gas pollution in the atmosphere; this constitutes a huge climate debt as far as third world people are concerned:

> So we are the ones who are supposed now to be mitigating. And I'm asking, what will a developing country, rural men or women—indigenous women in Bolivia doesn't even have electricity—will mitigate? And for what? So that developed countries can even still have two, three cars? ... We think that they are negotiating not an environmental agreement. They seem to be negotiating an economic agreement.[53]

First world leaders at Copenhagen ignored this point of view, insisting that everyone must sign up for cuts, especially China, already the world's largest current emitter, and India, which is steadily industrializing. In his address to the conference, Bolivia's President Morales spoke of "the use of atmospheric space by the developed countries. It's not possible that atmospheric space be the exclusive property of just a few countries ... that are *irrationally industrialized*" [my emphasis].[54] Rather, he argued, the atmospheric space needs to be fairly distributed.

As early as Rio, third world leaders and critics from the Left, such as Middleton, O'Keefe, and Moyo (1993), saw environmental concerns as merely tactics of the prosperous, designed to hamper the development of poor countries and thus their ability to compete with the West. At Copenhagen, as the climate crisis pressed on the world, many of the leaders of the poorest countries took the same view as the Bolivians—that the rich must commit to drastic contraction. The US, however, offered only minor immediate cuts (3 percent) to the agreed-upon 1990 baseline. Although "contract and converge" seems the only strategy capable of yielding climate stability in particular and ecosystem protection in general, there is no sign that the United States, one of the world's heaviest per capita polluters and responsible for some 30 percent of the existing atmospheric CO_2, will embark on drastic reductions in the near future. While this remains the case, China and India will also resist.

Thus, rather than a continuum, the relationship of wealth and poverty looks more like extraction. On the basis of *Forbes* magazine's global wealth figures, the French journalist Hervé Kempf noted that the 793 billionaires listed in 2005 possessed assets worth US $2.6 trillion, equal to the entire third world debt, while, as already noted, the income of the richest five hundred individuals equaled the income of the poorest 416 million.[55] Though most of the billionaires are American and many are European, China had eight and India ten. These latter billionaires are the men at the zenith of the elites who organize their nations' resources and security to suit their own business interests and those of the ubiquitous TNCs.

Transnationals own and control much of the world's mining, energy, transport, manufacturing, and even agriculture—a large part of the entire world's productive apparatus. These corporations are multifaceted organizations able to contract out elements of their chains of production in different countries to take advantage of low wages, for example, permissive regulation, or low taxes, and that shower riches on the men who run them. Although a number of women do feature on the Forbes list, most are beneficiaries of inheritance or marriage.[56] *Control* of global wealth is even more concentrated than ownership. Stefania Vitali, James Glattfelder, and Stefano Battison at the Swiss Federal Institute of Technology studied the connections between 43,000 TNCs in 2011, using topological analysis. This network study revealed that 747 of these corporations controlled 80 percent of all TNCs worldwide, and that a core "super-entity" of just 147 tightly interlocking TNCs controlled 40 percent of all global revenue, the majority being financial institutions.[57]

The "trickle-down" effect mooted in neoliberal rhetoric did not eventuate; indeed, wealth trickled up. The billions who still live without clean water, sanitation, or adequate food or the Bolivians who live without electricity are hardly running in the same race as those who manage these entities. Arundhati Roy's notion of the "secession" of the rich in India seems far closer to reality:

> You have one India which has now seceded into outer space and has joined the elite of the world and is looking down at the old India and thinking, "Why are these tribals living on our bauxite and why is our water in their rivers and why is our furniture in their forests and how do we get them off the land?"[58]

Industry Sent South

Much of the third world's growth in the past decades has occurred via industrial expansion in the numerous special economic zones (SEZs) and "free trade zones." In these enclaves, TNCs and local corporations have established what Western people call sweatshops, characterized by long hours, low wages, poor safety, abusive treatment, and so-called flexible work arrangements (where the workers can be dismissed if they get involved with a union).[59] Whether ownership is local or transnational, most of the output is bound for first world consumption. Corporate agriculture is also conducted in conditions Westerners would regard as totally unacceptable, from twenty-hour days on rubber plantations and the use of child labor to being sprayed with pesticides in the course of the working day.[60]

All this means that corporations can sell at prices that are very cheap from a Western consumer's point of view—running shoes and television sets cost much less than they did only twenty years ago. This is the reality that lies at the base of the "free trade" era of globalized production that we are still in the midst of. It suits the ultra-rich, the chief beneficiaries of the profits, and it suits the employed Western consumer, who can buy stuff beyond the wildest dreams of his or her grandparents. "Employed" is a key term in this scenario, which depends on the export from our own countries of millions of manufacturing jobs chasing cheap labor. Just as most of the entire first world population qualifies as "rich" on a global scale, most of the manufacturing class of the world is now removed from the consumers it supplies. Instead, raw materials and components are imported into the SEZs of distant lands and shipped out as cheap goods, which are commonly thrown away within days or weeks of purchase. After the Western working class fought its way over two centuries to reasonable wages and conditions, much of what was gained has now been nullified.

Not only were the gains of manufacturing workers reversed, but so too were the beginnings of the environmental regulation that had emerged in the late 1960s and 1970s. Heavy industrial production was moved away from the first world to China in particular, avoiding wage gains, safety rules, and the regulation of air and water quality. Massive pieces of equipment such as second-hand blast furnaces, which were becoming environmentally unacceptable in the West, were dismantled and transferred to China. Chinese corporations became the world's main makers of steel, coke, aluminum, cement, chemicals, leather, and paper—goods whose manufacture involves high wages and tough environmental rules elsewhere. The *New York Times* reported that an economist with China's Ministry of Commerce conceded that "the shortfall of environmental protection is one of the main reasons why our exports are cheaper," while a study conducted by the European Parliament found that "China's less efficient steel mills, and its greater reliance on coal, meant that it emitted three times as much carbon dioxide per ton of steel as German steel producers."[61] The Berkeley economist Richard Carson, co-author of a detailed assessment of China's CO_2 emissions,[62] told the Environment News Service that from about 2000, Chinese "government officials turned away from energy efficiency as an objective, to expanding power generation as quickly as they can, and as cheaply as they can.... Many of the poorer interior provinces replicated inefficient Soviet technology." Thus, China has been building power plants that are dirty, inefficient, and outdated at the outset and are intended to operate for another forty to seventy-five years.[63]

The transfer of heavy industry to China, with the concomitant rush to build the cheapest coal-burning power plants to service it, constituted a big step backward for the world as a whole as far as industrial efficiency, environmental protection, and climate safety are concerned. The manufacture of vast quantities of the Western consumer's gewgaws in these conditions has added immense environmental penalties to our consumption. Though the emissions are reckoned as China's, it is we who are the end consumers. Viewed through the corporate lens of price, profit, and growth, however, it rates as a grand success.

Livelihood or "Progress"?

The putative success of three decades of "development" and three more of "globalization" has always been measured in GDP and in GDP per capita, a technique that assumes, first, that GDP is a reliable gauge of national economic health, and second, that an average taken over any population will fairly represent the prosperity and well-being of most of its individual citizens. Neither is necessarily the case (box 9.4).

Vandana Shiva has pointed to values that are unreflected in GDP but are vitally important to people who still produce their livelihood in symbiotic cooperation with nature. Shiva describes the cultural and spiritual connection that India's farmers have with their land and celebrates the communal practices and technologies that have kept India's agriculture viable for millennia. She vigorously denies that the peasant life is "below dignity" and something from which to be delivered. "It is disposability that robs people of their dignity and selfhood," she argues, not the work of tilling the soil and lighting a wood fire. GDP-based notions of well-being are irrelevant in this world.[64]

Shiva believes that indigenous and peasant people are entitled to control over the fate of their ancestral lands—the opposite view from the one that accords all resources to the control of the nation state. She also draws attention to the madness of expanding hydrocarbon-based

Box 9.4
Some Problems with GDP as a Measure of Prosperity

GDP is made up of the value of "all the goods and services sold in a country in the course of a year ... supplemented by ... the cost of producing the non-market services provided by government."[a] The critique of GDP as the measure of a country's prosperity has existed since the concept's inception in the early 1930s, when the architect of national accounting, the Nobel laureate Simon Kuznets, told the US Congress that social well-being cannot be inferred from national income and that the question "growth of what and for whom?" would always need to be asked.[b] GDP remains, however, the accepted measure of wealth in mainstream economics and politics and is widely used as a proxy for well-being. Though economists sometimes retreat from such a claim, arguing that GDP is just a measure of cash flow and claims to be nothing else, the ills of the world recession that began in 2008 have been couched in terms of the failure of growth in GDP, confirming that GDP serves as the benchmark for things going well. But it is ill-fitted for that task.

First, the only items included in the calculations are market-based activities that can be quantified in monetary terms; this excludes the entire range of subsistence, household, and voluntary work described in chapter 6, which accounts for at least half the economic activity in the countries that still have a large rural sector and makes up a significant slice of economic activity in developed economies as well. Second, GDP does not subtract environmental or social costs from its total; these are negative externalities in mainstream accounting and rarely quantified in monetary terms. This is not to argue that environmental damage can or should necessarily be quantified in a monetized cost–benefit regime but to point out that GDP imposes a regime of measurement where items without a market price are excluded altogether, even when they are of immense value. Third, and compounding the second problem, many of the costs of ecological and social damage end up as *additions* to our supposed wealth. For example, the cost of cleaning up the Moreton Island oil slick of 2009 is counted on the plus side in Queensland's and Australia's 2009 GDP.

Under the guidance of Pan Yue as deputy director, China's State Environmental Protection Administration (SEPA) worked for years in the early twenty-first century to produce a system, known as green national accounting (or "green GDP"), that would factor the immense costs of pollution and ecological damage into GDP, and account for the depletion of land, minerals, forests, water, and fisheries.[c] This was intended to provide a more realistic assessment of the real progress involved in China's extraordinary economic growth since 1979. In 2005, Pan Yue spoke openly with the German press, criticizing "the assumption that the economic growth will give us the financial resources to cope with the crises surrounding the environment, raw materials, and population growth."

Box 9.4 (continued)

Spiegel: Why can't that work?
Pan: There won't be enough money, and we are simply running out of time.[d]

In this interview with *Der Spiegel*, Pan Yue outlined the extent of the destruction of China's agricultural land and the poisoning of its rivers in the pursuit of industrial growth up to that point. In June 2006, Zhu Guangyao, another senior official of SEPA, told *China Daily* that "damage to China's environment is costing the government roughly 10 percent of the country's gross domestic product."[e]

The first report quantifying China's green GDP was published in September 2006; it estimated that pollution alone was costing the equivalent of three percentage points of economic output while acknowledging that this was doubtless an underestimate.[f] The second report, due for release in March 2007, was never published—its lead researcher told the *New York Times* that provincial officials had killed it off. According to the *Times*, "the early results were so sobering—in some provinces the pollution-adjusted growth rates were reduced almost to zero—that the project was banished to China's ivory tower ... and stripped of official influence."[g] The work was revived, however, in 2008 and results since then suggest that 3.5 percentage points of China's GDP are swallowed up by environmental damage—but items such as depletion of groundwater and loss of arable land through erosion were still not fully included.[h]

GDP remains a worldwide convention that treats cash flow as if it represents genuine economic and social health. Numerous alternative measures have been put forward,[i] but none has even begun to rival GDP in daily discourse as an indicator of human well-being. Those who have

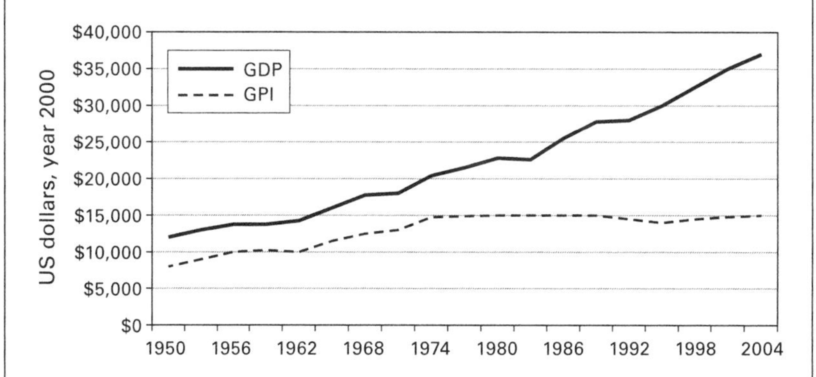

Figure 9.2
Gross Production vs. Genuine Progress, 1950–2004 (in 2000 dollars).
Source: Talberth, Cobb, and Slattery (2007, 19). Courtesy of Clifford Cobb.

(continued)

Box 9.4 (continued)

devised indexes to incorporate actual well-being have found that the growth of GDP exceeds any increase in genuine social progress. Proponents of the Genuine Progress Indicator, which emphasizes sustainability, note that, while GDP per capita has grown continuously, the index of genuine progress has stagnated since the 1970s.

Notes

a. Gadrey 2005, 263.

b. Gadrey 2005, 263, 319.

c. State Environment Protection Administration of China (SEPA) 2006.

d. Lorenz 2005.

e. *China Daily* 2006.

f. SEPA 2006.

g. Kahn and Yardley 2007.

h. O'Rorke 2013.

i. See Nordhaus and Tobin (1973) for Measures of Economic Welfare (MEW); Daly and Cobb (1989, 401–416) for the Index of Sustainable Economic Welfare (ISEW); Talberth, Cobb, and Slattery (2007) for the Genuine Progress Indicator (GPI); Abdallah et al. (2009) for the Happy Planet Index (HPI); and UN Development Programme (2010) for its Human Development Index (HDI).

infrastructure and technology as the climate crisis grows and the natural world is eroded: "chasing economic growth while ecosystems collapse is stupidity."[65] As noted in chapter 6, projected volumes of cars on the roads of India and China alone would require more than the entire current oil supply of the world (and that on top of the undiminished requirement for three-car garages in US homes). Even the most heroic and optimistic of the oil corporations admit that quantities adequate for such demand will never be pumped, while attempts to meet such demand will rely on ever more grossly polluting fossil alternatives such as tar sands (discussed in chapter 14).

India's "superhighway"—the fifteen-year project to widen and pave 40,000 miles of dusty roads—has, according to journalist Amy Waldman, "cut a swath of destruction, swallowing thousands of acres of farmland, shearing off the fronts of thousands of homes" and grafted "western notions of speed and efficiency onto a civilization that has always taken the long view.... Ever-flashier cars, evidence of a frenzied new consumerism, leave bullock carts in the dust."[66] Local people oppose the cutting of trees along the roadsides—neem, mango, sisam, sacred fig, all valuable for shade as well as timber, fuel, fruit, and medicine—and the bifurcation of villages, where women now have to detour through occasional cuts in concrete barriers and sprint across the motorway to get to the water pump or carry back fuelwood. Though Waldman is in basic sympathy with modernization, she lucidly portrays the same two Indias as Arundhati Roy.

In *Soil Not Oil*, Shiva also touches on the destruction of India's roadside trees in the service of the motorcar. She stresses, above all, the vital function of shade trees for the predominant foot and cart traffic on India's roads. Planted sequentially for millennia, hundreds of years old, these trees blunt the searing Indian sun in areas where temperatures frequently exceed 40°C (104°F). Though presented as benefiting the nation, the roads primarily benefit car owners, less than 1 percent of the population. Even the potential market for cars is only a fraction of the whole population—a middle class of 216 million out of 1.2 billion people, according to Shiva's 2008 estimate. What is more, many of the SEZs swallowing up the land of India's farmers are dedicated to automobile manufacture and associated industrial processes. Some cars are made for locals, but a large part of the production is intended for export, which increased at 39 percent annually in the five years to 2008, fueled by trade liberalization policies and the arrival in India of numerous first world car manufacturers.[67]

Since the days of Lewis and Rostow, it has been assumed that farmers would be better off "liberated" from the soil and given "employment." While the wealthy classes of third world countries, exemplified by India's finance minister, have wholeheartedly embraced this recommended development path, no doubt in many cases genuinely believing it to be the way to national prosperity, it has often seemed to the poor that the objective is merely to steal their livelihoods, culture, security, and land.

The drive to bring "development" to the third world by industrializing both farming and manufacture assumes the universal applicability of the Western template and rests on the faith that unlimited resources and sinks will always be available for ongoing expansion. The dominance of neoliberal ideology since 1980 has only intensified the push for "progress through profit" and deepened the pursuit of unending economic growth as the only answer to unrelieved poverty. In its 2012 *World Development Indicators*, the World Bank records nearly 2.5 billion people still living on less than $2 a day, only slightly fewer than in 1981. The bank also records nearly 1.3 billion people who remain in extreme poverty. As critiques of its methods indicate, these figures very likely underestimate the extent of both extreme and "non-extreme" world poverty. When attention is paid to the gross numbers in all the World Bank's categories of "poor" people rather than to its ever-improving percentages, lifting millions out of poverty is accompanied by leaving billions behind.[68] But neither minimal success in reducing the *numbers* of people in any category of disadvantage nor the obvious extension of injustice and dispossession in the service of cheap consumer goods for the first world has so far been able to cut through the generally accepted view that the growth strategy is making great progress in this respect.

This view is one of many that prevail despite the evidence. In part III, I examine the processes that make this possible, how ideas about the successes of growth and the triumphs of progress have been intentionally buttressed and propagated, so that they circulate as "common sense," and fortified so that they are able to sweep aside contrary evidence and alternative values. I begin with the emergence of the big corporation around 1900.

III

Persuading the People

Touch that most vital of all business matters, the question of general federal regulation of industrial corporations, and the people amongst whom I live my life become immediately rabid partisans. It matters not one iota what political party is in power, or what President holds the reins of office. We are not politicians, or public thinkers; we are the rich; we own America; we got it, God knows how; but we intend to keep it if we can by throwing all the tremendous weight of our support, our influence, our money, our political connection, our purchased senators, our hungry congressmen, and our public-speaking demagogues into the scale against any legislation, any political platform, any Presidential campaign, that threatens the integrity of our estate.

—Frederick Townsend Martin, 1911

10

Propaganda: "Business Finds Its Voice"

The engineering of consent is the very essence of the democratic process, the freedom to persuade and suggest.

—Edward Bernays, 1947

Corporate propaganda directed outwards, that is, to the public at large, has two main objectives: to identify the free enterprise system in popular consciousness with every cherished value, and to identify interventionist governments and strong unions (the only agencies capable of checking the complete domination of corporations) with tyranny, oppression and even subversion.... The subject embraces a 75-year-long multi-billion dollar project in social engineering on a national scale.

—Alex Carey, 1997

The Corporation and Persuasion

The big corporation entered the US economy at the end of the nineteenth century and soon began to adopt professional public relations in its struggle with its opponents: first to counter popular resentment at the destruction of pre-corporate patterns of ownership and everyday life, then to contain dissatisfaction with the cataclysmic failure of the corporate economic system from 1929 to the beginning of World War II. Corporations applied the new advertising techniques, so successful in the incitement of consumption as a guarantee of ongoing growth (see chapter 5), to the promotion of the capitalist system itself. They proposed what they called "free enterprise" as the natural path to the "good life," defined as a life of material comfort and abundance.

After the war, this practice continued uninterrupted, but a new centerpiece emerged, the ideology of the "bigger pie," where economic growth was given the role of spreading abundance without disturbing economic power. In this refinement, expansion could encourage a docile and compliant workforce and democracy could be made safe for corporations. Though productivity had been considered a promising substitute for redistribution back in the 1920s, it lost traction during the long years of the Great Depression. Now, however, the idea of redistributive justice was jettisoned by many in the unions and the Democratic Party who had been former opponents of business. Economic growth was embraced broadly, and the new corporate order could expand unchecked as the solution to social discontents. As growth became the self-evident, obvious, and inevitable basis for almost all government policy across the world, collision with planetary limits crept closer.

The Fall of Economic Growth?

H. W. Arndt in his *Rise and Fall of Economic Growth* (1978) outlined an abrupt emergence of economic growth as a preeminent policy objective for governments in the course of the 1950s. This approach appeared in tandem with the emergence of "development" as described in chapter 7 and was promoted by some of the same economists—W. Arthur Lewis, for example, and Walt Rostow.[1] Just as material development had not been a specific policy focus before the war, Arndt found that growth too had not been an object of government policy during the first half of the twentieth century. In the roaring twenties, for example, though growth was spectacular, "economic growth was hardly ever discussed by academic economists," and neither businessmen nor politicians thought its promotion a proper activity of government anyway. Examining the work of prominent academic economists and government officials, chiefly from the United States and the UK, Arndt identified several imperatives that underlay the new direction after 1950, all of them clustered around the idea that growth was the best solvent for class discontents aimed at redistribution and an essential prop to full employment. After the traumatic unemployment of the Great Depression, the latter had come to be considered an essential social objective.[2]

Arndt traced the various strands of the critiques of growth launched in the 1960s and regretfully concluded that the critics had succeeded and the push for growth had had its day; by the early 1970s, he argued, growth was an idea in decline. Yet, as is evident from public discourse in the years since then, the status of growth as the central objective of policy recovered from all setbacks. Arndt grossly overestimated the impact of the critique, or perhaps underestimated the ability of the promoters of growth to counterattack.

Emergence of the Big Corporation

The modern corporation began wrestling with issues of image and legitimacy soon after its rapid emergence in the United States in the late nineteenth century. In 1870, there were no national US corporations at all, and the average factory employed fewer than ten workers.[3] A mere thirty-five years later, by the early 1900s, the vital industries of the entire country were dominated by large corporations, with fully two-fifths of US capital in the hands of a few large new firms. The age of the giant "vertically integrated, bureaucratically managed corporation"[4] began rapidly, and Adam Smith's free enterprise economy, comprised essentially of a multitude of small and medium-sized businesses in healthy competition, would soon become a thing of the past.

This revolution took off in the context of a protracted series of depressions, stretching from 1873 through to the mid-1890s. A handful of significant mergers occurred from the 1880s on, beginning with the creation of the Rockefeller-dominated Standard Oil cartel in 1882.[5] Two further waves of mergers occurred in the period from 1890 to 1902, hundreds of them in the second tranche (1898–1902), when around two thousand independent businesses disappeared.

Alfred Chandler, historian of the rise of the modern corporation, argues that merger (or horizontal integration) sometimes enabled companies to fix prices and to control the level of production, but it was insufficient to establish decisive advantage and ongoing growth. In Chandler's view, what achieved this end was vertical integration, which was adopted by the newly merged entities in almost all the most successful enterprises.[6] In this adaptation, single multifaceted units controlled the entire chain of production, from the supply of raw materials to the marketing of finished products, overcoming vulnerability to market fluctuations in supply prices and, in fact, superseding the putative market role of allocation altogether by internalizing the whole process. As Chandler puts it, "the visible hand of management replaced the invisible hand of market forces."[7] This strategy cut costs and improved profits, which was useful as a counter to the price contraction during the long depression of the late nineteenth century; it extended through the first two decades of the new century. The combination of merger with vertical integration established the business structure of the US economy for the long-term future. Indeed, a great many of the corporate giants still familiar in the twenty-first century, such as Coca-Cola, DuPont, and Alcoa, had been established by 1917.[8]

By the turn of the twentieth century, these new behemoths had already begun to extend their consolidated activities internationally. In 1914, foreign direct investment already accounted for 7 percent of US GDP, and by the time of the 1929 Wall Street Crash, US private interests held over $8 billion in foreign loans and investments.[9] These developments, enhanced by the profits the United States reaped from its role as industrial supplier to the UK and its allies in two world wars, prepared US business for the internalization of much of the world economy within its giant corporations after World War II. These US companies were the prototypes of the modern transnational corporation (TNC) and were increasingly emulated by corporations based in Europe and Japan. By the 1960s, between 20 and 30 percent of world trade took place *within* transnational corporations, a proportion that had grown to some 40 to 50 percent by 1990. At that time, more than half of US exports and imports were internal transactions of TNCs.[10] In the twenty-first century, the Fortune Global 500 includes numerous TNCs based in China and several more scattered throughout the developing world, mainly in Asia.[11] The globalization era has favored the ascent of transnational capital, which now dominates world trade, foreign investment, and world production.

The emergence of these unfamiliar corporations was not at first welcomed by many Americans. According to the economic historian Naomi Lamoreaux, the United States—hitherto made up of small-scale, small-town communities—was "transformed overnight from a nation of freely competing, individually owned enterprises into a nation dominated by a small number of giant corporations."[12] There was great alarm over the wave of mergers around 1900: in one of these, for example, the International Harvester trust consolidated 85 percent of the manufacture of US farm machinery into a single organization. Even if, as Chandler holds, the key competitive advantage lay in vertical integration, the numerous horizontal mergers made the integrated corporations far bigger than they would otherwise have been and helped fuel the fear of the concentration of economic power in very few hands.

The Struggle for Acceptance and Legitimacy

These centralized organizations, integrated both horizontally and vertically—as predicted by Marx[13]—have persisted for more than a century as the dominant form of enterprise in the world economy, gaining international primacy in the second half of the twentieth century; their immense material superiority gave them the deep pockets to finance their multifaceted counterattack against environmental concerns and the critique of growth in the 1970s. The means by which the emerging corporations of the early twentieth century won over a reluctant population were varied, but all owed their success to the new profession of public relations, touched on in chapter 5. Blunt force retreated as an everyday option for US business and the professionalization of persuasion began; it would extend the principles of product advertising into every arena of life, thought, and politics.

At the beginning of the twentieth century—the era of the "robber barons" in the United States—conflict between capital and worker continued as it had throughout the nineteenth century, and intensified. Businessmen bitterly opposed any form of regulation and regarded labor unions as a part of the pernicious tendency to regulate conditions and wages. In 1895 the National Association of Manufacturers (NAM) was founded to lobby for business-friendly legislation and, under the leadership of David Parry, in 1903 it launched "a crusade against unionism."[14] George Baer, railroad president and spokesman for the mine operators, remarked during the 1902 anthracite coal strike that "the rights and interests of the laboring men will be protected … not by the labor agitators, but by the Christian men to whom god in his infinite wisdom has given the control of the property interests of the country."[15] Views such as these, accompanied as they were by corporate militias, swindles, and lockouts, underpinned panic about democracy. Gustave Le Bon, one of the earliest theorists of the danger of imminent democratic change, commented, "Today the claims of the masses are becoming more and more sharply defined and amount to nothing less than a determination to destroy utterly society as it now exists.… The divine right of the masses is about to replace the divine right of kings."[16]

Complicating the legitimacy problem for business, the middle classes—including the farming community—were unhappy with the concentration of economic power and the dissolution of the world they knew. As Stuart Ewen observes, it was no longer just the working class that opposed a world where

> a small number of powerful and disdainfully arrogant men were dictating the social circumstances and life rhythms of countless people throughout the United States.... Middle class people began to look toward the state and toward the device of regulation as necessary instruments for controlling the rapacity of those who held unrestrained private wealth.[17]

In the infancy of public relations, before World War I, the Enlightenment-based idea of disseminating factual information to a rational audience for sensible assessment persisted, but the concept of a "fact" had already begun to shift. Ivy Lee, one of the very first men to set up as a PR practitioner, started his practice by offering "facts" (known today as "factoids") to the public. In his PR activities for the Rockefellers after the Ludlow massacre of 1914, when a miners' camp was set alight, burning men, women, and children to death, Lee published a "witness report" from the wife of a railroad official who was not, in fact, present at all. When at the subsequent US Commission on Industrial Relations investigation Lee was asked what personal effort he had made to ascertain whether the facts given to him by the owners were correct, his frank reply was, "None whatever."[18] Thus, although Lee still embraced an ostensibly factual rhetoric, notions of evidence intrinsic to the concept of scientific knowledge were evaporating. Facts could be manufactured to suit the purpose. Lee also pioneered the presentation of publicity as news. In one early job in 1906, to help an insurance company improve its profile after a fraud conviction, he arranged for the company president to write letters to seven hundred newspapers across the country; Lee got most of them published, many in prominent slots masquerading as "news."[19]

The Manufacture of Consent

The government propaganda effort after the United States entered World War I in 1917 demonstrated to the corporate world just how persuasive the new techniques could be. President Wilson hired George Creel, a journalist and erstwhile trenchant critic of big business, to run his Committee on Public Information (CPI), designed to promote US involvement in a war that many Americans did not support. Creel, who later wrote about his work under the title *How We Advertised America*,[20] marshalled the advertising industry to lead the effort and demonstrated that advertising could be just as effective in the propagation of ideas and beliefs as it was at selling consumer goods. He noted in one article that "somebody once said that people do not live by bread alone; they live by catch phrases."[21] The "public information" Creel organized was favorable to big business. Creel's CPI recruited the Four Minute Men, community leaders who spoke briefly in thousands of local movie theaters—about 150,000 times a week. They preached the need to "make the world safe for democracy," branding as German sympathizers any citizens who thought the war was a "capitalists' war" or was being pushed by the captains of industry for its money-making potential, or who otherwise felt the United States had no interest in it. Loyal Americans were instructed to report antiwar sentiment, and by 1918, a Censorship Board was actively suppressing dissent.[22] Creel also used the full range of social institutions to spread the patriotic message—from churches and schools to Rotary Clubs and Chambers of Commerce, from cinemas to the print media.

It was in the wake of this immensely successful PR campaign that Walter Lippmann, a journalist and another early theorist of public opinion, coined the phrase "manufacture of consent," later revived by Noam Chomsky. "Increasingly," Lippmann wrote, people "are baffled because the facts are not available; and they are wondering whether government by consent can survive in a time when the manufacture of consent is an unregulated private enterprise." Lippmann, fearing that democracy was not actually possible under such circumstances, appeared to fault private ownership of the means of communication: "So long as there is interposed between the ordinary citizen and the facts a news organization determined by entirely private and unexamined standards, no matter how lofty … no one will be able to say that the substance of democratic government is secure."[23]

The prominent business analyst Roger Babson remarked in 1921 that "the war taught us the power of propaganda. Now when we have anything to sell the American people, we know how to sell it."[24] Edward Bernays, too, noted that the "astounding success of propaganda during the war opened the eyes of the intelligent few in all departments of life to the possibilities of regimenting the public mind." In the world Bernays aimed to foster, "propaganda is the executive arm of the invisible government."[25] As well as demonstrating how effective PR could be, the CPI showed that existing organizations of all kinds could be recruited to spread any message.

During the 1920s, US corporations were not only concerned about their image and legitimacy in the public eye, they were gravely fearful of overproduction and anxious that growth would be jeopardized if ongoing consumer desire was not systematically stimulated (discussed in chapter 5). Yet, even while consumer desire was being carefully constructed by advertising, one of the emerging PR gambits was the idea that the corporation existed primarily to serve the people's wishes.[26] Simultaneously, the cabal of wealthy individuals who had created the huge new companies was beginning to attract the first wave of middle-class investment, an initial small step toward the idea of the citizen shareholder class;[27] this trend helped solidify middle-class identification with the interests of big business. Corporations also recognized that joining regional and community organizations gave them "unusual opportunities for establishing contacts with the leaders in general public activities and those who are molding public sentiment." Two subsidiaries of AT&T alone paid dues of nearly $5 million for noncommercial memberships of every conceivable kind between 1925 and 1934.[28]

Republican presidents were returned throughout the 1920s on platforms similar to those of the late twentieth-century neoliberal era: tax cuts, the promise of universal prosperity through laissez-faire economics, the minimization of union power, and deregulation—"less government in business and more business in government," as President Harding put it during the 1920 campaign.[29] The decade also saw more consolidation, with the disappearance of some five thousand public utilities and more than eight thousand separate mining and manufacturing businesses.[30] Newspapers were among the consolidating enterprises,[31] and Roy Howard, who headed Scripps-Howard, one of the two largest print concerns (the other was Hearst), approved heartily of the "system of setting in action identical thought processes in all communities of the nation."[32] At the same time, commercial radio was granted the lion's share of the bandwidth; according to Ewen, "the public sphere itself was becoming the private property of corporate gatekeepers."[33] The concentration of both economic and ideological power proceeded unchecked. As was the case during the era of the mortgage bubble, so abruptly burst in 2008, the financial sector dominated the economy and debt fueled the fiesta of consumption. Most of the biggest US corporations at the start of 1929 were holding companies engaged in financial and speculative activities—only four out of the ninety-seven largest corporations were strictly operating companies doing real business in the productive economy.[34]

Bruce Barton, one of the foremost executives in the new PR industry and holder of the accounts of DuPont, US Steel, General Motors, and General Electric, told a convention of electric light chiefs in 1926, after two decades of the infant PR industry's efforts, that corporations were still "only tolerated" and that the people did not "fully understand" or "fully trust" them and did not yet "love them."[35] The 1920s had nevertheless seen the reputation of corporations improve greatly: in *Propaganda*, Bernays suggested that the big corporations were by then regarded as "friendly giants" rather than the "ogres" they had been thought to be, a change he attributed to "the deliberate use of propaganda."[36]

The Wall Street Crash of 1929 brought this success to an abrupt halt, delivering a serious reversal to the more positive image achieved by big business in the decade following the war. As the hard times dragged on, a new campaign would soon be launched to persuade Americans that, despite the trauma of the Great Depression, business would still serve them better than any government ever could.

Depression and War

For the Love of the Corporation

Along with the major corporations (which already had their own PR staff), NAM played a central role in the fight against the New Deal. The US Congress had investigated NAM's influence on legislation as early as 1913. It found some evidence of direct inducements to legislators but reserved its "severest censure" for what it called "coercion through propaganda," a disguised campaign to influence public opinion, conducted through newspapers, speakers, and the dissemination of literature to schools, colleges, and civic organizations across the country.[37] Robert Lund, who became NAM's president in September 1933, urged the corporate world to unite against Roosevelt under the NAM banner, and launched a campaign to dislodge him in the 1936 election; this was not successful, which "gravely shocked business men all over the country."[38] As PR man Bruce Barton told NAM's 1935 convention, "industry and politics ... are competitors for the confidence and favor of the same patron ... the public." If businessmen talked to ordinary people, he said, they would be astonished to find "how little we are liked, how much our incomes are resented and our motives misunderstood." They must, he told them, persuade people that "we are more reliable than the politicians [and] ... will work for them more cheaply and with more satisfaction." Barton echoed the exhortations of General Motors president Alfred Sloan Jr., who at the same meeting forecasted a great battle between "political management" and "private enterprise," a battle for the "very foundation of the American system." NAM's president declared that they would campaign to "sell the American way of life to the American people."[39]

This they did. NAM's public relations budget rose from $36,000 in 1934 to almost $800,000 in 1937.[40] The campaign yoked the New Deal notion of the greater good to the role and purpose of corporations and argued that private enterprise was the citizen's guarantee against a "collectivistic" future where people were mere cogs in the machinery of the state. In contrast to this specter, NAM offered the "freedom of opportunity for the individual to strive, to accumulate, and to enjoy the fruits of his accomplishments."[41] In essence, a wealthy elite set out to ensure that people would be convinced, as *Fortune* magazine put it, that "business is not just a phenomenon on the surface of American life ... it *is* American life."[42]

If the level of its ownership and domination is any guide, business was indeed the essence of American life. In a 1938 speech, President Roosevelt characterized the corporate world as "a concentration of private power without equal in history ... a cluster of private collectivisms masking itself as a system of free enterprise." One-tenth of 1 percent of corporations owned more than half of corporate assets and earned half of corporate income, he told Congress, while less than 5 percent owned 87 percent of the assets and earned 84 percent of the income of all of them. Three-tenths of 1 percent of the US population received 78 percent of all dividends—the equivalent, the president pointed out, of one person in every three hundred receiving 78 cents in every dollar of corporate dividends while the other 299 got to divide up the remaining 22 cents between them.[43] *Fortune* magazine, in contrast, advised business that it should present itself as "a public utility" rendering "well-defined public services" rather than as a money-making apparatus.[44]

The NAM campaign for the "American way" was massive. It replicated Creel's World War I model in establishing local cells, "Special Committees of Public Information," which enlisted local Chambers of Commerce, Rotary Clubs, and churches, as well as lawyers, teachers, and local dealers of the appliances and cars made by the corporations. These committees of influential people were responsible for the regional face of NAM's multifaceted "publicity program"; they funneled articles, features, and films to newspapers, radio stations, and movie theaters; they sent speakers to the theaters as Creel had done, as well as to every local group of any sort (including women's groups and what were then called "negro groups"); they distributed pamphlets and weekly bulletins to schools, clubs, and libraries.[45] Aware that the adult population was cynical about the corporate claim to "service," they aimed specifically at schools, where *Young America*, their weekly children's magazine that portrayed capitalism as dedicated to looking after them and their communities, was sent to thousands of teachers, who used them in classroom assignments. *You and Industry*, a series of booklets written in simple language, linked individual prosperity to unregulated industry, and was distributed to public libraries everywhere.[46] One million booklets were distributed every two weeks by the US Chamber of Commerce, which, along with the giant industrial corporations, was also involved in the campaign.[47]

The American Way campaign reflected the transformation of PR practice from a rhetoric of alleged facts, such as Ivy Lee had used in the early years, to a rhetoric of symbols and images that entertained the viewer (as recommended by Bernays) and, as Lippmann described it, "assemble[d] emotions after they have been detached from their ideas, ... [causing an] intensification of feeling and a degradation of significance."[48]

Images and symbols were deployed at all levels. Cartoons were distributed to more than three thousand weeklies; in one such cartoon, the "forgotten man," symbolic of people destroyed by the depression, is portrayed as the fleeced taxpayer, and the "fat cats" are not corporations but pro-welfare politicians.[49] Billboards looked down on every town of

Figure 10.1
Bread line under an "American Way" Billboard, Louisville Flood, Kentucky, 1937. *Source:* Margaret Bourke-White/Masters Collection/Getty Images.

twenty-five hundred people or more by 1937 (figure 10.1),[50] combining happy families with the slogans "There's no way like the American Way" and "What's good for industry is good for you." Serials such as *The American Family Robinson* were sent to radio stations without charge and broadcast across the country through more than two hundred local stations every week;[51] in this particular serial, a happy white family provided the setting for engaging stories that pitted their sensible, pro-business father against socialist troublemakers such as the benighted "Friends of the Downtrodden." Movie theaters screened feature films and "documentaries" depicting the upward march of America, "a tale of uninterrupted scientific progress … a history driven by the genius of American industry."[52] The proliferating avenues of mass communication were saturated with this message.

Just as NAM's activities had been covert before World War I, the avalanche of "American Way" propaganda in the 1930s continued the tradition. In its 1939 investigation into violations of labor rights, the

La Follette Civil Liberties Committee (US Senate) found that NAM had "blanketed the country with a propaganda which in technique has relied upon indirection of meaning, and in presentation upon secrecy and deception."[53] NAM rarely disclosed that it was the originator and distributor of the innumerable films, radio programs, news editorials, advertisements, and purported news it circulated. The La Follette Committee slammed NAM: "They asked not what the weaknesses and abuses of the economic structure had been and how they could be corrected, but instead paid millions to tell the public that nothing was wrong and that grave dangers lurked in the proposed remedies."[54]

The New York World's Fair of 1939, the "World of Tomorrow," was the business community's last great PR extravaganza of the decade; it took the production of theater and spectacle to a new level. Again, while the rhetoric was uncompromisingly democratic— "this is your fair, built for you, and dedicated to you," according to the official guidebook— space was sold to the corporate participants with the promise that this would be the "greatest single public relations program in industrial history." Participants were counseled that it would not do simply to display their products—they must sell ideas. They must propel the public into a "world of abundance, where consumers could be kings and queens."[55] NAM's pavilion, for example, aimed to fascinate and delight as it thrust private enterprise into the role of the true "friend of the downtrodden," which could fulfill the promises of the New Deal: opportunity, security, and the greater good.[56]

While opposing government relief for the 12 to 17 million US unemployed, blaming the destitute for their own failure to practice sufficient thrift, and characterizing old age pensions as "socially inadvisable" and an "unwarrantedly weakening drain on industry,"[57] the big-business people of the United States embraced a strategy of humanizing the corporation, positioning themselves as common folk cut from the same cloth as the ordinary rural citizen. US Steel adopted a "folksy classless persona," portraying itself as "a bunch of hometown boys" making a few nails, and Henry Ford enjoyed unusual popularity as "the friend of the common man."[58]

There were certain successes: the *Advertising Age* credited the National Chamber of Commerce with divorcing the word "big" from the word "business" in the public mind, and historian Wendy Wall identifies the rapid replacement of the terms "private enterprise" and "laissez faire" with the more appealing "free enterprise" in public discourse during the late 1930s, a change she attributes to the strategies of US corporate leaders in general.[59] Republicans also did better in the congressional elections of 1938 and 1940 than they had done since 1928, but the American people continued to support their interventionist president. The corporate entreaty to oppose "restriction and regimentation" and to allow "the true creators of wealth to serve their constituencies" was still viewed with skepticism.[60]

The coming of World War II, however, was to be a boon for the business of propaganda and the propaganda of business. The Roosevelt administration had no choice but to reconcile with big business in order to mobilize for war, and the corporations were increasingly able to present themselves publicly as serving "the nation and the world so well in this hour of peril."[61] The war in any case fostered feelings of national solidarity and finally brought the return of economic growth, which would ultimately make it far easier for big business to cast itself as the friend of the people. War, like consumerism, is a wonderful driver of growth; it induces massive production of ephemeral products that, in war, are soon exploded and, in consumerism, quickly superseded.

War and Aftermath

Throughout the war, NAM's rhetoric downplayed the role of government and touted the superiority of American corporate "know-how," manifested in the heroic contribution of industry to the war effort ("the initiative, ingenuity and organizing genius of free enterprise").[62] This was blended with visions of a cornucopian future of endless consumer products, in which families would fly their personal helicopters and planes to the shops "as readily as you drive a car." "Put my groceries in that blue helicopter," one customer tells the grocer in a DuPont advertisement for cellophane reproduced by historian Roland Marchand.[63]

Privately, however, most US businessmen remained apprehensive about the future. They still saw government involvement and trade union influence as enduring threats. When President Roosevelt delivered his "Four Freedoms" speech in 1942 (discussed in chapter 6), businessmen thought he was missing the essential freedom: the corporate vision of the American way of life defined freedom first and foremost in relation to enterprise. The business community and their PR operatives insisted that a "fifth freedom," the freedom of enterprise (or, as Noam Chomsky later put it, "the freedom to rob and exploit"),[64] needed to be added to the other four. As champions of the "fifth freedom," the corporations sought to wear the costume of the New Deal while undermining its values and priorities. Monsanto's chairman expressed the wartime view that the business world was "engaged in two wars," and Paul West of the Association of National Advertisers identified the greater of these challenges as that of winning "the war of ideologies" at home.[65]

Long-time California Chamber of Commerce officer Vernon Scott described the crusade that corporations needed to mount in this way: "We must constantly preach and prove that a planned economy destroys free markets, shackles free enterprise, reduces the standard of living for all, beats down private initiative, and cripples competition, the life blood of democratic capitalism."[66]

Henry Link of the Psychological Corporation[67] recommended a continuing strategy of manipulation that would emphasize the rights of all people, "a transfer in emphasis from free enterprise to the freedom of all individuals under free enterprise; from capitalism to a much broader concept: Americanism." This approach, Link said, was based on research showing that "Americanism" had a "terrific emotional impact" that "free enterprise" lacked.[68]

Just as fear of communism had been a compelling tool for the pro-business thrust after the Russian Revolution in 1917, so too was the Cold War suited to corporate interests. By 1946, just a year after the United States' wartime alliance with the Soviet Union, the Chamber of Commerce was already distributing millions of pamphlets with titles such as *Communist Infiltration in the United States*.[69] The second Red Scare of the 1950s played a key role in creating a climate conducive to the protection of traditional property relations and to splintering the labor movement.[70] While congressional committees terrorized intellectuals and film actors, and liberal radio broadcasters were fired, popular magazines ran lurid stories of imminent takeover, not unlike the sudden appearance, just before the invasion of Iraq in 2003, of the spurious but widely publicized "proof" that attack by weapons of mass destruction could be unleashed on Britain or the US mainland within forty-five minutes.

The Chinese revolution, the Soviet development of nuclear weapons, and the Korean War all intensified the fear of communism and promoted antisocialist sentiment, providing fertile ground for the ongoing multimillion-dollar selling of the American way. Assisted by both history and consummate PR machinery, pro-business Republicans won control of both houses of Congress in 1948, and Dwight Eisenhower took the presidency in 1952.

Television was also a potent new tool in the postwar onslaught. Not only was the medium visual, bearing all the potential of images and symbols to manipulate unconscious desires, but these images could now be beamed directly into people's homes. As RCA's TV manager T. F. Joyce told a media conference in 1944, "we will have thirty million showrooms where personal dramatized demonstrations can be made simultaneously."[71]

A number of conservative think tanks, characterizing themselves as nonpartisan, were founded in the late 1940s. By 1949, four thousand US corporations had set up dedicated public relations departments, and some five hundred independent PR firms were in business.[72] By late 1951, business-sponsored films were drawing one-third of the US movie audience, 20 million people a week. In factories across the country, employees were given time off to attend sessions on economics as business saw it and the corporate commitment to workers' welfare. By 1952, according to the editor of *Fortune* magazine, corporations were spending $100 million a year to sell "free enterprise," an outlay that rose tenfold to $1 billion by 1978.[73]

Growth Enthroned

In the early decades of the twentieth century, theorists such as Bernays warned that democratic processes would threaten corporate power and profits: elections could install governments that would tax away the profits of the "great corporations" and try to impose unacceptable regulation. The Roosevelt years seemed to realize these nightmares, when a government that was inimical to unrestrained private enterprise held unprecedented power. Roosevelt regulated the financial system, which had been out of control in the 1920s, then many aspects of industry itself, including prices, as well as establishing taxpayer-funded welfare programs to assist ordinary people ruined by the economic collapse. These developments, buttressed by Roosevelt's own ability to propagate his message, made it all the more necessary for business to perfect "the conscious and intelligent manipulation" of ordinary people in order to "bind and guide the world."[74] Bernays's warnings were also, of course, designed to encourage corporations to engage his services as a "propaganda specialist" and expand the influence of the new profession he had been instrumental in creating.[75] It was a lucrative pursuit yielding substantial fees, as journalists S. H. Walker and Paul Sklar noted in their *Harper's* series, "Business Finds Its Voice."[76]

In the late 1930s, and especially after the severe 1937 recession that followed a reduction in Roosevelt's stimulus spending, one prominent school of thought suggested that the party might already be over—that capital accumulation in the United States might have reached its limit. Western expansion had hit the Pacific Ocean decades before; the frontier lay only in unexploited gaps. These ideas, put forward by economist Alvin Hansen and akin to John Stuart Mill's "stationary state," suggested that the next great challenge for the United States was to find a way to live without growth or, as economists called it, with "stagnation."[77] The war brought such cautious views to an abrupt halt and heralded the unambiguous return of economic expansion on a previously unimagined scale. After its almost complete absence from the economic discourse of the English-speaking world, both political or academic, and with only sparse discussion in the decade leading up to 1945,[78] the status of economic growth in government policy was about to be comprehensively transformed.

A Bigger Pie

The ongoing business propaganda campaign played an indispensable role in the ascendancy of the values of private enterprise in US culture and prepared the ground for the gradual erosion of the New Deal in capitalism's heartland; these trends were enhanced by the postwar advent of the ideology of the bigger pie.

Studebaker chairman Paul Hoffman founded the Committee for Economic Development (CED) in 1942, proposing a partnership between business and government that would engage in fiscal activism along Keynesian lines. The CED embraced a business ethos devoted, like

NAM's, to minimizing union influence and fostering individualism, competition, and a respect for the necessity of profits, but it emphasized productivity as the route to these objectives, defining productivity as output per unit of labor: "Productivity is a vitally needed lubricant to reduce class and group frictions. As long as we can get more by increasing the size of the pie there is not nearly so much temptation to try to get a bigger slice at the expense of others. That applies particularly to the common and conflicting interests of labor and capital."[79] The CED also encouraged a role for government, envisaging vital input from businessmen, who would exercise determinate influence on actual policy as they had done in the years before the Great Depression. From 1942, leading businessmen had been recruited into Roosevelt's war cabinet and had begun to enjoy significant influence again.[80]

Working closely with the CED was the Advertising Council. A reincarnation of the War Advertising Council, it had sought and won tax deductibility for "institutional and public service" advertising during the war. The council's president explained the relationship as follows: "You might say that whereas CED is concerned with the *manufacture* of information in the public interest, the Advertising Council is concerned with the *mass distribution* of such information."[81] The council believed that Americans needed "economic education" to inoculate them against "foreign ideologies." To this end, it convened a Public Advisory Committee that enlisted representation from the American Federation of Labor (AFL), one of the two major national trade unions and, ultimately, secured endorsement from the presidents of both the AFL and the Congress of Industrial Organizations (CIO),[82] the other big union, for its economic education campaign. This allowed it to represent its program as "nonpartisan," rallying all groups, according to the chairman of the council's Public Policy Committee, "for the common effort to improve our system by constantly increasing productivity and a wide distribution of its benefits." While the education curriculum conceded rights to organize and to collective bargaining and a limited role for the state, it also stressed "freedom of enterprise and expanding productivity through mechanization and increased efficiency," focusing on the past benefits of capitalism and the great gains to come.[83]

By 1950, in just two years, the Ad Council had published 13 million lines of newspaper advertisements and 600 magazine pages, set up 8,000 billboards, and circulated 1.5 million pamphlets.[84] Like NAM, the advertising industry established networks of speakers, trained at "Freedom Forums," the first of which was described by one participant as "long sessions of indoctrination in the fundamentals of our economic system."[85]

Numerous other groups also purveyed economic "education," including NAM's new "Education Department," set up in 1949; the Foundation for Economic Education (FEE), founded in 1946 by leading corporate CEOs (chapter 11); and the American Enterprise Association, which became the American Enterprise Institute in 1962—one of the most prominent and influential US think tanks of later decades.[86] FEE's propaganda, fed to US schoolchildren and college students, advanced such notions as the abolition of the income tax, US withdrawal from the UN, and getting rid of government roles in education, roads, and the postal service.[87] Sharon Beder, a major Australian analyst of corporate power, describes a mushrooming network of organizations, funded by business and extending into every aspect of education. The Opinion Research Corporation (ORC) claimed that the education program had "check[ed] the school use of many materials that are anti-business in nature" and developed in teachers "a much friendlier attitude toward American business." Surveys showed that a staggering 89 percent of teachers were using the program's materials in their classrooms by 1951, and that by 1955, citizens were more worried about Big Labor and Big Government than about Big Business.[88]

The success of postwar business ideology in the United States also owed much to the shift in the position adopted by organized labor and favored by the Democratic Party. By the end of the war, as the historian Elizabeth Fones-Wolf argues, US liberals had already modified their expectations:

> They shifted from demanding that the state control the economy through social planning and extensive business regulation to advocating that the government promote economic growth while only occasionally compensating for the private sector's failures through social welfare and social insurance. An expanding economy, a demand that easily meshed with business's goals, rather than the reform of capitalism became the clarion call of American liberalism and the Democratic party.[89]

Thus, the business world's "moderates" (or "neo-Progressives") forged a common cause with liberals and noncommunist labor organizations and adopted the strategy of the bigger pie. Economic growth became the solvent that could neutralize old class conflicts and provide everything everyone wanted without disturbing the distribution of wealth or of power. The US Chamber of Commerce's 1964 newsletter, for example, claimed that health care programs were less necessary as incomes rose and that relief for the poor became redundant as GDP steadily grew.[90]

Productivity had been canvassed as a substitute for redistribution in the 1920s when it was championed by Herbert Hoover (US secretary of commerce at the time) as a means of transcending scarcity and spreading abundance without disrupting economic power. Though this vision disintegrated during the Great Depression, the productive triumphs of wartime encouraged its resurrection. In the words of the Harvard historian Charles Maier, "the mission of planning became one of expanding aggregate economic performance and eliminating poverty by enriching everyone, not one of redressing the balance among economic classes or political parties."[91] These postwar ideas heralded the first world's middle-class societies of the late twentieth century and are echoed in current visions, such as Rupert Murdoch's, of an entirely middle-class world.[92] Productivity became the crucial ethos of postwar business and, according to Maier, involved the gradual acceptance of business *as* society: "The manager or executive was the man fitted to run society as a whole." Despite intermittent hysteria from the muscular Right, the drive for productivity gave business the role it wanted, while avoiding head-on conflict with its opponents.[93]

Maier also traces the broader dissemination of the new growth ideology after World War II. Described as "peaceful intercourse and economic expansion," its main targets were West Germany and Japan, where the Americans had full command, then Europe more generally as the United States took control of the world economy. Part of what they required of Europe was an end to industrial strife.[94] Through the late 1940s, Europe's communist unions opposed the Marshall Plan and split away from the labor mainstream, rejecting US influence and the emphasis on private enterprise. By 1949, the AFL and CIO were meeting with British and French unions to found a new noncommunist international labor federation based on "the politics of productivity."[95]

For the next two decades, this strategy was immensely successful in economic terms, even if biologists progressively sounded warnings of its ecological costs. Osborn and Vogt (see chapter 3) were worried about the impact of such growth on the biosphere as early as 1948. Vogt explicitly identified the adoption of economic expansion as an "article of economic faith" that would exhaust minerals and jeopardize soils, water, forests, grasslands, and wildlife.[96] It was into this ebullient first world bonanza that *The Limits to Growth* was launched in 1972, so it is perhaps unsurprising that the reception from professional economists, described in chapter 4, was not favorable.

Having established well-organized and well-rehearsed PR machinery to sell the private enterprise system, the business community was equipped to confront the gathering resistance to widespread environmental damage and to counteract the debate about the risks of untrammeled growth. It moved on to this next battle by generating hundreds of free enterprise think tanks, staffed largely by economists. Alongside already entrenched patterns of buying influence over the public and the Congress, business interests now had their own alternative academia, well-funded institutions that could challenge and then counter the voices of scientists and concerned citizens.

11

Sleight of the Invisible Hand

Economic institutions designed to make money must search for philosophical legitimacy when they instead try to make policy.
—Mark Green and Anthony Buchsbaum, 1980

The Perfecting of Corporate Persuasion

In the first seven decades of its efforts at keeping the US public on its side during the twentieth century, capital pushed the concept of free enterprise as the very foundation of American prosperity. The market researcher Opinion Research Corporation (ORC) found in 1960 that "free enterprise" was a more persuasive and acceptable term than "capitalism."[1]

In the early years of the twentieth century, some of these propaganda campaigns met with success. The distaste for big business that arose as corporate forms of business transformed small-town life was effectively countered. After the setback of the 1929 crash, the task became more challenging, and ongoing multi-million-dollar campaigns were launched to convince a suspicious public that this event did not alter the fundamental truth: private enterprise was "the American way," the bedrock of everything that mattered to Americans and the best of all possible systems. After World War II, business extended the coordination of its efforts and systematically inserted its free enterprise ideology into the entire educational structure. Before 1970, the great majority of this corporate propaganda was funded and disseminated in the United States. From then on, the practice became more generalized throughout the world, especially in the other English-speaking countries.

Even though a free market economy is frequently characterized as inexorable and "natural," a force no mere human can affect, the level of funding and application devoted to it would indicate a different reality, one that the business world has known for a century. Neither natural nor inevitable, the free market needs massive advocacy to create, retain, and extend public acceptance. The new techniques of advertising had demonstrated techniques for selling products. It was evident that the same methods could sell anything. S. Walker and Paul Sklar, in their congratulatory series for *Harper's Magazine*, "Business Finds Its Voice," noted that "business has adapted a machine intended for distribution of products to distributing ideas, thus releasing a new social force in America."[2]

Before Rachel Carson published *Silent Spring* in 1962, the interaction of corporate business with the natural world was not much at issue—worries such as those of Vogt and Osborn (discussed in chapter 3) were marginal. By 1970, however, when the Cuyahoga River burned in Cleveland, Ohio, there were disturbing signs that the environment was in trouble. The establishment of environmental watchdogs and other government institutions to deal with pollution began in both the United States and the UK in the 1960s and proceeded apace. This process instituted a regime of oversight and regulation for business and rested on the confident expectation that the newly visible environmental problems were amenable to technological solutions. The growth of these institutions and the introduction of the environmental impact study as a precondition for most new developments drew significant sections of a critical scientific community into careers within the new structures.[3]

Some scientists, the MIT *Limits to Growth* researchers included, began to warn that economic expansion could not go on indefinitely. Around the same time, increasing numbers of citizens became concerned about pollution and the degradation of the natural environment. The coordinated apparatus of persuasion outlined in chapter 10 was at hand: spin skills, propaganda ploys, and PR professionalism developed over sixty-five years stood ready to respond to all perceived threats to corporate values and corporate control, buttressed from the 1970s by the proliferation of conservative think tanks.

The promotion of economic growth was largely implicit rather than argued. As a shared value of mainstream economists, it was a preanalytic assumption for all, and there was little need to defend it. It was and continues to be presented by media of almost all persuasions as the indispensable underpinning of all realistic solutions to social problems big and small. The immediate postwar years saw it embraced as the answer to every awkward question about distribution. We did not need to consider how to share the pie since, by making it bigger, the crumbs would also increase. Economic growth became the principal yardstick for success in economic policy.

While economic growth rarely received explicit endorsement or defense at this time, warnings about environmental dangers were systematically countered and an entire machinery of denial was established through the think tank apparatus. The attack included the demonization not just of environmental activists but of science and scientists as well—an unthinkable situation just a few decades earlier.

Business Raises its Voice

The Powell Memo Recommends Guerrilla Scholars

The great majority of the peer-reviewed condemnation of *The Limits to Growth* in the 1970s (chapter 4) came from economists, and many of them preferred ridicule and personal attack to a rational examination of the book's arguments. In the mainstream press, several reviewers touched explicitly on the necessity of economic growth to deal with third world problems, and some stressed the centrality of growth to the functioning of any capitalist economy. Both arguments were seen as decisive against limits thinking. Economists also supplied most of the reviews published in the popular press. There is little evidence of businessmen or corporate spokespeople making explicit attacks on the idea of limits to growth or on the book itself in the first fifteen to twenty years after publication, though some, such as the agricultural economist Karl Brandt, a founding member of the Mont Pèlerin Society, were already engaged in the MPS project to set the market free. The immediate attacks were conducted by members of economics faculties in mainstream tertiary institutions, in the English-speaking world in particular.

However, as part of a broader reaction to the introduction of regulation and to criticism of business practices, the 1970s saw the relaunch and concerted extension of a well-rehearsed campaign to sell and resell private enterprise. On August 23, 1971, not long before *Limits* came out, corporate lawyer Lewis Powell sent a memorandum to his friend Eugene Sydnor, the director of the US Chamber of Commerce, urgently recommending such a campaign.[4] The memo was symptomatic of a sense of threat among America's businessmen, and of a mood for the renewal and extension of ideological warfare.

Powell, who was nominated to the US Supreme Court by President Nixon two months later, identified an escalating "assault on the enterprise system" that was, he warned, gaining momentum. He claimed that half the nation's university students "favored socialization of basic US industries" and quoted Milton Friedman's contention that "the foundations of our free society are under wide-ranging and powerful attack … by misguided individuals parroting one other." Powell named Ralph Nader, the consumer advocate—an avocation never popular with business—and Charles Reich, the author of the 1970 pop counterculture bestseller, *The Greening of America*, as examples of people conducting the frontal assault on the private enterprise system. Even more dangerous, he believed, were the social science faculties in most universities; he decried the malign influence of Marcuse, who was at the University of California at the time, and went on to claim that "Yale is graduating bright young men who despise the American political and economic system." This is at best a partial truth; soon afterward, the majority of Yale's bright young graduates were seeking employment in merchant banking.[5]

The fear that business was losing its grip on power had already reappeared throughout the twentieth century, as outlined in the last chapter. Powell, recapitulating his predecessors' themes, painted businessmen as the backbone of US prosperity and as the well-meaning victims of sabotage by propagandists. "The time has come for the wisdom, ingenuity and resources of American business to be marshalled against those who would destroy it.… The ultimate issue may be survival—survival of what we call the free enterprise system." To this end, corporate chiefs must focus beyond the selling of their products; they "must be equally concerned with protecting and preserving the system itself." This demanded collective organization, and Powell thought the National Chamber of Commerce with its plethora of local cells could be an ideal vehicle for the project.

In order to address the priority task of countering "the campus origin of this hostility," Powell recommended that the Chamber of Commerce establish a staff of highly qualified and sympathetic scholars, especially in the social sciences, a stable of reputable speakers who would pursue "equal time" on the college circuit, and a program of monitoring and evaluating textbooks at all levels. These strategies would be aimed at "restoring the balance essential to genuine academic freedom" and ensuring "fair and factual treatment of our system of government and our enterprise system" and its accomplishments. Additionally, a "steady flow" of scholarly papers for journals should be provided, as well as articles for mainstream magazines like *Harper's* and the *Atlantic*; the "increasingly influential graduate schools of business" should be cultivated, and secondary schools also needed intervention. These activities should be embedded in a multifaceted media campaign that would include a pervasive demand for "equal time" and a program to monitor television and other media along the same lines as textbooks. Complaints should be lodged wherever necessary. This demand for "balance" and "equal time" was to play a key role in the ideological wars of the next forty years.

Politicians, Powell complained in his memo, stampede to support "almost any legislation related to 'consumerism' or to the 'environment.'" In this unhappy situation, "business must learn the lesson that political power is necessary; that such power must be assidously [*sic*] cultivated; and that when necessary, it must be used aggressively ... without embarrassment and without the reluctance which has been so characteristic of American business."[6] As outlined in chapter 10, however, there had been scant evidence of this alleged reluctance through the greater part of the twentieth century.

Billionaires to the Fore

Joseph Coors, the billionaire brewer, told the official historian of the Heritage Foundation that Powell's memo had "'stirred' him up and convinced him that American business was 'ignoring' a crisis." In 1973, Coors provided seed money to the Heritage Foundation (which became one of the most influential of the new breed of US think tank).[7] His $250,000 contribution was handsomely supplemented, in 1976, by $420,000 from another billionaire, Richard Mellon Scaife, an heir to the Mellon fortune, derived from banking, oil, and steel. Scaife's foundations gave $13 million to conservative groups and think tanks in 1980, out of their overall $18 million "philanthropic" budget; in 1998, the Heritage Foundation got $1.3 million from Scaife. It also received early contributions from the Olin and Noble Foundations, which were established by chemical and fossil fuel interests, respectively.[8]

The Heritage Foundation was one of dozens of free market think tanks to appear in the following decades, establishing a parallel academic universe funded by and sympathetic to private enterprise. Prominent additions were the Pacific Research Institute (1975), the Centre for the Defense of Free Enterprise (1976), the Cato Institute (1977), and the Manhattan Institute (1978),[9] followed in the early 1980s by Anthony Fisher's Atlas Foundation in London (1981), the Competitive Enterprise Institute (1984), and the Heartland Institute (1984).[10] The proliferation of such institutions continued rapidly in subsequent decades.

Familiar Tactics: Advertising, PR, and "Economic Education"

Family foundations have been a crucial element in the funding of the modern think tanks. Journalist Lewis Lapham lists the richest conservative US foundations (as of 2001), with assets approaching $2 billion. The "Four Sisters"—Richard Mellon Scaife's group of foundations, the Lynde and Harry Bradley Foundation, the Olin Foundation, and the Smith Richardson Foundation—figure in Lapham's top five, along with the Earhart Foundation, devoted to free market scholarship (Hayek was among its beneficiaries). Numbers six and seven are the Coors foundations and those of the Koch family, prominent in 2010 for their covert funding of the tea party movement through Americans for Prosperity.[11] Think tanks founded or funded by these foundations include the Cato Institute, the Heritage Foundation, the American Enterprise Institute (AEI) , the Manhattan Institute, and the Hoover Institution.[12] The donations of the principal foundations to the Hoover Institution, the AEI, the Heritage Foundation, and the Cato Institute during the period 1985–2002 give an indication of the extent of the funding (table 11.1).

While Coors and other wealthy individuals—and their family foundations—launched, so to speak, a thousand think tanks, Powell's ideas were also being implemented by the preexisting institutions of business propaganda described in chapter 10. From 1976, the Advertising Council, supported by the Chamber of Commerce, spearheaded the work outlined by Powell, just as it had in the postwar period. The National Association of Manufacturers (NAM) also played its part. S. Alexander Rippa, a historian of education in modern American society, described the entire

Table 11.1
Donations of principal US family foundations to key conservative US think tanks, 1985–2002

	Hoover Institution	American Enterprise Institute	Heritage Foundation	Cato Institute
Sarah Scaife Foundations	$7.6 million	$4.4 million	$17 million	$1.8 million
Lynde and Harry Bradley Foundation	$1.7 million	$15 million	$13 million	$560,000
John M. Olin Foundation	$5 million	$7 million	$8 million	$800,000
Koch Family Foundations	$5,000	—	$1 million	$12.5 million
Smith Richardson Foundation	$1.3 million	$4 million	—	—

Source: After Beder 2006b, 27. Courtesy of Sharon Beder.

campaign to disseminate the "free-enterprise creed" as the "most elaborate and costly PR project in American history."[13]

The Advertising Council, still funded and directed by corporations, bankers, and media chiefs, attracted immense amounts of free commercial time, worth an estimated $642 for every actual dollar donated to the council. As well as services donated by the advertising, media, marketing, and PR executives who sat on its board, the council also monopolized the time set aside by the government for public service announcements—about 3 percent of all airtime. It was the largest single advertiser in the country, taking up double the time and space of the top corporate advertiser, Proctor & Gamble. The commercial cost would have been about $460 million, but the Ad Council ran on a mere $2 million a year, thanks to the generous donations of airtime and services.[14]

The Ad Council ran twenty-eight "public service" campaigns every year. Some, such as the "Future Is Great in a Growing America" campaign, promoted business and growth directly. Others, such as the campaign to prevent forest fires, actually approximated public service, but most held the individual responsible for cleaning up any mess, and all put business in a positive light. In the series on pollution, the campaign pointed the finger at the public and away from industrial pollution: "People start pollution. People can stop it." The 1970s campaign against inflation held government regulation responsible. In the words of the political scientist Michael Parenti, "collectivist, class-oriented, political actions and governmental regulations are not needed in a land of self-reliant volunteers."[15]

The council also acted as the arbiter of the definition of public service and met no obstacle in running the business propaganda campaign under that rubric. In this situation, no attempt at "balance" or "equal time" was considered necessary. As Congressman Benjamin Rosenthal (D-NY) commented:

> The Ad Council and the networks have corrupted the original intent of public service time by turning it into a free bonus for the special interests. The Ad Council is a propagandist for business and government ... it not only makes sure its own side of the story is told, but the other side isn't. The public has no meaningful access to the media.[16]

Since public service work was by definition nonpartisan, the council justified the propaganda campaign by arguing that the American public, tragically ignorant of its own economic system, needed "economic education."

Under the leadership of Barton Cummings, head of Compton Advertising, market research was conducted that "revealed" an ignorance that required remedy. In *Free Market Missionaries*, Sharon Beder assembles questions (with their "correct" answers) from Compton's nationwide 1976 survey of US public attitudes to the economic system. For example, the approved response to the proposition that "when business profits are up, times are good for more people" was "Agree."[17] Although more than 80 percent of the respondents *did* agree on this point and on many others, suggesting that previous "education" efforts had succeeded handsomely, this level of "knowledge" apparently rated as unsatisfactory. Such claims justified the economic education program as public service. In addition to media advertising, the Ad Council distributed "millions of booklets to schools, workplaces, and communities."[18] Individual corporations launched their own campaigns alongside the Ad Council, and numerous business coalitions sprang to assist—the US Industrial Council, representing 4,500 companies, ran its own multifaceted media blitz. This avalanche of advocacy for private enterprise, disguised as an educational service to the community, was described in *Fortune* magazine as "a study in gigantism, saturating the media and reaching almost everyone."[19]

At the same time, corporations also began to fund specific institutions inside existing universities and to withdraw funding with no strings attached. The Joint Council on Economic Education "funnel[ed] money from business to some 155 university centers and 360 school district programs that helped teachers give instruction on the free enterprise economy." Between 1974 and 1978, corporations endowed more than forty academic centers and chairs of free enterprise in American universities.[20] Business was not only establishing its own dedicated ideological apparatus via think tanks, it was colonizing universities with similar institutions.

Littering the World with Think Tanks

Forerunners

Think tanks were not first invented at this time. Nor were they all dedicated to the ascendancy of radical neoliberalism. The great majority, however, were advocates for private enterprise. One typology of think tanks distinguishes between the "old guard," founded before the 1970s, and the "new partisan."[21] Many of the old guard, however, were part of precisely the same movement as the later efflorescence. This is particularly true of Anthony Fisher's Institute of Economic Affairs in London. Other old guard think tanks such as the AEI (US) and the Institute of Public Affairs (Australia) are argued to have been less doctrinaire in their earlier days. Most caught the neoliberal wave in due course, however. Of eighty-three think tanks operating in Australia in the 1990s, only five had any kind of "wet" or leftist leaning, confirming that the great majority were linked to the conservative side of politics.[22] The AEI was an early US example, organized during World War II by "a partnership of top executives of leading business and financial firms (Bristol-Myers, General Mills, Chemical Bank) and prominent policy intellectuals."[23] Its program at this time involved opposing the continuation of government involvement in the economy after the war. The partnership of CEOs and free market theorists that characterized the beginnings of the AEI would prevail throughout the think tank proliferation. Like the new think tanks, the AEI dramatically expanded both its budget and its staff during the 1970s.[24]

Aims of Industry[25] was founded by the UK Chairman of Ford in 1942 to foster appreciation of private enterprise in the UK and, after the war, to resist the nationalization of industry and curtail planning by government. Anthony Fisher launched his Institute for Economic Affairs (IEA) in London in 1955, another pioneer think tank predating the 1970s economic crisis and an early outgrowth of the missionary zeal of Hayek's Mont Pèlerin Society (discussed in chapter 6). Fisher, who funded the venture with the considerable fortune he made from introducing the factory farming of chickens into the UK, was also instrumental in founding the Manhattan Institute, the Pacific Research Institute, the Fraser Institute (Canada), and the Atlas Foundation (1981), committed to "littering" the world with think tanks dedicated to these ideas.[26] Fisher met Hayek just before the first MPS meeting in 1947 and attended his first MPS meeting in 1951. He acted as entrepreneur, giving organizational form to this aspect of the project Hayek sought to initiate.

In Australia, think tanks and libertarian rhetoric were rare before the 1970s, but not entirely unknown. The Institute of Public Affairs (IPA) was founded in 1942 at the instigation of the Victorian Chamber of Manufacturers to "combat socialism"[27] and was an integral part of the conservative side of Australian politics: its branches from both Victoria and New South Wales participated in the conference that gave birth to the Liberal Party in 1944.[28] In 1956 the IPA was urging Australian businesses to fund "economic education" along American lines, and in 1950, W. D. Scott, a prominent accountant and management consultant, claimed that "the system of free enterprise is at stake" and recommended a wide-ranging public relations program that would "educate" the public and sell private enterprise along the lines he had observed in the United States.[29] But it was not until the 1970s that the long-standing US corporate project to naturalize and extend the idea of the free market was ignited in Australia.

"Independence"

While their funding, when revealed, reflected its corporate sources, most think tanks claimed to be "nonpartisan" or "independent." From the US Ad Council's emergence in the 1940s to the IPA in Australia today, this self-styled dispassion has always been questionable, since they are almost all linked to the conservative side of local politics and were developed with an explicit propaganda agenda. In Australia, dense interconnections exist between influential think tank operatives and conservative politics; of twenty-three key neoliberals listed by Bette Moore and Gary Carpenter, eleven are members of or staffers for the Liberal Party, Australia's conservative party.[30] Liberal Party members, including many in senior positions, appear with regularity among their personnel. The political economist Damien Cahill identifies think tank affiliations for thirty-seven free market campaigners from the conservative Coalition, many of them members of parliament (MPs).[31]

These links are not trumpeted. In the UK, Fisher's cofounder at the IEA warned back in 1955 that it was "imperative that we should give no indication in our literature that we are working to educate the Public along certain lines which might be interpreted as having a political bias."[32] Similarly, in Australia, political connections were concealed. Greg Lindsay himself, the founder of the Centre for Independent Studies (CIS), told the CIS magazine that his Liberal Party allies "were very conscious of my unwillingness to be seen as being involved in party politics and they were careful not to compromise us."[33] Their supposed independence helped to qualify many think tanks for the tax exemptions that go with charitable status, as well as giving an appearance of political neutrality. Almost all the free market think tanks worldwide are tax exempt.

The Foundation for Economic Education (FEE), dedicated to "individual economic freedom, private property, limited government and free trade through 'economic education,'" is, with the AEI, among the earliest of the free market think tanks and provides another US example of corporate origins and influence. Its seven 1946 creators included senior executives from Goodrich, GM, the *New York Times*, ORC, and a former top manager with both the Chamber of Commerce and NAM. Initial funding came from GM, Chrysler, Edison, DuPont, and several oil and steel companies; forty-six corporations had made million-dollar contributions by the end of 1949.[34]

The erstwhile workers' parties in the two-party systems of the English-speaking world have increasingly embraced the business agenda, partly in an alliance similar to that between business and the parties of the Right and partly as a response to the propaganda campaign that engulfed policymaking from the mid-1970s on. Once financial deregulation had taken hold, what Thomas Friedman called the "golden straitjacket" cramped governments' freedom of movement. Government policies that offended the "electronic herd" of international financial traders and speculators brought swift retribution in the form of flights of capital and credit downgrades from international ratings agencies.[35] The Australian Labor Party pioneered the adoption of the neoliberal program in Australia, though it tried to maintain government regulation in many areas—more than was comfortable from a free market think tank viewpoint. Its senior MPs did not found neoliberal think tanks and only occasionally participated in them. Nonetheless, it laid the groundwork for the Howard government—when the conservative Coalition governed from 1996 to 2007.

At the libertarian extreme, some think tank staffers have defined independence as *freedom from government funding*. Myron Ebell of the Competitive Enterprise Institute (CEI) told John Humphrys on the BBC's *Today* program that, while "we still have independent scientists in this country [US] ... virtually all scientific funding in Europe and Japan and Australia is held directly from government." According to Ebell, governments are biased, whereas business is "independent."[36]

While the Australian political scientist Diane Stone argues in favor of this putative independence,[37] and most think tanks insist on it, others freely admit their dependence on corporate finance. As one American think tank vice president remarked, "there is no such thing as a disinterested think tanker. Somebody always builds the tank, and it's usually not Santa Claus or the Tooth Fairy [and] unfortunately, many of these folk are often interested in satisfying the requirements of whoever pays the tab."[38] The prominent conservative columnist Irving Kristol expressed similar views in a 1977 essay in the *Wall Street Journal*. Kristol confirmed that the funding of think tanks such as the AEI, where he was then a resident scholar, is not and should not be disinterested:

> When you give away your stockholders' money, your philanthropy must serve the longer-term interests of the corporation. Corporate philanthropy should not be, cannot be disinterested.... Most corporations would presumably agree that [any donation] ought to include as one of its goals the survival of the corporation itself.... And this inevitably involves efforts to shape or reshape the climate of public opinion.[39]

"Capitalism Fights Back" in Australia: Centre for Independent Studies and Enterprise Australia

The efforts of US and British think tanks reverberated as far away as Australia, where think tank activity escalated in the mid-1970s. In 1976, the IPA helped bring Hayek to Australia for a lecture tour, and Greg Lindsay, dubbed the man who "controls your future" by Diana Bagnall in her *Bulletin* article,[40] began building the innocuously named Centre for Independent Studies in a shed in his Sydney backyard. Lindsay was initially inspired by Ayn Rand, in particular the cinematic version of *The Fountainhead*, a classic of American pro-capitalist ideology. Pursuing the trail through Rand, he joined forces with the libertarian Workers Party for a time and began importing books from FEE; in 1975 he traveled to the US to visit relevant think tanks in person.[41] Lindsay also met Anthony Fisher in 1976 when Fisher visited Australia and began convening meetings and conferences with local academic libertarians and economists. Lindsay attended his first MPS meeting in Hong Kong in 1978, where he met Hayek and Friedman and numerous members of overseas think tanks.[42] The CIS got its first big financial boost—some $40,000 a year for five years—from a group of prominent businessmen led by Hugh Morgan, CEO of Western Mining Corporation.[43] Lindsay was also later assisted financially by Fisher's UK-based Atlas Foundation. By 2004 the budget of the CIS had grown to $1.55 million.[44]

Enterprise Australia (EA) was set up in 1976, the same year as the embryonic CIS, and was funded by six major corporations, most of them with links to US-based multinationals.[45] It was expected to be "by far the most important group in the propaganda warfare for capitalism," according to Geoff Allen,[46] a lecturer in business administration at Melbourne University who went on to lead the Business Council of Australia when it was formed in 1983.[47] EA saw itself as responding to two supposed threats to "free enterprise": government encroachment and "public misconceptions" about "the size of profits and who benefits." EA claimed that the main beneficiary was the community rather than the owner of the business.[48] Both these concerns had been exacerbated by the election of the first Labor government in over two decades, that of Gough Whitlam (1972–1975), which "shocked Australian capitalism."[49]

Modeled on the AIMS template, EA introduced Enterprise Week in 1977. Overseas speakers were imported regularly, a media campaign was launched, and schools were targeted for "economic education." EA's first guest was Phil Gramm, notable for his subsequent role in the US Congress in abolishing the firewall the Roosevelt administration had placed between retail and investment banking—with consequences well known after the 2008 financial collapse when, to prevent a retail "run on the banks," governments were obliged to pay for the improvident speculative losses of the banks' investment arms.[50] EA also arranged for Hayek to conduct seminars on the market economy at universities in Sydney during his 1976 visit. Like its predecessors, EA presented itself as nonpartisan, but its response to the Whitlam dismissal[51] suggests otherwise—EA's launch was deferred until 1976 to avoid its being seen as a political front for the Liberal Party.[52]

Softening up the Enemy

The purpose of all this think tank work, according to Ralph Harris of the IEA in London, was "the conversion of the brighter and younger intellectuals," to provide an "intellectual artillery to soften up the entrenched enemy strong points."[53] Greg Lindsay echoed Harris's metaphor when he told Bagnall, "We always saw ourselves as the artillery, namely, firing shells into the distance, trying to soften up the ground."[54] Lindsay, like Hayek and Friedman, wanted to transform the policy agenda of government. For both Harris and Lindsay, the idea was to recruit people who were or would be in a position to influence the climate of public opinion and the actual policy of governments.

The great innovation of the ideological onslaught of the 1970s, which distinguished it from the foregoing century of propaganda, was the parallel business-funded research school—both "independent" think tanks and the "free enterprise" chairs and centers established in universities. In line with Powell's warnings that the mainstream universities were preaching revolution and that big business needed its own staff of writers, speakers, and intellectuals, scores of such private schools were funded by the wealthy, providing an institutional apparatus that has been expanding ever since. Think tanks everywhere sought influence on the policies of governments and targeted those who might provide it. They addressed themselves to opinion-makers: "editors, columnists, commentators, MPs"—a new terrain that Alex Carey, Australia's pioneer propaganda analyst, called "tree tops" propaganda, as opposed to that directed at the grass roots.[55]

Self-styled as independent, think tanks conveyed the business view as objective research, and this required a relatively qualified staff. Their tactics paralleled those developed by the tobacco industry from 1953, when researchers first demonstrated a direct link between tobacco and cancer. After a period of "frantic alarm," the industry hired the PR firm Hill and Knowlton in 1954 to disseminate more acceptable information to the public. At the same time, reputable scientists were hired, research was commissioned, immense sums were spent, and the idea of "balance" was artfully deployed to persuade the media to air their position. The tobacco corporations were later found guilty of fraud and conspiracy to suppress the truth.[56]

The rise of the think tanks embodies an alliance forged between corporate leaders and neoliberal intellectuals wedded to the same economic theories and the same business agenda. Throughout the development of the neoliberal movement, one of its hallmarks has been intense networking among a relatively limited cadre of players. Anthony Fisher contributed to numerous new think tanks worldwide and set up his Atlas Foundation for that express purpose. Meetings and conferences gathered together the cream of the global neoliberal intelligentsia, and international lecture tours were mounted for prominent propagandists. In 1978, EA's chief executive, Jack Keavney from Australia, participated in a typical gathering, an international conference in London organized by AIMS and attended by Ralph Harris, director of the IEA in London, and Edwin Feulner, president of the Heritage Foundation in the US.[57] Keavney was an important part of the network at the time, touring the US on a number of occasions and reporting to NAM and the US Chamber of Commerce on the Australian situation.[58] Richard Cockett, historian of the neoliberal movement in the UK, estimated that only about fifty people were involved in the British branch of the endeavor.[59]

The Australian think tanks' meteoric rise to such influence, like that of their US analogues, relied on the funding provided by the leaders of key segments of the business class—including transnational corporations based or operating in Australia. The most prominent of these were mining interests, finance capital, and manufacturing, with the manufacturers clustered in industries that served the miners and energy producers or that came under direct threat from regulation—tobacco, chemical industries, cement and lime, and construction suppliers such as James Hardie, which manufactured asbestos board. A small group of militant business leaders, Hugh Morgan prominent among them, who shared the century-long American agenda set out by publicist Hofer in the 1920s, allied themselves with the intellectual heirs of Hayek, Milton Friedman, and the MPS.[60]

Business and Environmental Crisis

Corporate Campaigners in the 1970s

In the early years of the twentieth century, the establishment of a machinery of pro-business propaganda was the open objective of men such as Edward Bernays, who was perfectly comfortable outlining his strategy for maintaining the influence of the business elite and sparing them the power of organized labor as the imminent democratic order took hold. He was unapologetic. At the beginning of the twenty-first century, the existence of such propaganda—and its history—is largely occluded.

The first rush of environmental awareness in the United States and the UK peaked briefly around 1973 before falling away. The new institutions of regulation, established in seventy countries by 1976, brought substantial improvements in environmental quality and may partly explain this decline in concern.[61] Moreover, segments of the environmentalist and scientific communities were subsumed into the policy apparatus of the new regulatory regimes, blunting the locally based momentum of environmental activism.[62] In Sharon Beder's view, however, the decline in concern about environmental problems from the mid-1970s is largely attributable to corporate campaigning, an effort more thorough and multifaceted than ever before.[63] Anti-union sentiment was a perennial aspect of this activism, but the emergence of consumer and environmental groups and the regulatory initiatives of the Nixon administration plunged corporate America into the panic implicit in the Powell Memo. In their study of US business coalitions in the 1970s, Mark Green and Andrew Buchsbaum interviewed one corporate lobbyist who told them that "the free enterprise system was in danger because you have the Ralph Naders of the world and the environmentalists."[64] In response, by 1978, US business was spending around $1 billion every year on a variety of propaganda campaigns intended to persuade Americans that their interests were the same as those of business.[65] They had some success. Just before Ronald Reagan was elected president in 1980, the Ad Council's annual poll, designed to monitor the effectiveness of the business campaign, found that people who thought there was "too much government regulation" had risen from 42 percent in 1976 to 60 percent just four years later.[66]

Alongside the advertising blitzes, the massive dissemination of "economic education," and the elaborate think tank apparatus outlined in the previous section, the direct political influence recommended by the Powell Memo was also pursued. It was thought expedient to intervene directly in the legislative arena to combat the specific dangers posed by consumer rights and environmental protection. New coalitions were forged, and lobbyists arrived in Washington in unprecedented numbers. In 1971, only 175 firms had political representation there; by 1982 the number had risen more than tenfold, to 2,445. In 1980 there were 15,000 business lobbyists in Washington spending $2 billion each year, a sharp contrast to the roughly fifty genuine public interest lobbyists, who spent $3 million per year.[67]

In tandem with the upsurge of lobbyists came the rise of the conservative political action committees (PACs), which in the two years from 1976 to 1978 doubled in number and quadrupled in terms of candidate donations. The corporate funding of these PACs, according to chemical industry documents published in 2001, was a long-term strategy designed not simply to influence current congressmen but to "upgrade the Congress"—to engineer the election of a Congress that would sympathize more fully with their objectives.[68]

Of the new industry coalitions that were formed at the time, the Business Roundtable was the most significant. Founded in 1972, it provided a boost to the existing peak bodies, the US Chamber of Commerce and NAM, in their long-established work for the business point of view. It focused on direct political influence, and by 1976 the Roundtable had eclipsed its predecessors; *Business Week* rated it "the most powerful voice of business in Washington." In 1978 the revenues of its members—a mere 192 in number—equaled half the GDP of the US, a sum greater than the GDP of every other nation on earth. Though it claimed to represent business both big and small, the Roundtable's Policy Committee, which determined its positions and assessed the input of its fifteen task forces, was a big-business entity. The Roundtable CEOs, drawn from the upper echelons of the Fortune 500, came personally to Washington to court senators, congressmen, administration officials, and presidents. Members of the Policy Committee had close personal ties to President Ford, for example, and though access to President Carter was less automatic, several Roundtable CEOs were well connected to his top officials.[69] Key congressmen and top officials were personal colleagues or ex-members of the dominant Roundtable clique or conspicuous members of antienvironmental pressure groups. In the late 1970s, more than a quarter of the Roundtable's governing Policy Committee were members of the Federal Reserve.[70]

Among the Roundtable's early successes was its PR campaign against the proposed consumer protection agency. The Roundtable's tactics included the promotion of a deceptively "framed" poll[71] "revealing" that 81 percent of Americans were against the creation of such an agency, and a concerted advertising campaign that emphasized the costs of bureaucracy and the perils of regulation.[72] After numerous close calls, the US Congress abandoned the formation of a consumer protection agency in 1978, a "signal victory" for the Roundtable, according to *Fortune* magazine.[73] Among other congressional retreats on environmental regulation were reductions in vehicle fuel economy standards, delays in the implementation of emissions standards for US motor cars, and relaxation of the nitrogen oxide standard. By 1978, business had "defeated much of the legislative program of both the public interest movement and organized labor."[74]

In Australia, too, new business coalitions were formed. The Confederation of Australian Industry was founded in 1970, the National Farmers Federation in 1977, and the Australian Business Roundtable, modeled on the US version, in 1980. The Business Council of Australia, founded in 1983 with a larger representation of the biggest corporations, now represents big business.[75]

A Second Wave of Corporate Antienvironmentalism

Despite the ascendancy of deregulation under Reagan and Thatcher—and the Hawke-Keating Labor government in Australia a little later—and despite the ideological crusade for unfettered enterprise emanating from business coalitions and the elaborate think tank apparatus outlined in the previous section, there were signs at the end of the 1980s that people in the Western world were still concerned about environmental degradation, perhaps even more so than before. A *New York Times*/CBS poll in1989 found that 80 percent of respondents thought that "protecting the environment is so important that standards cannot be too high and continuing environmental improvements must be made regardless of the cost." In Australia, a 1990 Saulwick poll found that 67 percent of people thought Australia "should concentrate on protecting the environment even if it means some reduction in economic growth," a finding echoed in a Gallup poll the following year.[76] In these years, ordinary people in the first world valued the environment ahead of the economy—or said they did—and told pollsters that they were prepared to pay a price for their preference. However, such views did not accord with the priorities of business, in particular those of corporations linked to the fossil fuel industries powering the ever-expanding industrial apparatus.

These changes in public perception were shaped by new sources of concern about environmental decline. The US climatologist James Hansen addressed the US Congress in 1988, the same year that the International Panel on Climate Change (IPCC) was set up to review world research into global warming. The IPCC's first report was released in 1990, and the UN conference in Rio de Janeiro, known as the Earth Summit, followed in 1992, attempting to address not only global warming but the destruction of the diversity of life on earth, the pollution of the oceans, and the threat from toxic waste. Despite the Earth Summit's capitulation to market solutions and its extremely modest results, the first framework agreements on control of carbon emissions and biodiversity protection were put in place, with ongoing negotiations scheduled. In achieving these embryonic accords, however, the Rio summit raised the specter of environmental regulation on an international scale, an even greater threat to corporate freedom than existing national efforts to control the ill effects of ongoing economic expansion.

To oppose these trends, business adopted a multiplicity of tactics in its second wave of opposition to environmental values. Many elements are familiar from the foregoing history; some of these were intensified at this time, while new strategies also emerged. Apart from the role of the think tanks, which is explored in detail in chapter 12, the major initiatives were the PR-based creation of fake grassroots organizations and of a legal infrastructure to intimidate environmental protestors by taking them to court.

Greenwashing and Front Groups

Public relations companies conducted numerous campaigns on behalf of corporations, the most novel being the "greenwash" exercise, whereby the public was to be convinced that polluting companies were "going green." When British Petroleum set out to rebrand itself "Beyond Petroleum," the advertising campaign cost as much as or more than BP's actual investment in solar technology.[77] Nonetheless, an impression of green credentials was successfully created. Sponsorships and "green partnerships" were established, such as one between the clear-cut logging and paper mill company Georgia-Pacific and an organization for injured animals, and another between Chevron and National Geographic. Public relations firms continued their well-established function of damage control but were also paid to create specialized front groups, such as the Global Climate Coalition (GCC), dedicated to minimizing concern about climate change. The GCC represented NAM and automotive, coal, and oil corporations, and shared personnel with industry associations and think tanks, including the American Petroleum Institute and the George C. Marshall Institute.[78]

Seeking to replicate the authenticity of citizen participation, business began to finance putative grassroots campaigns, forming organizations with innocuous (or totally misleading) names such as the Environmental Conservation Organisation, Citizens for Effective Environmental Action Now, established by the chemical industry, and the National Wildlife Institute. These organizations, funded by corporate interests and often set up by PR firms, mobilized discontented citizens (often unwittingly against their own beliefs and interests) in campaigns designed to ensure corporate access to resources such as forests and minerals. It was industry insiders who first dubbed them "astroturf" organizations, after the synthetic grass known as AstroTurf. Although citizens were enlisted in these entities, they did not arise as grass roots groups but were instigated from above by corporate interests for propaganda purposes. The "wise use" umbrella organization, founded in the United States in 1988, was one of the most successful of innumerable such groups and had links to many corporate bodies, including the Heritage Foundation, logging companies, resource trade organizations, and off-road vehicle manufacturers; the CEI sponsored their first conference.[79] Ron Arnold, who helped to organize the gathering, acknowledged the underlying agenda: "We don't even care what version of Wise Use people believe in, as long as it protects private property, free markets, and limits government."[80] Thus, although aggrieved or politically conservative citizens were attracted to aspects of its program, its intention was to advance corporate goals: to free private property from all regulation and open public lands to unrestricted commercial exploitation. Influential US legislators are also linked to an array of bodies like the National Wildlife Institute, founded by timber interests, and the Environmental Conservation Organisation, funded by the trade associations of earthmoving contractors and farmers.[81]

Astroturf methods facilitated the camouflage of corporate values and priorities and advanced a corporate version of facts without business actually seeming to be involved. Corporations could enroll private citizens and masquerade as "the ordinary person," just as the US tea party movement, though attacking the big banks in its rhetoric, has been funded and facilitated by the billionaire Koch brothers through their own astroturf organization called Americans for Prosperity. The same Kochs funded the launch of the Cato Institute in 1977 and have spent hundreds of millions over the years on such conservative think tanks as the Economic Education Trust, the Mercatus Center at George Mason University, and the Heritage Foundation.[82]

Funding Antienvironmental Research

Think tanks play a similarly covert role on behalf of business. Instead of corporations putting research forward transparently as their own, think tanks, with the studied appearance of independence, purport to be supplying research comparable to peer-reviewed work from the academic world. Vested interests can be concealed in this way and media organizations encouraged to air think tank scholars as if they were in fact independent.[83] Indeed, in the United States, research by Fairness and Accuracy in Reporting (FAIR) showed mainstream media outlets (major newspapers, radio and TV stations) had quoted, hosted, or published Heritage Foundation staff 2,268 times in the study year (1995), AEI staff 1,297 times, and Cato Institute staff 1,163 times.[84]

In 1998, as corporations faced the prospect of the Clinton administration signing on to the Kyoto Protocol, adopted on December 11, 1997, John Cushman of the *New York Times* revealed that "an informal group of people working for big oil companies, trade associations and conservative policy research organizations ... have been meeting recently at the Washington office of the American Petroleum Institute." Their plans encompassed a media program, with $600,000 in funding, to recruit, train, and finance a team of credible scientists who would question and undercut the "prevailing scientific wisdom" on radio talk shows and in opinion pieces in newspapers. They also planned a Global Climate Science Data Center with a budget of $5 million over two years, which would again recruit credible scientists and act as a "one-stop resource" for members of Congress, the media, and industry.[85] The document Cushman obtained stated that "victory will be achieved when ... recognition of uncertainties becomes part of 'conventional wisdom.'"[86]

Industry sources claimed that the *Times* publicity had forced them to abandon that particular plan, but people involved in the meeting have been prominent in climate change denial work ever since—including ExxonMobil lobbyist Randy Randol, "junk science" proponent Steve Milloy, Myron Ebell from Frontiers of Freedom, now with the CEI, and representatives from the American Petroleum Institute, Chevron, the Marshall Institute, the Science and Environmental Policy Project, and the Committee for a Constructive Tomorrow.[87] As negotiations for a treaty beyond Kyoto drew closer, the AEI offered $10,000 to any scientist who would write articles emphasizing shortcomings in the IPCC's 2007 draft assessment report.[88]

In these documented cases, vested interests planned to pay individual scientists to present an industry-friendly opinion in the public sphere as if they were unconnected to industry. Though it is often difficult to link specific individuals to precise corporate donations, some evidence does exist: in the early 1990s the coal conglomerate Western Fuels revealed in an annual report that it was enlisting prominent scientists Patrick Michaels, Robert Balling, and Fred Singer as spokesmen. The coal industry paid these and a handful of other self-styled skeptics $1 million over a three-year period;[89] Michaels admitted at a 1995 hearing in Minnesota that he had received more than $165,000.[90] Evidence that the Heartland Institute has spent over $20 million since 2007 funding scientists and "skeptical" bloggers was leaked in early 2012.[91] Among the recipients were the Australian geologist Bob Carter and the US weatherman Anthony Watts. Even where proof of direct funding is lacking, there is ample evidence of corporate donations to think tanks and corporate involvement in their boards, while think tank relationships with self-styled contrarians are openly disclosed. Think tanks constitute a go-between that sanitizes industry propaganda and turns it into "independent research."

Mother Jones journalist Chris Mooney has documented connections between ExxonMobil and various think tanks and front groups. He found forty organizations with close ties to climate change denialists that were funded by the petroleum giant, which spent more than $8 million on them between 2000 and 2003. The AEI received nearly $1 million while ExxonMobil chairman Lee Raymond served as vice president of its board of trustees. The CEI got $1.38 million, Frontiers of Freedom $612,000, and the Committee for a Constructive Tomorrow $252,000. Smaller sums were disbursed to many other entities, including the Cato Institute, the Center for the Defense of Free Enterprise, where Ron Arnold is based (discussed in chapter 12), and the Advancement of Sound Science Center, registered at Steve Milloy's address.[92] *Mother Jones* has compiled a table of think tank funding by ExxonMobil,[93] and the Greenpeace investigative website exxonsecrets.org provides extensive information on the connections between dozens of think tanks and their funding sources.[94] In their open letter to ExxonMobil in 2006, Republican senator Olympia Snowe and Democrat senator Jay Rockefeller pointed out that "since the late 1990s, ExxonMobil [alone] has spent more than $19 million on a strategy of 'information laundering,' enabling a small number of professional skeptics, working through so-called scientific organizations, to funnel their viewpoints through non-peer-reviewed websites."[95]

Intimidating Citizens with Lawsuits

In a 1971 speech to the US Chamber of Commerce, Lewis Powell, of Powell Memo fame, recommended that business set up its own law firms, call them "public interest" firms, and prepare to fight for the business agenda in the courtroom. The Chamber of Commerce established its own litigation center, one of many such corporate interest law firms.[96] These provided a weapon later used widely to threaten individuals involved in protest or activism against polluters and developers; this sort of intimidating litigation was dubbed "strategic lawsuits against public participation" (or SLAPPs) by the University of Denver academics Penelope Canan and George Pring. Canan and Pring had observed an upsurge in civil damages suits mounted against citizens.[97] In court, the pockets of corporations were too deep for ordinary citizens to oppose. In the United States, many activists were scared off and silenced. The Melbourne barrister Brian Walters has documented a number of cases in which Australian businesses—often developers—used the defamation or trade practices laws to sue citizens who expressed concerns about environmental and community issues, sometimes by merely writing to the paper.[98] Many of these suits succeeded in silencing the dissent and, even when people chose to fight, the risk of losing everything was high and led others to fear the consequences of public participation.[99]

As well as discrediting, bankrupting, and scaring off private individuals, the neoliberal Right and its think tank infrastructure went on to accuse scientists of distortion and bias while fostering the denial of environmental problems with its own distortion and bias. The next chapter examines these developments and looks at the rhetorics that have undermined environmental science and prevented or delayed action on environmental issues.

12

The Free Market Assault on Environmental Science

The environmental policies of the most powerful and gluttonous nation on the planet are being written by the world's most powerful oil company.
—Mark Morford, 2005

Climate change policy in Canberra has for years been determined by a small group of lobbyists who happily describe themselves as the "greenhouse mafia."… This cabal consists of the executive directors of a handful of industry associations in the coal, oil, cement, aluminium, mining and electricity industries. Almost all of these industry lobbyists have been plucked from the senior ranks of the Australian Public Service.… The revolving door between the bureaucracy and industry lobby groups has given the fossil fuel industries unparalleled insights into the policy process and networks throughout government.
—Clive Hamilton, 2007

Science Loses Favor

In chapters 10 and 11, I traced the step-by-step creation of channels of propaganda and direct influence by corporate America, and their spread to other countries. I have also indicated the process whereby pro-corporate ideology was internalized in popular belief and became the commonsense way to see the world. Economic growth is intrinsic to the corporate system so that, even when growth itself is not the overt topic of the propaganda, it remains an underlying objective. This is particularly true of the battle to continue burning the fossil fuels on which the entire productive apparatus currently depends.

The core rhetorical task for nearly a century was to persuade ordinary Americans—and then others round the world—that their interests were identical to those of big business and best served by keeping the government out of economic decision making. By the late twentieth century, there were signs that corporate interests wished to subsume under their own control not only strictly economic decisions but much of the entire range of government policy as well. One manifestation of this broadening intent was the attack on the credibility of science itself.

Science had been regarded as an indispensable ally of capital throughout the successive technological transformations of the late nineteenth and twentieth centuries. Though neutral in theory, scientific knowledge fuels technical innovation, an indispensable aspect of a regime of "creative destruction" and growth in GDP. However, as scientists began to document the ecological damage caused by the surge of economic growth after World War II, American business turned against them. During the 1960s, biological scientists in particular started studying the evidence of unintended, often noxious, consequences of production; in response, the status of science in the industrial system began to plummet. From this time on, pro-business interests perceived science in two categories: "productive science," the natural support of industry for a century or more, and "impact science," the unwanted study of industrial outcomes. This involved the rise to prominence of the ecological and health sciences; according to the sociologists Riley Dunlap and Aaron McCright, these biologically oriented sciences presented a "fundamental challenge to the dominant social paradigm" of endless growth and ever-mounting prosperity.[1] This was a challenge that business was anxious to suppress since, as has been argued throughout, expansion is crucial to the profit system. The rise of the biological sciences also diverted part of science funding into the new disciplines, a trend that some of America's elite physicists criticized forcefully.[2]

In a radical departure from their long-term alliance with science and technology, industrial enterprises and free market theorists found it expedient to attack science itself and the scientific community that produced it. A number of analysts argue that this strategy was adopted after the American public objected to the Reagan administration's attempts to roll back environmental protections in the 1980s.[3] If the public was inclined to insist on remedies for environmental damage, it was preferable that they saw no damage in the first place.

On the question of global warming, David Goldston, Republican chief of staff for the House of Representatives Science Committee until 2006, told *Newsweek* that opponents of greenhouse curbs had "settled on the 'science isn't there' argument because they didn't believe they'd be able to convince the public to do nothing if climate change were real."[4] Since scientific studies gave substance to many calls for environmental action, science and scientists would have to be challenged: the best tactic was to cast doubt on the seriousness of environmental problems and depict environmentalists—and environmental scientists—as extremists who would be willing to falsify evidence so as to exaggerate the problems.[5] Although denial of the reality of global warming is the most palpable manifestation of this strategy, the entire gamut of environmental science is subject to it. First, environmental skepticism denies that environmental problems are serious and dismisses scientific evidence; second, it denies any need for policies to protect the environment or remedy damage; third, it opposes regulation and corporate liability; and last, it attacks environmentalism as a threat to Western progress.[6]

Along these lines, the commentary of the mainstream economists when *The Limits to Growth* was first published in 1972 featured out-and-out denial of limits to economic growth. Such denial became a key tool for the opposition to environmental protection. Free market advocates had always rejected regulation, seen as an unwanted imposition on profitability. This was even more imperative for business when the pollution of air, water, and soil increased dramatically in scale after World War II and as clearing, logging, and mining decimated human communities and natural habitats worldwide. If this was the price of economic growth, however, it was a price that corporations wanted people to pay.

Environment and the Neoliberal Think Tanks

The "New Right" Skeptics

In 1985 the Australian journalist Tim Duncan, who has since worked for the Business Council, Rio Tinto, and the PR company Hinton, published his sympathetic portrait of the Australian "New Right." Duncan tabulated the political and social beliefs that have buttressed neoliberal economics in Australia. Alongside "the rescue of Australian history" and "reasserting traditional social values," a detailed table spelled out the New Right's approach to "the future of mankind." In response to environmentalist attitudes, which he termed the "orthodoxy," Duncan set out the beliefs of the free marketeers about resources, growth, and progress, uncanny in their resemblance to the economists' attack on *Limits* back in 1972. According to Duncan, the New Right believed there were "more resources available now than ever before" and that "energy [is] cheaper now because there is so much of it. If oil gets more expensive it will become redundant." As for economic growth, "[a] sustainable society is an ecological dreamtime—something which has never happened in human history," and "there are no physical or economic constraints to continued progress." In Duncan's picture, Australia's New Right was not only dismissive about limits, it rejected notions of sustainability as well.[7]

In the United States, free market think tanks promoted environmental skepticism almost single-handedly. Of the 141 books published between 1972 and 2005 that denied or downplayed environmental problems, more than 90 percent had a clear link to one or more think tanks. The bipartisan conservationism of the Nixon era had been abandoned and the prevalent cultural practice of accepting the scientific integrity of the academy discarded. In 2005, 90 percent of think tanks that addressed the environment issued policy statements that embraced a skeptical view of environmental damage and all eight of the climate-focused entities contested the reality of global warming.[8] These US think tanks went on to lead the climate "skeptics" worldwide.

The gathering torrent of antienvironmental literature began slowly in the 1970s when the first neoliberal think tanks were being established. In the 1980s, fourteen books that dismissed environmental problems were published and all were linked to conservative think tanks. The 1990s saw a fivefold increase over the previous decade, spiking at the time of the Rio Earth Summit in 1992 and again around Kyoto in 1997.[9] The production of such books continued to build in the first decade of the new century, when sixty-four books characterized by climate change denial were published.[10] Ongoing attempts at global regulation and reduction of greenhouse gas emissions play an important part in the intensity of this backlash, which clearly aims to obstruct, if not prevent, any such action.

Explicit Political Connections in the United States

The substantial interpenetration between free market think tanks and US administrations, Republican in particular, is an open secret. The Heritage

Foundation provided the newly elected Reagan administration with its *Mandate for Leadership*, a massive guide to free market policies. Think tank influence was similarly reflected in the appointments George W. Bush made to his new administration in 2001. He told the American Enterprise Institute (AEI) that "some of the finest minds in our nation are at work on some of the greatest challenges to our nation. You do such good work that my administration has borrowed 20 such minds."[11] Vice President Cheney's wife Lynne was and remains a Senior Fellow at the AEI. Edward Feulner, president of the Heritage Foundation, claimed in 2001 that the Bush administration had dipped "deep into the Heritage pool of talent."[12] According to the journalist Sharon Begley, Myron Ebell from the Competitive Enterprise Institute (CEI) arranged to apply pressure on the administration when rumors circulated that the new president was going to announce he would honor his campaign pledge to cap carbon dioxide emissions. The CEI's president told Begley that the CEI had alerted anyone who might have influence to get the line out of the speech—if it was, in fact, in there.[13] By March, not only had Bush abandoned his pledge to cap carbon dioxide emissions, he had withdrawn from the Kyoto treaty altogether.

In addition to more circuitous avenues of persuasion through its funding of think tanks, industry exerted a direct influence on the appointments of the 2001 Bush administration through its lobbying apparatus. The ExxonMobil lobbyist Randy Randol asked the newly inaugurated president to remove the British scientist Robert Watson, then chairman of the International Panel on Climate Change (IPCC), from his post.[14] Watson's acceptance of the consensus position on global warming, that humans were very likely responsible, seems to have troubled Exxon. Watson was denied a second term and replaced in 2002 by Rajendra Pachauri; though Pachauri's scientific views were similar, the Bush administration backed his declared "political neutrality."[15] Randol also recommended that a fierce opponent of climate action, Harlan Watson (unrelated), who was trained in physics and economics but was not a climate specialist, be seconded to help the United States with its climate negotiations; he became the US lead negotiator at subsequent post-Kyoto climate conferences.[16] Philip Cooney, a former attorney for the American Petroleum Institute, was appointed to head the Council on Environmental Quality, where he later came to prominence for his role in unilaterally amending the reports of the actual scientists who worked there. On resignation, he joined ExxonMobil.[17]

The major industry associations also opposed action on global warming directly. The National Association of Manufacturers wrote to President Bush in 2001 congratulating him on reversing his pre-election promise to cap carbon emissions:

> Dear Mr. President: On behalf of 14,000 member companies of the National Association of Manufacturers (NAM)—and the 18 million people who make things in America—thank you for your opposition to the Kyoto Protocol on the grounds that it exempts 80 percent of the world and will cause serious harm to the United States.[18]

The US Chamber of Commerce was on much the same track, telling the president, "Global warming is an important issue that must be addressed—but the Kyoto Protocol is a flawed treaty that is not in the US interest."[19]

Think Tanks, Front Groups, and Corporate Money

The establishment of a business-friendly reservoir of alternative scholarship in the free market think tanks was to serve American business well, just as Lewis Powell had foreseen. Alongside AEI and the Heritage Foundation, openly favored by President Bush, the Cato Institute and the CEI also commanded extensive media attention.[20] The same family foundations and the same corporations, mainly linked to fossil fuels or motivated by regulatory issues, also funded a number of smaller think tanks and front groups dedicated to antienvironment and counterconsumer work.

The American Council on Science and Health (ACSH), founded in 1978, still claims to promote "coverage of health issues ... based on scientific facts—not hyperbole, emotion, or ideology."[21] While presenting itself as a consumer advocate, the group has consistently supported industry, playing down risks from DDT, dioxin, and asbestos,[22] supporting bovine growth hormone for dairy cows, and opposing EPA regulation of the defoliant 2,4,5-T in 1981 and the fumigant ethylene dibromide in 1984. The ACSH has acknowledged receiving "40 percent of its money from industry, particularly manufacturers in the food processing, beverage, chemical and pharmaceutical industries, and much of the remainder from industry-sponsored foundations." Monsanto was a prominent contributor, as were Exxon and several other chemical industry giants such as American Cyanamid, Dow, Union Carbide, and Uniroyal.[23]

Fred Singer, a physicist who characterized himself as a contrarian and whose activities are explored below, founded the Science and Environmental Policy Project (SEPP) in 1990 with money from Monsanto, Texaco, and the Bradley, Forbes, and Smith Richardson Family Foundations;[24] Bradley and Smith Richardson are listed as the top two in Lapham's 2004 list of the richest. Oil companies such as Arco, Exxon, Shell, Sun Oil, and Unocal also funded Singer's research.[25] Fred Seitz, another of the contrarian physicists, served as chairman of SEPP's board, and yet another, William Nierenberg, was on SEPP's Board of Science Advisers.[26] Singer has held positions with many think tanks and front groups, including Steve Milloy's Advancement of Sound Science (see below), the Cato Institute, the Heritage Foundation, the Environmental Conservation Organization (ECO), and Elizabeth Whelan's ACSH.[27]

Almost all of the free market think tanks are staffed by economists, policy analysts, and lawyers rather than scientists,[28] so it is all the more extraordinary that their dismissive views on environmental science have been taken so seriously by so many. The cultivation of an appearance of credible scholarship has assisted them in this but, while their staff members are frequently qualified in economics, their claims to legitimate expertise in science are often dubious.

Ronald Bailey is typical. Affiliated with both the Cato Institute and the CEI, Bailey has written numerous antienvironmental books, including *Eco-scam* and *Global Warming and Other Eco-Myths*,[29] yet his BA is in philosophy and economics. Myron Ebell from CEI appears frequently in the US and UK media attacking mainstream climate science. Trained in philosophy, history, and political science,[30] Ebell is another example of a nonscientist who portrays himself as a climate expert. In his BBC interview with *Today* anchor John Humphrys, Ebell attacked the UK's chief scientist, Sir David King, as "ridiculous," "alarmist," and having "no expertise in climate science," and went on to pontificate on the faults of climate modelers. In a subsequent BBC interview with Jeremy Paxman, however, he was obliged to admit that, unlike King, he is not a scientist of any sort.[31] Julian Simon, associated with both the Cato Institute and the Heritage Foundation,[32] made the first concerted denial of the environmental crisis in *The Ultimate Resource*, and followed up with *The Resourceful Earth*—a reply to President Carter's *Global 2000* report—which he coedited with Herman Kahn.[33] Simon was an economist with dense think tank affiliations. The Danish statistician Bjørn Lomborg, who cited Simon as his inspiration, is another nonscientist who has had an enormous impact on the public assessment of the scientific evidence of environmental damage.

Fostering Doubt

Global Warming: Science versus Fiction

Scientists have been almost unanimous about global warming since the 1990s, and have endorsed the consensus position embraced in the IPCC's Third and Fourth Assessments of 2001 and 2007, respectively—that the climate is warming and human greenhouse gas emissions are the likely explanation.[34] Conservative think tank commentators, however, maintained their dissent throughout the same period, and as late as 2007 Vice President Cheney, while conceding that warming was happening, still claimed there was no consensus about whether it was part of a "normal cycle" or not.[35]

The denial of global warming continues to affect public opinion and political processes, especially in the United States and Australia. A poll in the United States showed a steady rise, between 2006 and 2010, in the belief that global warming is exaggerated and a corresponding rise in the belief that increases in temperature are due to natural causes rather than human activities.[36] A similar though less pronounced trend was seen in Australia, where a poll in late 2009 found a decline in community concern over climate change.[37] These trends in public opinion ran counter to the increasing confidence expressed by the IPCC in its endorsement of anthropogenic global warming.

In the US Congress, which has resisted measures to control greenhouse gas emissions, the novelist and global warming skeptic Michael Crichton was the star witness at a 2005 Senate hearing before the Committee on Environment and Public Works, prompting Senator Jim Jeffords[38] to ask, "Why are we having a hearing that features a fiction writer as our key witness?"[39] Crichton's techno-thriller, *State of Fear*, depicts ecoterrorists plotting to fake weather catastrophes and fuel unfounded fear, something the committee must have regarded as qualifying Crichton for his appearance. Across the Pacific, the Australian conservative politician Tony Abbott, who had declared climate change science to be "absolute crap" only a few months earlier,[40] won the leadership of the Australian Liberal Party in 2010. He became prime minister after the conservative coalition won the September 2013 election with the slogan "Axe the tax" at the center of his campaign—aiming to abolish the Gillard Labor government's carbon price.

"Doubt Is Our Product": Tobacco, Acid Rain, and CFCs

In *Merchants of Doubt* (2010), Naomi Oreskes and Erik Conway examine in detail the networks that have fostered environmental denial from the late 1970s on—and that continue in the current century. They identify the key players as the conservative think tanks; a handful of self-styled contrarian scientists, including a core group of physicists who worked on the US weapons program during the Cold War; vested interests such as tobacco corporations, electric utilities, and chemical industry front groups; a naive and sometimes partisan press; and free market fundamentalists of all shades, whether academics, fellows at conservative think tanks, or bureaucrats and politicians.

The tobacco industry pioneered the deployment of doubt. In 1969, a notorious industry memo maintained, "Doubt is our product, since it is the best way of competing with the 'body of fact' that exists in the mind of the general public. It is also the means of establishing a controversy."[41] Big tobacco had realized that doubt could undermine fact. Two of the prominent physicists who disagreed with the science of environmental impacts, Fred Seitz and Fred Singer, had already worked for the tobacco industry. Seitz joined R. J. Reynolds Tobacco in 1979, supervising a massive research program designed to establish alternative causes of diseases attributed to smoking. Another physicist, the rocket scientist Fred Singer, was involved in the work of APCO, the PR firm hired by Phillip Morris to counteract the evidence of passive smoking risks. Seitz, a Cold War physicist, was one of the founders of the George C. Marshall Institute in 1984, along with his Manhattan Project colleague William Nierenberg and fellow physicist Robert Jastrow.[42]

All three were men with towering reputations. They had held prestigious posts as heads of America's premier scientific institutions and had enjoyed long careers in military and security arenas. They had excellent connections to the highest echelons of government and personal involvement in matters of national security and in weapons-oriented solutions to Cold War conflict. Though it may seem odd that such eminent scientists would contest the scientific work of climate specialists, Oreskes and Conway suggest that these physicists, veteran Cold War hawks with deep roots in the Republican Right, transferred their hostility from socialism to the "new great threat" of environmentalism. As the perceived Soviet threat disappeared, the physicists' distrust of regulation and "government interference" was transferred to a new target. Green became the "new red," "the last vestige of communism's collectivist, one world government plot to subjugate the planet." Along with many other conservatives, they saw people expressing environmental concerns as "watermelons"—thinly disguised communists.[43]

These same influential physicists reappeared regularly in the campaigns against action on acid rain and ozone depletion. In 1982 the Reagan administration appointed Nierenberg and Singer to a peer review panel on the connection between acid rain and emissions from coal-fired power stations. In the 1950s, coal-fired electric utilities had declined to remove sulfur dioxide and nitrogen oxides at source and instead had installed tall chimneys with scrubbers to diminish local effects. The removal of particulates (soot), however, led to increased acidity. When it became clear that these emissions were the main cause of acid rain, the utilities had resisted regulation, arguing that the science was "fuzzy" and the costs of removing sulfur dioxide so "prohibitive" that installing suitable new scrubbers would "break the economic backbone of the Midwest." The Nierenberg report emphasized uncertainties and recommended more research, delaying action until after Reagan left office.[44]

The campaign against regulation of chlorofluorocarbons (CFCs), designed to stop the destruction of the ozone layer in the stratosphere, reveals similar dynamics. CFC manufacturers reacted to early concerns in 1971 by setting up their own research organizations, which attacked the academic science as disaster-mongering and blamed volcanic eruptions for ozone loss. Singer attacked the "ozone scare" as an overreaction of panicky scientists, claiming that fluctuations in the ozone layer were part of "a natural cycle." Even if there was a problem, he insisted, solutions would be too difficult and expensive to pass a cost–benefit test.[45]

Singer's predictions of the catastrophic costs involved in remedies for power plant emissions and CFCs fit the "Chicken Little" and "Doomsday" description better than any of the measured warnings of ecological disruption that came from the scientific community. Business has a track record of resistance to ameliorative change, despite its supposed immersion in a heroic narrative of "creative destruction." The American Automobile Manufacturers Association resisted seat belts in the 1960s, fuel economy standards on numerous occasions, and the 1970 Clean Air Act, claiming that manufacturers "would be forced to shut down." Industry officials insisted in 1990 that further reducing auto emissions "is not feasible or necessary and that congressional dictates to do so would be financially ruinous." Despite such claims, the car industry boomed in the decade after the reductions went into force.[46] Countless environmental and health regulations, including the removal of lead from petrol, the introduction of catalytic converters, and the transition to alternative refrigerants, were all accomplished without precipitating the predicted economic apocalypse.

Global Warming: Creating Doubt

By the late 1960s, some climate scientists had already begun to warn about a potential major problem. Though it was not yet possible to predict the timing, climate experts declared that, given the "great and ponderous flywheel of the climate system," the process might well be irreversible before its seriousness became obvious.[47]

The Reagan White House asked the National Academy of Sciences (NAS) to clarify the timing. A panel chaired by the economist and game theory specialist Thomas Schelling—and including Nierenberg, who had headed the 1982 acid rain panel that recommended no concrete action—produced a brief response, recommending more research and preferring adaptation to mitigation. Mitigation involves reducing greenhouse gas emissions by modifying such core economic activities as the burning of fossil fuels, while the option of adaptation has been consistently preferred by business interests and economists. In this scenario, no great changes need be undertaken in the operation of the fossil-fuel-based world economy; growth will not be jeopardized. Human technological ingenuity is predicted to supply ongoing solutions to climate problems—and people will have to migrate to higher ground or more temperate areas if technology fails to develop quickly enough.

A further NAS panel followed, with Nierenberg in the chair and both Schelling and another economist, William Nordhaus, on the panel, alongside a number of America's top climate scientists. The Nierenberg panel produced an extensive report, of which nearly half was written by Nierenberg and the economists, including the executive summary and policy recommendations. They stressed the need for more research, arguing that many uncertainties were associated with projecting the future, whether scientific, technological, or economic. The last chapter, on policy implications, rejected any preference for prevention rather than amelioration: "It would be wrong to commit ourselves to the principle that if fossil fuels and carbon dioxide are where the problem arises, that must also be where the solution lies." It was suggested that building defenses against sea level rise was a straightforward matter and that selective retreat was "inevitable."[48]

The brief executive summary referred to "irreducible" uncertainties, flawed models, and unpredictable outcomes. It recommended a "balanced research program," an openness to making "adjustments," and no change to "current fuel-use patterns."[49] As is common, the executive summary approach was the one reflected in the press. While accepting that climate change could be drastic, the *New York Times* credited the NAS with ruling that "there is no politically or economically realistic way of heading off the greenhouse effect," so we would have to adapt. It was neither the Academy nor even the entire panel that made this judgment, but only Nierenberg and the economists. An EPA report issued at the same time called for immediate action. President Reagan's science adviser dubbed it "unnecessarily alarmist,"[50] and the Reagan White House embraced the Nierenberg strategy: research, delay, and perhaps migration to higher ground if all else failed. Though it did not deny the warming trend, the Nierenberg report demonstrated that a counsel of delay could crush action just as effectively as outright denial.

Five years later, in the summer of 1988, James Hansen delivered his famous testimony to the US Congress, stating baldly that global warming was already happening, no longer merely an imminent risk, and that he was 99 percent certain it was caused by human emissions. In the same year, the IPCC was set up to review all climate research and collate its findings, to assess impacts both environmental and socioeconomic, and articulate possible responses.

These developments provoked a surge of environmental denial from nonscientists associated with think tanks in the early 1990s. As well as the physicists at the Marshall Institute, a few other scientists also joined the effort. According to internal strategy papers obtained at the time by the journalist Ross Gelbspan, the coal industry's generous funding of Patrick Michaels, Robert Balling, and Fred Singer was intended "to reposition global warming as theory (not fact)."[51]

Denial and Delay in Australia: The Greenhouse Mafia

In Australia, the conservative Howard government presided over denial and delay on global warming in much the same manner as the Reagan and Bush administrations had done. Guy Pearse, speechwriter for Robert Hill, the environment minister in the Howard government's early years, writes of his dawning realization at the time of the Kyoto conference: "I started to think the unthinkable—the Liberal Party was taking the country in precisely the wrong direction on climate change. It had been captured by a cabal of powerful greenhouse polluters and had no intention of reducing Australia's greenhouse pollution, ever."[52]

This realization emerged from his work for the Liberal Party—senior partner in the conservative coalition that Howard led from 1996 to 2007—and his PhD research, where he sought to discover why denial was so prevalent in business circles when many industries such as insurance and tourism seemed to have so much to lose from global warming.[53] Pearse interviewed members of the Australian Industry Greenhouse Network (AIGN), "a highly influential collection of Australia's biggest greenhouse polluters" with an alarmingly blunt title for their organization. Several told him they privately called themselves, even more bluntly, the "greenhouse mafia."[54]

Part 2 of *High and Dry* (2007), the book that resulted from this work, lays bare the dense web of connections linking industry funding, business coalitions and interest groups, a few key think tanks, public relations companies, supposedly independent economic modelers, lobbyists, media skeptics on both sides of the Pacific, and senior federal bureaucrats and ministerial staffers in the departments handling industry and trade—not to mention several senior cabinet ministers led by Prime Minister Howard himself. A flavor of musical chairs pervades the narrative, as the key players jump from the Canberra bureaucracy to the boards of mining and energy corporations, and on to CSIRO advisory bodies, free market think tanks, or industry peak bodies. These were the determinate influences on the climate policies of the eleven years of the Howard government, which sent AIGN representatives along on government climate negotiation teams. In their interviews with Pearse, members of this greenhouse mafia admitted they were involved in the writing of actual cabinet submissions and ministerial briefings for the government, activities properly conducted by the public service under the Westminster system rather than by lobbyists.[55] In the Howard government, the polluters wrote the government's climate policy.

Pearse exposes Howard's links to the world of global warming denial and his long-standing doubt about the science. His decision to embrace the IPCC's science and commit to an emissions trading scheme (ETS) in late 2006 was a political decision, forced on him by public concern in Australia. His ultimate successor at the helm of the Liberal Party and now prime minister, Tony Abbott, is another self-confessed denier; although he adopted a climate policy after gaining the leadership, he told a forum in March 2011 that the science on climate change is still not settled. "Whether carbon dioxide is quite the environmental villain that some people make it out to be is not yet proven," he told his audience.[56]

With a mandate for action, the incoming Labor government of 2007, headed by Kevin Rudd, ratified the Kyoto Protocol and set out to put a price on carbon via an ETS. But the carbon lobby was already entrenched inside the institutions that were to be responsible for the curtailment of its emissions, as well as enjoying the formidable organizational assets built up over decades in think tanks and industry associations. It continued to dominate the policy process, artfully confusing the national interest with its own.[57] Pearse points out that Australia's reliance on coal-fired electricity had deepened since the 1970s under governments of all stripes, fostering an alliance between key unions and the corporations who employ their members. He identifies an alliance between polluters, elite politicians, and elite bureaucrats, the so-called "iron triangle," one that persisted, he argues, when the government changed in 2007.[58]

While power generators warned of blackouts and price spikes and industry associations forecasted deep job losses,[59] the actual details of Rudd's scheme were extremely friendly to fossil fuel interests. A modest reduction target of 5 percent from 2000 levels was adopted. Electric utilities, which had known for decades that carbon would one day be priced, were to be absolved of responsibility for their decisions and paid nearly $4 billion in compensation for the reduced value of their businesses.[60] Transport emissions were to be excluded, at least initially. On average, heavy polluters would get four out of five pollution permits for free, while the heaviest would receive up to 90 percent for free. "Clean coal" would also be generously funded, without reference to the actual scale required to make a tangible difference, a scale that the Canadian scientist Vaclav Smil has quantified. In order to sequester *just 10 percent* of current global CO_2 emissions, the world would need to force underground annually a volume of compressed CO_2 comparable to the volume of crude oil extracted annually by the global petroleum industry—whose infrastructures and capacities have been put in place over a century of development. Achieving even this limited degree of sequestration appears to be highly unlikely.[61]

What most of the public was not clearly told, or did not really grasp, was that the Rudd trading scheme would not launch new, cleaner industries at home and would not reduce actual emissions from Australian industry at all in the first few decades. The reduction, already very modest, would be accomplished by buying low-cost permits from countries such as Indonesia and Papua New Guinea, mainly in exchange for forest preservation. The Australian Treasury estimated that the country would be emitting 585 million tonnes (Mt) of greenhouse gases in 2020 (about 560 Mt excluding land-use change and forestry),[62] 34 percent more than comparable emissions in 1990. Not until 2050 would the actual emissions approach the 1990 level. Australia's apparent compliance with Kyoto during the Howard years was achieved by reductions in land clearing, while its industrial emissions actually rose 38 percent. Under Rudd's scheme, the country's ongoing compliance with greenhouse targets was to be achieved largely by paying others to reduce land clearing. Thus, the Australian economy was not slated to emit much less carbon in the next few decades and would not commence a transition to low-carbon alternatives.[63] In addition to the coal burned at home, Australia provides 25 percent of the entire world's supply, including more than half of its high-grade metallurgical coal,[64] but this is ignored under current accounting rules, as is the carbon emitted in the production of most of Australia's consumer durables by countries such as China.

As the ecological economist Michael Jacobs has observed, many governments have resisted the fossil fuel industry's attempts to crush their moves to reduce emissions. According to Jacobs, only in the United States, Canada, and Australia have the arguments of climate change deniers weakened the government's resolve to act on climate change.[65] The approach of the conservative coalition in Australia is typical of the politicization of environmental issues described above, where science—and even fact—is irrelevant and the objective is denial, delay, or the transfer of costs. These tactics have been less effective in Europe and other parts of the world. Why this is so merits further research, but it may be connected to the influence of fossil fuel interests and conservative think tanks over the governments and media of the three rogue nations Jacobs cites.

While Abbott pledged a "citizen's revolt" against carbon pricing, a tactic along US tea party lines, a broad coalition of business interests began an advertising campaign to oppose the Gillard government's "carbon tax" (though it was merely a lead-in phase to the ETS framework that business had always favored).[66] The campaign relied on fostering fear and doubt and circulating half-truths and distortions. It included the spurious idea that Australia should not act if countries such as China did not, and asserted that Australia was acting ahead of other major industrialized nations.[67] The government's Productivity Commission report found this not to be the case: Australia ranks behind Germany and the UK and is in the mid-range of countries when it comes to addressing carbon emissions.[68]

Rhetorics of Environmental Denial

Rhetorics of Obfuscation

The master propagandist of Nazi Germany, Joseph Goebbels, is widely credited with the observation, "If you repeat a lie often enough, it becomes the truth." Closer to the present time, Wendell Potter, a US insurance executive who resigned from his post in 2008, has revealed a similar duplicity involved in his prior PR work. Speaking of the deployment of the term "socialized medicine" to demonize the public option during President Obama's health care reform efforts, Potter commented, "When you say something like that, it's not true, but it doesn't have to be true. You just say it, and you say it over and over and over again. You get your allies to say it over and over again, to the point that Americans believe it."[69]

Frank Luntz epitomizes this type of PR practice. Pollster, focus group guru, and Republican political consultant, Luntz is notorious for writing a memo that was used by the George W. Bush administration in framing its messages in the early 2000s.[70] The climate section of Luntz's memo constitutes an astute compilation of phrases that manipulated debate in the interests of delayed action and the rollback of regulation, while appearing to seek a "cleaner," "safer" environment through a more "commonsense" approach to the problems. He advised that "climate change" is less frightening than "global warming," a proposal derived from focus groups, who thought it sounded less "catastrophic." It is perhaps an indication of the ubiquitous influence of right-wing PR that "climate change" quickly became the accepted terminology in public discourse.

Fostering doubt about the validity of the climate science was a central Luntz recommendation, calculated to justify delay:

> The scientific debate remains open. Voters believe that there is no consensus about global warming within the scientific community.... You need to make the lack of scientific certainty a primary issue in the debate.... Emphasize the importance of "acting with all the facts in hand." ...
>
> The scientific debate is closing [against us] but not yet closed. There is still a window of opportunity to challenge the science.[71]

It should be remembered that the *scientific* debate did not, in fact, remain open at all.[72]

Luntz summarized tactics typical of the overall neoliberal assault on environmental action, centered on opposition to regulation and what was called "bureaucracy." He also alluded to the fear of a conspiracy to create world government, common on the fringes of the American Right:

> Give citizens the idea that progress is being frustrated by over-reaching government.... Emphasize how voluntary innovation and experimentation are

preferable to bureaucratic and international intervention and regulation.... Unnecessary environmental regulations hurt moms and dads, grandmas and grandpas. They hurt senior citizens on reduced incomes. They take an enormous swipe at miners loggers truckers farmers—anyone who has any work in energy intensive professions.[73]

Luntz also proposed to emphasize reciprocal restrictions and penalties on the developing world in any climate negotiations. The third world had been excluded from the commitments of the Kyoto Protocol, much to the chagrin of the US negotiators and fossil fuel representatives.[74] According to Luntz, "The international fairness issue is the emotional home run. ... Americans will demand that all nations be part of any ... treaty. Nations such as China, Mexico and India would have to sign.... Every nation must do its part."[75]

With his grasp of the importance of emotion, Luntz is a direct heir to Bernays. "My job is to look for the words that trigger the emotion.... We know that words and emotion together are the most powerful force known to mankind." He also noted that "a compelling story, even if factually inaccurate, can be more emotionally compelling than a dry recitation of the truth."[76] This attitude often defeats the scientific community, whose loyalty is to the accumulation of provisionally valid beliefs based on all the evidence.

Reverse Rhetoric: Sound Science/Junk Science

The terms "sound science" and "junk science" illustrate well what might be called "reverse rhetoric," where the target is accused of the most obvious weaknesses of the attacker. While advocating emotional triggers and encouraging strategic factual inaccuracies, Luntz also stressed that "the most important principle in any discussion of global warming is your commitment to sound science."[77] This would, of course, be a contradiction in terms if taken at face value, but Luntz reflects the quest to corner the label "sound" in order to sanitize denial and delay. In Luntz's view of reality, like Ivy Lee's in 1915, facts and factoids are equally valid.

The use of the term "sound science" as an antiregulatory slogan has been in circulation since at least 1982, when Dow Chemical claimed to be using sound science in its $3 million program to reassure Michigan residents about local dioxin pollution.[78] To counter the move toward banning smoking in public places, Philip Morris was instrumental in setting up The Advancement of Sound Science Coalition (TASSC) in 1993, through its PR agent APCO Associates. TASSC, describing itself as "a not-for-profit coalition advocating the use of sound science in public policy decision making," set to work to recruit representatives from other industries subject to regulation, such as food, plastics, chemicals, and packaging, so as to blur any focus on its campaign against smoking restrictions; the dangers of passive smoking would be obscured among numerous examples of alleged "unsound, incomplete, or unsubstantiated science."[79]

Conflicts over fact had emerged as the industrial economy developed through the late nineteenth and early twentieth centuries; efforts to regulate toxic substances often met with denial from manufacturers. Adulteration and poisoning of foods was controlled from 1906 in the United States, but the burden of proof lay with the government, which had to show that the additive was dangerous for humans. In 1938, a modification allowed poisonous substances in food, within a "safety" margin set by the government, if they were "essential" to its production.[80] These food rules illuminate the main territory of the conflict between industry and citizens over environmental pollutants. Industry always sought a high bar when it came to proof, advocating absolute certainty. Where small populations with scattered incidences made such proof difficult or impossible to establish, people could be disregarded—even when pollution was "easy to smell" and they could not breathe or developed cancer.[81] In pushing absolute proof as the only valid standard for assessing pollution, regulation of toxic substances could be delayed. Industry characterized precaution as a recipe for economic disaster, equating it with "economic and social stagnancy.... [It was] an unnecessary interference with the scientific advances essential to progress."[82]

The pursuit of such objectives earlier in the century became intrinsic to the idea of sound science as it was pushed by antienvironmentalists in the late twentieth century. What was always at stake here was the right of enterprise to pollute or poison rather than risk a reduction in its profit. In the case of lead, for example, despite evidence of its dangers dating back to the nineteenth century[83] and grim warnings from public health officials at the time, the petroleum and automobile industries succeeded in their bid to add tetraethyl lead to ordinary petrol in the 1920s. From that time, lead was gradually dispersed through every street in the world where people drove cars. Lead manufacturers insisted that there was no danger and, above all, that there was no *absolute* proof of danger. At the same time, they suppressed evidence that indicated risk and funded research that would treat it as minimal, as other manufacturers did with numerous suspected toxins.[84] Lead in petrol visited fifty years of poisoning on the entire developed world and parts of the rest before regulation began in the 1980s. Computer monitors containing an average of four pounds of lead are nonetheless disintegrating in countless landfills and leaching into groundwater at the present time. In *Deceit and Denial*, their book on industrial pollution, the historians Gerald Markowitz and David Rosner target the lead and vinyl industries, yet they make it clear that "lying and obfuscation were rampant in the tobacco, automobile, asbestos, and nuclear power industries as well."[85]

Steve Milloy, associated with the Cato Institute and the AEI, became executive director of TASSC in 1997[86] and, though TASSC ceased operations in the late 1990s, Milloy continues to run a renamed Advancement of Sound Science Center (ASSC) and the website JunkScience.com, also funded by Philip Morris through the PR firm APCO.[87] These associated sound science front groups define junk science as "faulty scientific data and analysis used to advance special and, often, hidden agendas."[88] Here they are referring to mainstream academic science and attributing their own hidden agendas to their target, another example of reverse rhetoric. Though initiated by Philip Morris in its attempt to position tobacco among less life-threatening products, TASCC and its offspring went on to champion the denial of global warming.

Claims to sound science are ubiquitous on think tank websites. The ACSH says of itself that it "was founded in 1978 by a group of scientists who had become concerned that many important public policies related to health and the environment did not have a sound scientific basis. These scientists created the organization to add reason and balance to debates about public health issues and bring common sense views to the public."[89] The Marshall Institute also claimed the label, stating its mission was "to encourage the use of sound science in public policy,"[90] though that particular formulation has been dropped from the website, which now simply claims that the institute promotes "accurate and impartial technical assessments."[91] Fred Singer, in almost the same words as TASSC, claimed that SEPP is "a non-partisan, nor-for-profit, privately funded research organization, devoted to the use of sound science in public policy."[92]

The Democratic congressman George Brown, in his 1996 report to the Democratic caucus of the Science Committee, unmasked the sound science mantra for what it was—an attempt to redefine science as absolutely certainty and thus create an apparent lack in the knowledge, a gap where doubt could multiply and almost all the findings of the environmental and health sciences could be delegitimized.[93] In their survey of the use of the term "junk science" in the US popular press, environmental consultant Charles Herrick and environmental philosopher Dale Jamieson found that it appeared in connection with a vast range of environmental issues, from global warming, sea-level rise, and species loss to risks from pesticides, dioxins, air pollution, and endocrine disruptors. They identified what they called the junk science "trope": "a punchy, dazzling, but highly misleading" rhetorical device designed to enlist emotional responses. The vast majority (84 percent) of the articles they examined for the years 1995–2000 ran an antiregulatory message, contending that the policy or regulation at issue was based on junk science.[94] This correlation reflects the close relationship between the term *junk science* and antiregulation advocates.

Claims to speak for sound science were accompanied by attacks on professional scientists and their work. In 1986, for example, Frederick Seitz's cousin, Russell Seitz, another physicist affiliated with conservative strategic policy institutes, attacked the concept of "nuclear winter," calling the modeling involved a "series of coin tosses." He went on to claim that scientists are guided by such "non-rational factors as rhetoric, propaganda, and personal prejudice" and asserted that "politically motivated" scientists had come to dominate "matters of science and public policy."[95] The climatologist Ben Santer was attacked for "scientific cleansing" and "secretly altering" the 1995 IPCC report when he was merely carrying out routine amendments according to IPCC protocols. His attackers, including Fred Seitz and Nierenberg, accused him of fraud and conspiracy using the *Wall Street Journal*, which declined to publish his rebuttals in full.[96] On the one hand, "contrarians" wrapped themselves in the sound science mantle and sidestepped the peer review regime by publishing their work through think tank channels and right-wing newspapers rather than in the usual scholarly journals; on the other, they targeted the scientific mainstream as leftist and unreliable. In this way they aimed to undermine the credibility of mainstream natural scientists in academic settings.[97]

In their survey of documents on global warming circulated by the major think tanks from 1992 to 1997, McCright and Dunlap found that 71 percent attempted to discredit the evidentiary basis of the consensus science. These think tank documents claimed that climate models were biased, alleging that the IPCC's consensus was "manufactured" or "doctored" and its peer review process "corrupted" and "thoroughly politicized." A significant minority of the documents (18 percent) engaged in simple abuse of the kind leveled at *The Limits to Growth* in the early 1970s, calling scientists "modern-day apocalyptics," part of "the doomsday crowd," "prophets of doom," and purveyors of "myth" or "scare tactics." Many of the documents (13.4 percent) labeled the work of the scientific mainstream as "tabloid science" and "junk science," a feat of cross attribution in which amateurs and nonscientists labeled as junk the work of the professionals.[98]

More Reverse Rhetoric

This kind of blatant projection is mirrored in other cases of reverse attribution. An even more astonishing swap has been the recasting of public interest groups as "special interest groups." This tactic shifts the accusation away from the actual vested interests and onto the community groups and citizen activists, who, with little to no economic interest, are struggling to curb their excesses.

The world government scare, touched on by Luntz in his memo on how to incite emotion in the service of environmental delay, carries a similar irony. While the business-friendly Right condemns the alleged theft of American sovereignty by the UN and its processes of negotiation, the now-entrenched coercive apparatus of the free market *actually* obstructs democratic governments from deciding about many matters affecting their citizens and environment. The international financial system threatens capital flight if governments fail to supply the requisite business-friendly measures, and the WTO rules explored in chapter 13 preclude many kinds of health, safety, and environmental protection measures. The South Korean prime minister who was responsible for implementing neoliberal policies in the 1990s remarked that "we did not realise that the victory of the Cold War was a victory for market forces above politics."[99]

Similarly, opponents of the business takeover of political life are popularly branded as "elites," "latte lovers" and "Chardonnay sippers"—effete, arrogant enemies of the "common man." In this trope, the actual ruling elites who shaped the world in the past forty years are made to disappear—or, where visible, to masquerade as the humble servants of all.

Attack Stance of the Free Marketeers

As previously discussed, the overall neoliberal movement owed its emergence to an alliance between the free market intelligentsia and sectors of capital with profits threatened by regulation or taxation. Similarly, the pioneers of the scientific denial machinery were the right-wing think tanks allied with industries anxious to avoid regulation and a widening group of politicians and policymakers who were influenced or paid outright. The mushrooming of conservative think tanks did indeed provide an "artillery," as Ralph Harris of London's IEA put it, capable of softening up any opposition. First in the United States, then gradually across the rest of the world, the think tank system took on environmentalism as a central target in its war to protect corporate dominance and profits.

Global warming is at the center of the campaign of doubt and delay because fossil fuels have been intrinsic to the very fabric of industrial growth and prosperity. As outlined in chapter 2, the exploitation of oil and the expansion of its use through the twentieth century was "something new under the sun"[100] and vital to the unprecedented growth of that century. So central is the role of petroleum that, as production of conventional liquid oil reached a plateau in the first decade of the twenty-first century, oil corporations scrambled to expand options such as tar sands and shale oil, with scant regard for the higher greenhouse gas emissions per gallon of liquid oil. Similarly, the use of coal, wound back in the middle of the twentieth century, rebounded after the oil shocks of the 1970s. While repeated warnings from climate scientists that emissions need to start trending downward by 2020 and that around 80 percent of known reserves of fossil fuels need to remain in the ground if we are to avoid catastrophic climate change, the US Energy Information Administration's 2013 *International Energy Outlook* predicts ongoing reliance on fossil fuels, especially coal, and a 46 percent increase in energy-related carbon dioxide emissions by 2040.[101] If this actually occurs, as the growth economy requires, runaway global warming is very probably inevitable.

Some industry representatives, PR professionals, and operators of think tanks and front groups acknowledged an explicit aim to destroy environmentalism altogether. Bob Williams, an oil industry journalist, wrote in 1991:

> What is the goal the petroleum industry should strive for in the Decade of the Environment? To put the environmental lobby out of business.... There is no greater imperative.... If the petroleum industry is to survive, it must render the environmental lobby superfluous, an anachronism.[102]

Ron Arnold, one of the initiators of the "wise use" umbrella organization and vice president of the Center for the Defense of Free Enterprise (CDFE) since 1984, is recorded in interviews with numerous journalists declaring war on environmentalism. For example:

> Our goal is to destroy, to eradicate the environmental movement.... We're dead serious—we're going to destroy them.[103]

> People in industry, I'm going to do my best for you. Environmentalists, I'm coming to get you.... We're out to kill the fuckers. We're simply trying to eliminate them. Our goal is to destroy environmentalism once and for all.[104]

Arnold saw himself and the CDFE as the forward assault team in this war against environmentalists: "We [CDFE] created a sector of public opinion that didn't used to exist. No one was aware that environmentalism was a problem until we came along."[105] In his interview for *Outside* magazine, Arnold revealed an appreciation of Bernays's technique—he had exploited people's "fear, hate and revenge," he told John Krakauer. "Wise use" itself was "a marvellously ambiguous expression.... Symbols register most powerfully in the subconscious when they're not perfectly clear.... Facts don't really matter. In politics, perception is reality."[106] "Wise use" was perfect. It smacked of good judgment and responsibility and could have meant almost anything.

Arnold cited two core conservative objections to environmentalism: It would "drastically reduce or dismantle industrial civilization" and it would "impose a coercive form of government on America."[107] Fred

Singer harbored the same twin fears, and explicitly named the second as the fear of socialism. In his paper about the alleged ozone exaggerations, he wrote:

> And then there are probably those with [a] hidden agenda—not just to "save the environment" but to change our economic system. The telltale signs are the attack on free enterprise, the corporation, the profit motive, the new technologies. Some are socialists, some are Luddites.... To them global regulation is the "holy grail."[108]

Whatever the danger of the regulatory holy grail supposedly sought by rabid environmentalists, Frank Mankiewicz, senior executive at the prominent PR firm Hill and Knowlton, was nearer the mark. He did not envisage any real threat:

> The big corporations ... are scared shitless of the environmental movement. They sense that there's a majority out there and that the emotions are all on the other side.... They think the politicians are going to yield to the emotions. I think the corporations are wrong about that. I think the companies will have to give in only at insignificant levels. Because the companies are too strong, they're the establishment. The environmentalists are going to have to be like the mob in the square in Romania before they prevail.[109]

Balance as Bias: The Role of the Press

The idea of balance, canvassed by Powell in his memo, has been deployed as a vital tool in the tactics of the think tanks throughout these crucial decades. The Marshall Institute used it from the outset, arguing that journalists were obligated to present "both sides" of the Star Wars controversy, effectively giving a tiny minority position equal weight to that of the large majority of the scientific community.

In matters of opinion, the notion of balance in media reporting is often appropriate, since conflicting views warrant adequate space in a democratic system. But, in matters of fact, there is little or no role for "balance." What is required is accuracy. Indeed, as *New York Times* public editor Daniel Okrent remarked, "The pursuit of balance can create imbalance" when something is obviously true.[110] In matters of science, as with any discipline pursuing evidence-based facts, "equal time" is only apposite to the extent that the scientific community is somewhat evenly split and consensus is unstable. In the case of global warming, the degree of agreement on human-induced climate change is almost total and has been for a decade or more;[111] it has been endorsed by all the premier institutions of science, even in the United States. As the physicist James D. Baker commented, "There is a better scientific consensus on this than on any other issue I know—except maybe Newton's second law of dynamics."[112] In such a situation, the maintenance of an even balance is misrepresentation.

The much-touted practice of "balanced" reporting, epitomized by the *Fox News* motto "Fair and Balanced," has displaced the commitment to accuracy in journalism that permeates Walter Lippmann's *Liberty and the News*. Lippmann quotes an editor in 1690 already disturbed by lies in the press and appealing to the principle that "nothing shall be published except what we have reason to believe is true." In 1920, Lippmann argued that a crisis in democracy inevitably flows from a crisis in journalism; he predicted disaster for any people "which is denied access to the facts" and not "protected by the rules of evidence."[113] For Lippmann, these were the necessary values: evidence, accuracy, truth; serious critical inquiry based on the available evidence. The norms Lippmann sought for journalism are analogous to the norms Australian scientist Peter Cullen defined for science. The objective of science is to "find a truth," and this is done "by gaining a consensus which becomes the orthodox view." In politics, the process is not about truth, but is one of bargaining and negotiation to find "an outcome acceptable to relevant interest groups."[114]

The BBC's benchmark also dismisses the notion of balance, as explained by Peter Horrocks, director of the BBC's Global News Service: "We don't use the term 'balance.' We talk about impartiality.... We clearly reflect the range of views, and we certainly do not exclude views.... However that doesn't mean that everyone has an equal voice and, in factual reporting by our specialist journalists ... they absolutely make clear where a consensus of views lies."[115] The clamor for balance was a catch-cry of the free market think tank culture, recommended by Powell as a tactic against the claims of environmental science and on behalf of the corporate sector in its quest to neutralize the opposition. Balance, like wise use, sounds unimpeachable but is in fact a bogus alternative standard to supplant accuracy and evidence as the basis of news reporting.

In the press coverage of global warming, the US media point of view diverges sharply from the actual consensus in the scientific community, where even the elite US press represented "debates" about global warming as "evenly divided." Maxwell and Jules Boykoff analyzed a random sample of news stories (not opinion pieces) from four prestige newspapers in the United States—the *New York Times*, the *Los Angeles Times*, the *Washington Post,* and the *Wall Street Journal*—for the period 1988–2002. On the issue of whether humans are contributing to global warming, over half their sample represented this question as an evenly divided debate. Fewer than 6 percent focused exclusively on human input, and only one-third approximated the actual scientific consensus and treated the human contribution as dominant.[116] When the Boykoffs turned to the issue of whether global warming warranted immediate action, they found a pattern that was even more skewed. Nearly 80 percent of all articles throughout the period (1988–2002) took a balanced view on the need for action, with the same surges of "extra-balanced" coverage in the key years 1992, 1997, and around 2000.[117]

The incidence of such balanced coverage waxed and waned. In the late 1980s, when George H. W. Bush campaigned on a platform of dealing with global warming, an "even" balance was not so prominent, but by 1990 it was on the increase. It spiked at the time of the Rio Earth Summit (1992), the Kyoto conference (1997), and the US election in 2000. At these times there was an increase in formulations such as "some scientists believe" followed by "but skeptics contend"; suggestions that global warming was a "hoax" or "gaffe" were juxtaposed with the findings of the IPCC.

The provision of a false balance is in fact so reflexive now that even reporters for Australia's ABC (public radio) feel constrained to provide it. In the *AM* report on the World Meteorological Organization's announcement that 2010 was one of the three hottest years on record, half the report was given over to the prominent Australian climate change "skeptic" Bob Carter, who claimed this was not related to global warming.[118] Whether the conduct of ABC reporters was affected by the views of Maurice Newman, its chairman at the time, is unknown. Newman is a stockbroker and merchant banker with links to both the CIS and the IPA, and was an appointee of the denialist Howard government. He told ABC staff, "Climate change is a further example of groupthink where contrary views have not been tolerated, and where those who express them have been labelled and mocked."[119] In his speech to the staff, Newman went on to recycle the standard "skeptic" attack on scientists over emails stolen at East Anglia University, claiming "sensational revelations of unprofessional conduct," although numerous independent investigations, including those by the British House of Commons and the US EPA, found no evidence of any kind of impropriety.[120] Newman also used his speech to accuse the IPCC of "dubious research," "politicised advocacy," "scientifically unsupported claims and errors," "questionable methods of analysis resulting in spurious temperature data," and a "lack of moral and scientific integrity." The incoming conservative Abbott government has appointed him chairman of its Business Advisory Council.

As Sharon Beder has argued, "Think tanks have more in common with interest groups or pressure groups than academic institutions. Nevertheless employees of think tanks are treated by the media as independent experts" and are in some cases preferred to scientists working in universities.[121] In the United States, reports prepared by think tanks routinely turn up in Senate hearings and congressional committees on science and the environment.

Industry has poured large sums of money into the rebuttal and denial of a wide range of environmental problems, from the damage caused by local toxins in soil, air, and water to the effects of substances with global impacts, such as CFCs and greenhouse gases. Whole sectors of industries have suppressed the known risks of their products in order to maintain and sometimes extend their business and its profits. Among these were the lead, asbestos, PVC, and tobacco manufacturers, as well as many branches of the chemicals and plastics industries. The fossil fuel sector and its dependents, such as the utilities and aluminum smelters, continue to expect society to bear the costs of their industries' side effects today.

Yet business has gone well beyond influencing public opinion to ensure the support of democratic governments, venturing into a domain that is not accessible to any electorate. In the next chapter I turn to the creation of the WTO and the tactics big business has used to cement its international trade agenda, which often nullifies national environmental legislation.

13

International Brakes on Environmental Priorities

These deals [free trade agreements] aren't about free trade; they're about the right of these guys, the US multinationals, to do business the way they want, wherever they want.

—Eugene Whelan, former Canadian minister for agriculture, quoted by Elizabeth May, Canadian Parliament, 1999

By the 1970s, neoliberals had accepted the Chicago school's repudiation of antitrust policies and its willingness to allow monopoly and cartel to operate without regulation or constraint.[1] Ideological support for such entities facilitated the emergence of the global financial market and integrated global production, in which dispersed elements can be designed, made, and assembled separately and incorporated into vast global chains of production and distribution. These in turn demanded a modified international rulebook.

Thus, while free market think tanks were cornering the policy debate inside the various nations of the developed world—and making inroads in a few others, such as India—a parallel shift was occurring in the international institutions that govern world economic policy, ensuring that global policy would be conducted through a lens of business prescriptions, without particular regard to the planet or the powerless.

Bretton Woods and Beyond: The World Bank and the IMF

The World Bank and the IMF were already in existence, and had operated under the influence of the US from their inception. Despite attempted reforms, voting rights in both institutions are based broadly on a country's wealth, reflected in its ability to contribute shareholdings, and, in the case of the IMF, on a country's GDP and level of "openness."[2] Thus, decisions are dominated by the US in particular and the developed world in general, and the heads of these organizations have traditionally been appointed by the US (World Bank) and the UK or Europe (IMF). More than half the World Bank's economists, who enjoy determinate influence at the bank, are trained in the UK or US. There is also a revolving door between the multilateral banks and the big international financiers such as Chase Manhattan, Goldman Sachs, Deutsche Bank, and JP Morgan.[3] Even before market fundamentalism gained ideological ascendancy, the US government and the international financiers objected to any whiff of socialism. In the five years before Salvatore Allende's election, Chile received $100 million in loans, while not a single loan was made during his term of office. Once the Pinochet dictatorship took over, it got $100 million in World Bank loans during the first two years after the 1973 coup, as well as $680 million from the US.[4]

As neoliberal economics took hold, the bank's economists inevitably reflected the new orthodoxy. Access to its loans was reserved for countries adopting free market policies and permitting unrestricted foreign investment. For countries aiming at equity or redistribution, prospects remained bleak. An "invisible blockade" on loans, analogous to the treatment of Chile, was imposed on Nicaragua during the socialist Sandinista regime, alongside covert military intervention via the Contras.[5]

The IMF was originally charged with maintaining financial stability in the world economy, a task that was seen to include responsibility for exchange rate stability and short-term lending to countries in trouble. In the late 1970s the IMF imposed "structural adjustment programs" (SAPs) as part of the conditionality of its loans. These SAPs demanded neoliberal restructuring in exchange for financial rescue, forcing privatization of state enterprises, reductions in government welfare spending, balanced budgets, and the abolition of barriers to foreign investment. The withdrawal of government agricultural programs that followed the imposition of the new rules on debtor nations deprived small farmers of whatever government support they had formerly enjoyed. As Jeffrey Sachs, a prominent US economist and former front-line neoliberal operative in Russia and Latin America, explains, "During the debt crisis of the 1980s and 1990s, the International Monetary Fund and World Bank forced dozens of poor food-importing countries to dismantle these state systems. Poor farmers were told to fend for themselves, to let 'market forces' provide for inputs. This was a profound mistake: there were no such market forces. Poor farmers lost access to fertilizers and improved seed varieties."[6] The IMF helped dismantle modest welfare measures in dozens of third world countries while also urging neoliberal changes on the developed world, including exhorting Australia to cut taxes for those in the upper income brackets, tighten welfare compliance, and lower the minimum wage.[7]

Labeled the "Washington Consensus," a term coined but disliked by economist John Williamson,[8] the neoliberal suite of policy modifications was adopted—or enforced—almost everywhere. In Australia, for example, the Hawke-Keating Labor government began the process in 1983, and it was comprehensively implemented by the subsequent Howard-Costello Coalition government (1996–2007). Both sides of Australian politics pursued similar objectives: reducing tariffs, lowering taxes—especially on the rich, who were assumed to be anxious to invest the proceeds productively—and imposing privatization and deregulation. As Sharon Beder notes, "These were measures that would expand business opportunities, reduce the costs of doing business and minimise the regulations that business would have to abide by."[9]

The driving force behind these changes, whether in Australia or elsewhere, was corporate interest, especially the newly mobile international financial markets. It was TNCs trading across borders and big business in national settings that stood to gain the most from the Washington Consensus. Their company taxes were slashed, progressive taxation scales were decimated, and consumption taxes were substituted. Obligations owed by foreign firms to specific nations were weakened or abolished, and regulation was eased. Gradually, capital was permitted to move across the world with little hindrance. Economic goals took precedence over all other priorities.

Draconian measures were imposed on the world's most vulnerable nations, beginning with Mexico in 1982 and extending to nearly eighty developing countries by the early 1990s and to East Asia after the 1997 financial crisis. The IMF's "cash for austerity" deals were indeed miserable arrangements for their recipients; in 1997, most of the cash secured at such cost was passed on to first world creditors in acquittal of loans, while the debtor countries still owed the money to the IMF or the banks that had provided the funds.[10]

International Lobby Groups and Think Tanks

New international organizations came into being alongside the World Bank and IMF after the war and especially in the 1970s as the CEOs and chief ministers of the developed economies grappled with stagflation.[11] The World Trade Organization was also gradually assembled, over the entire postwar period, coming into operation in 1995. These new institutions assisted the expanding TNCs in the prosecution of their specific interests.

The OECD, the World Economic Forum, and the G7, G8, and G20

The Organisation for Economic Co-operation and Development (OECD) grew out of the group of seventeen West European countries, plus Turkey, which administered the postwar Marshall Plan. In 1961, Canada and the US joined these countries to form the OECD. Later additions included Japan in 1964, Australia in 1971, South Korea, Mexico, and the eastern European countries in the 1990s, and Israel and Chile in 2010. The OECD functions primarily as a peak think tank for the "economic giants" of the world. In its own words, it "uses its wealth of information on a broad range of topics to help governments foster prosperity and fight poverty through economic growth and financial stability."[12] The OECD monitors performances, collects data, and publishes hundreds of surveys and reports every year. It may be regarded as the pioneer of the new elite international organizations.

The World Economic Forum (WEF) started out as a forum on management practices convened by Klaus Schwab, a European business school professor, with funding from what was then the European Community and European industrial associations. Schwab invited more than four hundred CEOs from European corporations to what was originally called the European Management Forum in the summer of 1971, shortly before the 1970s crisis emerged. His initial objective was to spread US management theory to Europe, but the events of the next few years triggered a broadening of Schwab's scope. Political leaders were included at the annual conference in Davos in 1974. A membership system was instituted in 1977, aimed at enlisting the "1,000 leading companies in the world," not just Europe, a goal apparently reached in 1992. The US began participating in 1983 with a satellite message from Ronald Reagan. In 1987 Schwab changed the organization's name to World Economic Forum.[13]

In addition to the annual Davos meeting, the WEF convenes numerous specialized meetings every year, including the Informal Gathering of World Leaders, which began in 1982 (and exemplifies the cozy "getting-to-know-you" atmosphere the WEF cultivates). The WEF also functions as an international think tank; it has produced an annual "global competitiveness" report since 1979, as well as regular reports on risk, the gender gap, information technology, and regional issues.

Although the WEF characterizes itself as a forum for "business, government and civil society," the last term of this trio is largely empty. The membership is composed of the one thousand corporations already mentioned, and the meetings are by invitation. Few community and citizen-based organizations attend—and only if the WEF invites them, as it does from time to time. Labor representatives such as Australia's union leader Sharan Burrow and Britain's Labour Party leader Arthur Scargill; Petra Kelly, then leader of the German Greens; and Muhammad Yunus, founder of the Grameen Bank, have attended at various times. Since about 2000, Burrow has attended most Davos meetings with a small team of labor delegates and Kumi Naidoo, the Greenpeace activist and currently its international executive director, has also accepted regular invitations in the past decade—a "pinch of public participation" as Naidoo describes it, among 2,500 CEOs, presidents and prime ministers.[14]

The actual membership includes most of the 737 TNCs that Vitali and colleagues found to be in control of 80 percent of global revenue.[15] The WEF is run by these corporations and functions as their own elite club into which select political leaders are invited, as well as reserve bankers, top personnel from the IMF and World Bank, and chosen UN officials. In this way, top global CEOs get access to the policymaking of the world's elected representatives, facilitating the privileging of business agendas in every corner of the world, especially the interests of transnational capital; input from citizens can be rationed and bypassed.

Socially sensitive phrases permeate the WEF's self-portrait: "committed to improving the state of the world," "a human face to the global market," "entrepreneurship in the global public interest." "Civil society" also gets a number of mentions on the WEF website,[16] and Schwab himself has spoken of "responsible globality" and of making "prosperity inclusive rather than exclusive.... Chief executives do not come to Davos to learn how to make money.... They come to improve the world." Nonetheless, in reality, there is no democratic element; Schwab told *Fortune* magazine in 1999 that the "sovereign state has become obsolete."[17] Only business, whose leadership is unelected, enjoys representation. The occasional invitation to citizen and labor bodies does not nullify this flaw since the board that controls the organization consists of Schwab himself, corporate chairmen and CEOs, a sprinkling of heads of prestigious academic centers, and, in many years, the head of the IMF.[18] Even if some degree of genuine concern for the public interest pertains, perhaps in the case of Schwab himself, the philanthropy of the WEF is paternalistic.

The difficulties of the 1970s, especially after the 1973 oil crisis, also precipitated the formation of the Group of 5. It consisted of the finance ministers of the US, UK, Japan, France, and Germany, who first met in 1974. Italy and Canada were invited in subsequent years, making up the G7. From 1976, the heads of these G7 governments began to meet annually and the finance ministers less regularly. A representative of the European Community was included from 1977. In 1997, Russia was admitted, to form the G8, which, with something like 15 percent of the world's population, produces 60 percent of world economic output in GDP terms. The members of the organization are the richest countries on earth, and many had been the chief colonial powers in previous centuries.

After the Asian financial crisis of 1997 a broader group, known as the G20, was founded. To the original G8 membership it added the finance ministers and central bank governors of major emerging countries: Brazil, India, and China, as well as South Africa, Mexico, Argentina, Turkey, and Saudi Arabia; also in the G20 are South Korea, Australia, and the EU as a formal member. The global financial crisis of 2008 triggered the elevation of the G20 as the premier economic council of the richest nations, displacing the G8 in that role.[19] Obviously more representative of the world's population (two-thirds) and the world's wealth (80 percent), the G20 nonetheless remains a collection of invited guests; a half dozen excluded countries, including Iran and Spain, have larger economies than some G20 members.[20] The Norwegian foreign minister, Jonas Gahr Støre, pointed out that the group is "self-appointed," or appointed by the major powers of the old G5, and thus lacks legitimacy. Støre argues that such a group undermines the agreed-upon Bretton Woods institutions (the World Bank and the IMF), which in his view should be reformed rather than supplanted.[21]

Other Elite Clubs

Members of the global elite meet in other exclusive clubs. The Bilderberg Group, a private organization initially aimed at toning down anti-American feelings in Europe, first met in 1954. It has met annually ever since and, though a list of its participants—tycoons, bankers, politicians, and a few select academics—is made public, the meetings are held in secret and no text is ever released.[22]

The Trilateral Commission is another such club. It was instigated in 1973 by David Rockefeller, heir to the Standard Oil fortune, president of Chase Manhattan Bank, and a founding member of the Bilderberg Group. The Trilateral Commission brought Japan into an elite club of the Bilderberg kind, alongside the US and Europe, for the first time. The commission's 1975 report, *The Crisis of Democracy*, echoed the fears of Bernays fifty years earlier; it warned of "an excess of democracy" in the US and yearned for the time when "Truman had been able to govern the country with the cooperation of a relatively small number of Wall Street lawyers and bankers" with input from "the law firms, foundations and media which constitute the private establishment."[23] In 1999, Rockefeller told *Newsweek* that the recent trend toward "democracy and market economies ... has lessened the role of government, which is something business people tend to be in favor of. But the other side of that coin is that somebody has to take governments' place, and business seems to me to be a logical entity to do it."[24] In other words, democratic representation can be discarded and government handed over to business.

The Club of Rome

The personnel of groups such as those described above differs little in class and gross economic interest from the members of the Club of Rome, which had formed just a few years before the WEF. All share an invitation-only membership. The elite origins of most of the Club of

Rome's members illuminate the suspicions of the Left (noted in chapter 4).[25] Numerous sources on both left and right have observed a certain amount of overlap between the personnel of the Club of Rome and that of the Bilderberg Group and the Trilateral Commission.[26] Yet in one crucial respect, the Club of Rome differed radically from its contemporaries: Peccei and his colleagues recognized a looming environmental crisis and were concerned enough to address it. Above all, they identified the cascade of environmental damage as a corollary of the growth of industrial prosperity itself. They could be characterized as intellectuals who saw beyond their narrow class interests and attempted, unsuccessfully, to change the course of industrial development. In contrast, most of the CEOs who meet at Davos and the politicians who frequent the G8 and G20 meetings share an unquestioned faith in the continuation of the industrial model of progress and in the notion that the international market will provide solutions to the problems that it has produced.

The World Trade Organization

Perhaps the most radical of all the entities invented since the 1970s, the World Trade Organization (WTO) has become what activists Debi Barker and Jerry Mander describe as a worldwide "invisible government." Built gradually over the entire postwar period, the WTO was officially launched in 1995 with the power to "strike down the domestic laws of its member nations and to compel them to establish new laws that conform to WTO rules."[27] Unlike UN organizations and forums, where environmental accords such as those on biodiversity and climate are wrestled through in conference sessions and always need ratification at home before coming into effect, the WTO gave itself coercive powers. Very few members of the public were aware that their trade negotiators were granting extensive powers to an unelected body working behind closed doors. What has been created at the international level epitomizes the ideology that puts economic priorities before all else and seeks to free corporate self-interest from democratic constraint.

From GATT to WTO: The Coming of "Free Trade"

When the World Bank and the IMF were designed in 1944 at the Bretton Woods meeting, the US government had wanted an international trade organization (ITO) that would work to remove trade barriers as the third part of the new architecture.[28] An ITO charter was drawn up at the outset and, though a generalized world trade organization did not materialize at that point, the US trade representative negotiated a tariff agreement with twenty-seven countries in 1947.[29] Called the General Agreement on Tariffs and Trade (GATT), it came into force on January 1, 1948, and became the foundation on which the WTO was built over the next forty years. The free traders tapped into anticommunist sentiment to advance their case, arguing that free trade would "immeasurably strengthen us and other freedom loving nations" in the struggle against communism. Although it dealt only with tariff reductions on goods and mentioned neither services nor investment, GATT became the vehicle and foundation for the extension of the free trade regime, functioning as rule book and negotiating forum.[30] Eight rounds of negotiations were undertaken over the next fifty years as more countries were gradually persuaded to join.

The eighth round of GATT negotiations (the Uruguay Round) gave birth to the WTO. It began in 1986 with furious lobbying from the representatives of international capital—the WEF, the Bilderberg Group, the Trilateral Commission, the International Chamber of Commerce (ICC), and the European Round Table of Industrialists.[31] In the US, more than five hundred corporate representatives had adviser status with US negotiators, and industry trade groups such as the Chamber of Commerce and the Business Roundtable had direct access to the actual negotiations; public interest NGOs representing environment, social justice issues, consumers, or labor were all excluded, along with the press.[32] Representatives of transnational capital were adamant that the WTO was essential to a bright future, and dark warnings of "chaos and impoverishment" (if it was not established) emanated from the Eminent Persons Group, composed of first world ex-politicians, bankers and trade bureaucrats.[33]

Reflecting the centrality of the US-based corporations in world capitalism, the negotiating positions of the US government and transnational capital have often been closely aligned, especially since World War II, even though the headquarters of TNCs are no longer located exclusively in the US. As Giovanni Arrighi has outlined in detail, the center of world capitalism has shifted over the centuries. Arrighi argues for its beginnings in the alliance between Genoa and Spain, followed by a long period based in Amsterdam, a British era, and the rise of US capital early in the twentieth century. Though it is clear that, if the system persists, the center may shift to China in this century, the US remains at its heart so far. This is evidenced by US power in the institutions outlined in this section—bolstered by the UK, which still harbors one pole of international finance, the City of London.[34]

Before the Uruguay Round, 108 countries had already joined GATT and tariffs had been slashed by 75 percent. The aims of this round were to integrate agriculture and textiles into the tariff-free regime and to set up the WTO, which would formalize an agreement on services. Business organizations in the US and across the world launched a massive lobbying effort to secure the US congressional vote, and when they succeeded, the WTO came into existence in early 1995.

As Sharon Beder points out, "given the thousands of pages of rules that the WTO now presides over, 'free' trade is not about doing away with rules altogether, but rather with replacing rules for companies with rules for governments, and replacing rules that protect consumers and the environment with rules that protect and facilitate traders and investors."[35] This is analogous to the construction of the "free market" in nineteenth-century England by edict of Parliament. The word "free" may be appended to the phrase, but markets and trade are always institutions conducted under sets of rules and regulations, imposed by human will. The freedom referred to here is the freedom of business interests to operate without hindrance.

The free trade narrative held that trade unimpeded by tariffs and other barriers would produce economic growth and cause all nations to prosper. Yet, over the past sixty years, the greatest rates of economic growth were seen before 1974.[36] The average annual growth of GDP per capita in the twenty years before 1980 is shown in the dark gray bar of figure 13.1,[37] outshining anything that came later. Since much higher tariffs predominated at that time, growth is obviously not determined by unrestricted trade alone.

Disputes and Environmental Protection

The environmental standards put in place in the late 1960s and early 1970s were never acceptable to a business class that desired free rein,

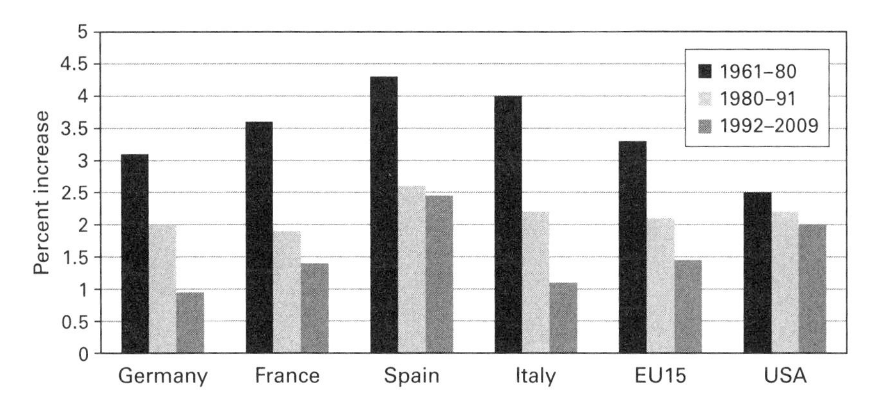

Figure 13.1
Average Growth of GDP per capita, EU and US, 1961–2009.
Source: Tridico 2011, 27. Data from Eurostat. Courtesy of Pasquale Tridico.

and the WTO has become a significant weapon in the hands of the corporate opposition to these and other regulations. As noted in chapter 2, the WTO views environmental regulation through the lens of trade. Its coercive apparatus enforces thousands of pages of rules dedicated to the compulsory pursuit of free trade. In the process, countries are stripped of the right to make democratic decisions that conflict with these rules. The central institution in the WTO's coercive capacity is the dispute panel, which settles conflicts in secret. Panel members are "trade bureaucrats, usually corporate lawyers," with no particular expertise in specific scientific issues or specific countries, no process for avoiding conflicts of interest, and no forum for review or appeal.[38] The panels are hardly impartial arbiters of the issues. The food safety standards observed by WTO panels, for example, are largely written by the food industry—and are far weaker than US standards.[39]

When a WTO dispute panel rules against a country, the country must either change its domestic laws, pay penalties representing "lost profits" to the aggrieved corporation, or face unilateral trade sanctions.[40] The US was obliged to weaken its air pollution legislation when the WTO ruled that it could not exclude petroleum of poorer quality imported from Mexico and Venezuela.[41] Japan has been obliged to accept more pesticide residues in food than its own regulations demanded.[42] In the dispute between Europe and the US over growth-promoting hormones in beef, the WTO panel found against the Europeans. (The dispute continues, however, as Europe has produced evidence showing that at least one of the hormones in dispute, estradiol 17β, is connected to an increased risk of cancer, and there is as yet insufficient evidence to determine the risk from five others.) The WTO reverses the burden of proof and requires objectors to prove harm rather than industry to prove safety.[43] Europe, on the other hand, applies the "precautionary principle" whereby substances are not permitted until the product is demonstrated to be safe on the basis of reliable scientific assessment of risk.[44]

A 1991 GATT panel also ruled against the US ban on imports of tuna caught with collateral slaughter of dolphins. The WTO's own report on this case states:

> What was the reasoning behind this ruling? If the US arguments were accepted, then any country could ban imports of a product from another country *merely because the exporting country has different environmental, health and social policies from its own*. This would create a virtually open-ended route for any country to apply trade restrictions unilaterally—and to do so not just to enforce its own laws domestically, but to impose its own standards on other countries. The door would be opened to a possible flood of protectionist abuses [emphasis mine].[45]

Here the WTO states baldly that trade has priority over environmental, health, and social justice considerations, regardless of the wishes of a government and the people it represents. To enforce trade obligations, the rules penalize countries if they choose to assess risk and protect citizens under their own standards.

More Than Trade

Capital began seeking greater global mobility from the 1960s on (discussed in chapter 6), something financial corporations had always favored—though early globalization was interrupted by World War I, the 1929 crash, and the financial regulation that came with the New Deal. Cross-border financial flows began to increase in the 1970s and exploded in the 1980s with the deregulation of domestic banking in many countries and the beginnings of electronic trading. By the end of 1991, annual international financial flows had increased to almost $200 trillion, fifty times greater than the volume of actual trade in concrete goods and services.[46]

Even mainstream economists acknowledge that such extreme capital mobility can have a destabilizing effect—for example in the East Asian financial crisis of 1997. The 50:1 ratio between money flows and actual goods gives some indication of the degree of speculation embodied in the activities of Thomas Friedman's "electronic herd," the traders tapping away on the computer terminals of international finance. The daily tsunami of speculative capital also reflects the financialization outlined by Woolley and Phillips, where the financial sector expanded its share of corporate profits from 10 percent in the 1950s to almost 40 percent by 2004.[47] The financial sector does not make anything tangible; theoretically, it exists to serve the funding requirements of productive enterprise. Nevertheless, financial institutions came to dominate the US, UK, and global economy in the last decades of the twentieth century, diverting capital from productive investment into the arcane arena of "innovative" but opaque financial "products," with an almost inconceivable face value of $640 trillion in September 2008—fourteen times the GDP of all the countries on earth.[48] It was claimed that these products would disperse credit risk and ensure resiliency, in line with the sanguine idea that a new business era of "the great moderation" had been engineered by the neoliberal seers.[49]

In reality, the entire financial system was on quicksand. Whether the crisis has been resolved remains to be seen (at the time of writing in 2013), since taxpayer-funded bailouts have transferred unpayable liabilities to nations and created or exacerbated immense sovereign debt, not just in countries such as Ireland and Greece but also in the US itself. Immense debts still underpin the entire financial system, and business confidence appears to remain dependent on the indefinite injection of US dollars via "quantitative easing."

Clearly, financial deregulation played a crucial role in this chain of events. In the words of the Indian economist Prabhat Patnaik, it detached the local "financial sector from its anchorage in the domestic economy to make it part of the international financial sector; to make it operate according to the dictates of the market which means the end of ... the distinction between productive and speculative credit needs; and to remove it from the ambit of accountability to the people."[50]

A return to greater financial regulation seems logical and necessary to militate against a further financial collapse. However, notwithstanding the trillions of dollars' worth of assistance they have received from US and other first world taxpayers, multinational bankers and financiers are unenthusiastic.[51] Now that transnational finance has become securely established, it is extremely difficult for national governments to exert discipline on it. To gales of objection from French financiers, President Sarkozy announced that France would unilaterally implement what is known as a Tobin (or "Robin Hood") tax, which would shave a minuscule amount from every financial transaction. The governments, banks, and business organizations of the US and UK, which are home to two of the foremost centers of transnational capital, London and New York, have declined to support anything of the kind.[52]

The agreements forged in the WTO at the peak of the deregulation drive during the 1990s are also implacably opposed to all regulation. WTO rules are, in fact, intended to be impossible for national governments to reverse. As US Treasury official Barry Newman told Congress in relation to the similar financial services provisions of the North American Free Trade Agreement (NAFTA): "Future Mexican governments may change and they may not have the same attitudes of the current government. The benefit of NAFTA is that it will lock [them] into an internationally legally binding and enforceable agreement."[53] The antiregulatory ideology of the neoliberal intelligentsia and the business interests it serves has been written into binding "free trade" contracts that democratic governments will struggle to repeal. UN treaties on climate or biodiversity, even though they require legislative ratification at home before implementation, are rejected by US think tanks on the grounds that they encroach on national sovereignty; the same entities ignore the very real curtailment of sovereign rights actually enforced by the free trade agreements and the WTO.

While the WTO operated as GATT before 1995, its central objectives revolved around reducing tariffs on concrete material goods. From the 1980s, however, the International Chamber of Commerce was already demanding new investment rules aimed at free movement for capital around the world.[54] The OECD attempted a comprehensive deregulation of foreign investment flows in the late 1990s when it launched the Multilateral Agreement on Investment (MAI), but the MAI failed in 1998 in the wake of an unprecedented groundswell of opposition from consumer, environmental, and labor groups and the defection of France. Despite such setbacks, the advent of free trade in capital and services would soon be upon us.

Once the WTO came into force at the end of the Uruguay Round, additional "trade-related" areas were added to the trade agenda, in particular services such as investment. The WTO's General Agreement on Trade in Services (GATS) launched a process of investment liberalization, and its Financial Services Agreement (FSA), adopted in 1999, has already imposed rules that limit national options to regulate. The "standstill" provision in the FSA forbids any limitations not already specified—no rollbacks permitted; Article B7, which permits foreign bankers to provide "any new financial service," prevents regulation of opaque or risky derivatives; Article B10 binds signatories to "remove or limit" existing measures that are adverse for foreign investors.[55] One of the "basic principles" of the GATS agenda is "progressive liberalization," now part of the Doha Round, which followed the Uruguay Round in 2001, after delays caused partly by vigorous opposition in Seattle. The goal of the Doha Round is to extend the scope of free investment flows so that governments will have no rights at all to regulate the entry, behavior, and operations of foreign-based corporations. The Doha Round is incomplete at the time of writing but is being implemented piecemeal in bilateral and multilateral trade agreements such as the Trans-Pacific Partnership and the Transatlantic Trade and Investment Partnership. In essence, the neoliberal agenda is being cast in iron: according to the WTO itself, "because 'unbinding' is difficult, the commitments are virtually guaranteed."[56]

This spells a particularly dismal outcome for developing countries should they wish to foster stability by restricting the stampedes of capital in and out of their countries, or by nationalizing services. A UN panel on financial reform, chaired by Joseph Stiglitz, pointed to numerous problems with the free trade regime. Any nationalization of services such as banking would be likely to incur compensation penalties. Capital controls are forbidden. The UN panel of experts recommended that "agreements that restrict a country's ability to revise its regulatory regime—including not only domestic prudential but, crucially, capital account regulations—obviously have to be altered, in light of what has been learned about deficiencies in this crisis."[57] In other words, they were warning that WTO rules prohibiting national capital controls are detrimental and should be dropped.

Despite the role of the bloated financial industry in the near collapse of the global capital market in 2008, the industry has not been reregulated, nor have its ideological preferences been widely challenged. It had already enshrined these preferences in WTO agreements and continues to enjoy determinate influence in the corridors of government, especially in the English-speaking world. The G20 and President Obama have attempted reregulation of finance with, at best, partial success. G20 meetings have canvassed the issue, made proposals, and promised action, but have achieved little so far.[58] The US legislation, passed by Congress in July 2010, was drafted with thousands of lobbyists in attendance on behalf of banks, hedge funds, and organizations such as the Chamber of Commerce and the Business Roundtable.[59] While it offers some consumer protections, the new law does not reinstate the firewall between commercial and investment banking; trading of the derivatives that played a key role in the collapse, though subjected to some disclosure, will still not be conducted openly in the same manner as share trading; and corporate size will not be limited to avoid the risk of being "too big to fail"—or "too big to bail," as economist Max Fraad Wolff put it in an interview with Amy Goodman.[60]

From the time of the economic crisis of the 1970s, the construction of an alternative business-friendly and business-funded intelligentsia in a host of think tanks enabled the precepts of neoliberalism to flourish in individual countries. Corporate leaders simultaneously established international lobbies and institutions to serve the same interests, often bypassing regulatory measures enacted by democratic national governments. By the new century, business priorities were entrenched in public discourse, government policy, and international institutions. Economic growth was established almost everywhere as the only way to solve anything. Environmental protection and social justice, both national and worldwide, were deemed to depend on it. But while these are the widespread beliefs, reality lies closer to the reservations expressed forty years ago by the MIT team.

In chapter 14, I return to the MIT researchers' projections and the way their work was distorted to create the prevailing popular view of them as sloppy peddlers of doom. I compare their actual output against recent statistical analysis, to check the plausibility of the much-repeated claim that the Club of Rome got it wrong.

IV

In Conclusion

A sustainable society would be interested in qualitative development, not physical expansion. It would use material growth as a considered tool, not a perpetual mandate. It would be neither for nor against growth.... Before this society would decide on any specific growth proposal, it would ask what the growth was for, and who would benefit, and what it would cost, and how long it would last, and whether it could be accommodated by the sources and sinks of the planet.

—Donella Meadows, Dennis Meadows, and Jørgen Randers, 1992

14

The Limits to Growth after Forty Years

Nowhere in the book was there any mention of running out of anything by 2000. ... The Club of Rome got the whole picture right. It was the rest of us who missed the mark!

—Matthew Simmons, 2000

When *The Limits to Growth* first appeared, its message was taken seriously for several years. However, the efforts of the US presidents Nixon and Carter and the Canadian prime minister Pierre Trudeau died away with their tenure in office. Carter's *Global 2000* was ignored by President Reagan and played no part in anyone's planning for the future. The UN environment agencies that were founded in the 1970s, though they have since assembled immense amounts of data and plugged away with conferences, projects, and reports, were gradually overshadowed by the trade-oriented institutions of the neoliberal "revolution" that took hold in the 1980s. More growth was confidently adopted as the solution to the problems growth had generated, while decision making about the means to that end was removed from democratic institutions and handed over to the "free market."

In the scientific community, concern did not abate. The team at MIT went on to publish *Beyond the Limits* in 1992, arguing that the human impact on the natural world was by then exceeding the earth's replacement abilities. In 2004, they published *Limits to Growth: The 30-Year Update,* which continued to argue that the general shape of their thesis remained correct. Each of the UNEP GEO reports warned of a deepening crisis, and GEO-4 suggested that numerous geophysical processes were approaching irreversible "tipping points." During the 1990s the human ecologist William Rees and his doctoral student Mathis Wackernagel, an engineer, developed ecological footprint theory, which aimed to quantify the precise extent to which economic processes are extracting more than the natural world is renewing. Their findings underscored the message of the MIT team.

Ronald Bailey's Attack on *Limits to Growth*

As outlined in part III of this book, think tanks have played a significant role in campaigning against environmental concern, and this has included direct attacks on the idea of geophysical limits. In the first decade after its publication, fierce criticism of *The Limits to Growth*, largely by economists, was accompanied by widespread interest and emulation, with record sales and the commissioning of parallel studies by President Carter and Prime Minister Trudeau. By the early 1980s, however, Julian Simon, associated with the libertarian think tank the Cato Institute, was prominent in amplifying the initial attacks on the concept of limits. The Cato Institute published three of Simon's books and dozens of his articles. Simon saw the human intellect as the "ultimate resource." He told interviewer William Buckley that "in the end, copper and oil come out of our minds."[1] Simon saw no need to limit the growth of population or production and, like Kaysen and the other economists discussed in part I, believed resources for expansion to be infinite.

It appears to have been Ronald Bailey who initiated the specific distortions of the *Limits* work that furnished the popularly accepted formula for contempt. Not a scientist, Bailey was a science and technology reporter for *Forbes* magazine from 1987 to 1990 and has been a science reporter for the libertarian magazine *Reason* since 1997. Like Simon, Bailey was closely associated with the Cato Institute, but it was another libertarian think tank, the Competitive Enterprise Institute (CEI), that appointed him its Warren T. Brookes Fellow in Environmental Journalism in 1993 and published several of his antienvironmental books. Bailey is still a fellow with the Cato Institute and an adjunct analyst at the CEI.[2]

In 1989, Bailey launched his attack on *Limits* and the Club of Rome with "Dr. Doom," a scathing portrait of Jay Forrester, in *Forbes*. Forrester was a pioneer of the new discipline of systems dynamics and the architect of the initial models from which World3, the computer model used by the *Limits* researchers, was developed. In the same issue, Bailey attacked Bill McKibben's new book, *The End of Nature*, as "garbage" in an article titled "Hi There, Bambi." The titles of these articles point to the contrast between the advocacy writings of Bailey and similar writers and the peer-reviewed work of serious scholars.[3] Bailey is emblematic of many opponents of *Limits*: a think tank fellow and a publicist for the free market viewpoint rather than an evidence-based analyst. Astonishingly, his distortions were adopted by many more reputable economists.

The Meadows team at MIT published their first sequel, *Beyond the Limits*, in May 1992, just before the Rio Earth Summit in June.[4] Bailey published *Eco-scam* in February of the following year, devoting a chapter to the question of geophysical limits. Here he distorts what Meadows and colleagues had claimed in 1972 about resource constraints and establishes the version that has become the widely accepted reading. According to Bailey, "they have been proven spectacularly wrong. In 1972, [they] predicted that at exponential growth rates the world would run out of gold by 1981 ... petroleum by 1992 and copper, lead, and natural gas by 1993."[5] None of this was claimed by the Meadows team, but the precise error has been repeated endlessly by many commentators, including the writers of UNEP's *Global Environment Outlook 3* (GEO-3) report.[6]

Bailey had engaged in sleight of hand. The figures he quotes in *Eco-scam*[7] are based on column 5 of table 4 of *The Limits to Growth*, "Nonrenewable Natural Resources." This contents of this table did not form part of the World3 model and did not contribute to the trends discerned from World3 output; its aim was to demonstrate how exponential growth in demand quickly diminishes supply (see table 14.1 for representative entries from the original table 4).[8] Column 5 states the years taken to use the *known reserves* of key resources if continued average growth in demand is assumed. The figures are those of the US Bureau of Mines. Bailey ignores the subsequent column, which shows how, in an exponentially growing system, a fivefold increase in the resource base does not multiply years of supply by five. This column, too, is merely illustrative of the unexpected effects of exponential growth and makes no claim the world will "run out" of anything. The figures in the table are not, in any case, part of the World3 model.

Bailey also ignores the overall analysis of potential resource scarcity expressed a few pages further on: "*Given present resource consumption*

Table 14.1
Selected nonrenewable natural resources

1	2	3	4	5	6
Resource	Known global reserves[a]	Static index (years)[b]	Projected rate of growth (% per year)[c]	Exponential index (years)[d]	Exponential index, 5 times known resources (years)[e]
Aluminum[f]	1.17×10^9 tons	100	5.1–7.7	31	55
Coal	5×10^{12} tons	2300	3–5.3	111	150
Copper	308×10^6 tons	36	3.4–5.8	21	48
Gold	353×10^6 troy oz	11	3.4–4.8	9	29
Iron	1×10^{11} tons	240	1.3–2.3	93	173
Lead	91×10^6 tons	26	1.7–2.4	21	64
Natural gas	1.14×10^{15} cubic feet	38	3.9–5.5	22	49
Petroleum	455×10^9 bbls	31	2.9–4.9	20	50
Tin	4.3×10^6 lg tons	17	0–2.3	15	61

Source: After Meadows et al. 1972, 66–68. Courtesy of Dennis Meadows.

a. *Source:* US Bureau of Mines, *Mineral Facts and Problems*, 1970.

b. The number of years known global reserves will last at the current global consumption rate. The calculation is made by dividing known reserves (column 2) by current annual consumption.

c. *Source:* US Bureau of Mines, *Mineral Facts and Problems*, 1970. The average is used to calculate columns 4 and 5.

d. The number of years known global reserves will last with consumption growing exponentially at the average annual rate of growth.

e. The number of years that five times known global reserves will last with consumption growing exponentially at the average annual rate of growth.

f. Bauxite expressed as aluminum equivalent.

rates, the great majority of the currently important non-renewable resources will be extremely costly 100 years from now" (emphasis in text).[9] This refers to the year 2072 and, in the second decade of the century, seems a plausible suggestion. None of the *Limits* books warns of resource exhaustion by the end of the twentieth century, or even at any specific date in the twenty-first. All of them offer a range of options about what society might do—from business as usual to strategies involving different degrees of "a deliberate turnaround, a correction, a careful easing down."[10] Various scenarios were run through the computer to reflect these different options. "It is possible to alter these growth trends and to establish a condition of ecological and economic stability," the *Limits* authors wrote in the introduction to the first book.[11] Although they were very clear that such options existed, it appears that most of their critics thought any interruption or reduction in economic growth was out of the question. Thus, models with scenarios that moderated rates of growth were sometimes totally ignored, as in an early review in the *New York Times Book Review*, which claims that the *Limits* researchers "jigger the assumptions just enough to eliminate non-catastrophic possibilities" to ensure that the world economy "obligingly" collapses.[12] This review's reference to a "false inevitability of doom" is based on ignoring the scenarios that do not lead to collapse; these are the subject of the entire fifth chapter, which discusses the models that yield a stabilized world. In a 2007 book chapter, Dennis Meadows notes that this is only one among many false and misleading statements made by the reviewers and suggests that they did not finish the book. Meadows enumerates several other false statements, including the claim that *Limits* states reserves of vital materials will be exhausted within forty years, an error similar to Bailey's. Meadows concludes that the reviewers were engaged in deliberate misrepresentation.[13]

Bailey's claims were widely adopted. The *Economist*, for example, stated that "the Club of Rome ... said total global oil reserves amounted to 550 billion barrels ... [and] made similarly wrong predictions about natural gas, silver, tin, uranium, aluminium, copper, lead and zinc."[14] The figure 550 billion barrels of oil does not appear at all in table 4, where the figure for *known reserves* (not total reserves) is 455 billion, a figure supplied by the US Bureau of Mines in any case, not the Meadows team (see table 14.1).[15] As far as I can ascertain, the *Economist* was responsible for this specific mistake. An article in *Foreign Affairs* lampooned *Limits* as part of "the 'sky-is-falling' school of oil forecasting [which] has been systematically wrong for more than a generation," and also reproduced the *Economist*'s error.[16] The British journalist Matt Ridley quoted the *Economist* verbatim in his Prince Philip Lecture of 2001.[17] Soon after, the Australian economist Chris Murphy remarked in an interview on ABC television's *Four Corners* that "they were claiming by now we would have run out of aluminium and we would've run out of natural gas about 10 years ago."[18] Another well-respected Australian economist, John Quiggin, claimed that "the Club of Rome model … predicted that reserves of most minerals would be exhausted before 2000."[19] These generally reputable but mistaken commentators might have misread the table in *Limits* themselves, or perhaps they were simply reproducing Bailey's error, having encountered his claims in the vast think tank echo chamber.

Throughout his chapter on what he calls "the depletion myth," Bailey makes much of the falling prices of commodities and food, as did his colleague, Julian Simon, and most of the other economists who disparaged the *Limits* work. Most rely on the seminal 1963 research by economists Barnett and Morse,[20] which found that resources became ever cheaper through the twentieth century, as their extraction costs in capital and labor terms declined; this was due, they thought, to technological advance. Simon and others who denigrated the limits work were confident that resources would remain cheap as technology continued to improve. The energy analyst and geographer Cutler Cleveland, however, has argued that Barnett and Morse overlooked the role of energy in general, and cheap oil in particular, during the period of their study (1900–1960). For Cleveland, cheap energy is the crucial factor in the declining price of resources since all resource extraction depends on the application of energy; energy is not just one resource among many but the "master resource" essential to the recovery and production of every other commodity. Cleveland's view is consistent with John McNeill's thesis that cheap and plentiful fossil fuels, especially cheap oil, underpin the unprecedented economic growth of the twentieth century.[21]

Other than the spike in the oil price during the 1970s, energy prices remained flat until the early twenty-first century. So too did those of key minerals. The belief in never-ending cheap resources appeared plausible.

But from 2003 on, commodity prices began to soar—in five years the price of copper tripled and that of zinc doubled.[22] Despite the downturn that followed the global financial collapse of 2008, the upward trend in prices of food, oil, and several other commodities had resumed by 2010. Economists insist that price signals "solve" all scarcities: as the price goes up, new options become affordable and substitutes are developed. The evidence regarding prices in the early twenty-first century indicates that they are at best an imperfect and tardy method of detecting scarcity. Until a resource is close to actual depletion, market forces do not recognize its decline.

In the second half of the 1990s, the petroleum geologist Colin Campbell began publishing analyses of world oil production that indicated it was likely to "peak" sometime in the first decade of the new century.[23] Not equivalent to running out, the term peak oil indicates that production has reached a level beyond which it can no longer expand. Notwithstanding drilling in deep water on the edge of technological feasibility and rash plans for the Arctic, the production of conventional petroleum reached a plateau in recent years and seems unlikely to rise significantly in the future. Though the OECD's International Energy Agency has resisted the idea of peak oil for many years, Fatih Birol, its chief economist, conceded in 2011 that crude oil production had peaked in 2006.[24] Campbell and Laherrère's figures show that the *discovery* of liquid petroleum peaked in 1964.[25] Increases in oil production in the second decade of the twenty-first century will depend on less accessible sources, a reality that also exacerbates global warming since emissions from the application of energy during production must be added to emissions when the oil is burned.

The recent US boom in shale gas and shale oil,[26] for example, captures hydrocarbons from source rocks rather than from reservoirs where the material has migrated over millions of years. It is no longer a matter of drilling into a cavity and recovering the liquid or gas, which gushed out under pressure in early wells. Instead, extraction relies on the newly developed technology of large-scale multistage hydraulic fracturing (fracking) of the rock, through wells drilled horizontally. Energy and capital have to be applied continually to keep the shale producing. Although there is considerable exuberance about "the end of peak oil," "a new age of plenty," and even "energy independence" for the United

States,[27] outcomes are unlikely to be so positive. Production from fractured shale falls away dramatically over the first and second years, demanding the ongoing drilling of new wells. Major new investment is constantly required to do this, so that the price required for profitable production is high and likely to rise further as the best well sites are exhausted.[28]

In addition to these difficulties, opposition on environmental and social grounds is increasing. The multiplicity of wellheads have to be connected by a lattice of pipelines and roads, which transforms rural or urban landscapes into industrial ones. The process also requires the injection of massive quantities of water and sand and generates a great amount of polluted wastewater which has to be disposed of. Water supply is already stressed in almost half of the areas where shale energy development is most intense. Although it is denied by industry sources, the contamination of groundwater has been linked to fracking in several studies (including one by a team of researchers at Duke University and two by the EPA).[29] The level of methane leakage from the whole production and distribution cycle for shale gas is also unsettled. Researchers at Cornell University have found that, because of methane leakage, greenhouse gas emissions exceed those from conventional gas and oil and are comparable with—or greater than—those from coal. Unburned, methane has far greater greenhouse potential than CO_2.[30] Although shale gas is argued to be a temporary answer to the decline in cheap oil and conventional gas, perhaps enough to buy time if environmental side effects can be controlled, the fact that the industry is predicated on releasing hydrocarbons from solid rock is emblematic of ever-increasing degrees of difficulty or, as some might say, desperation.

Bailey's claims about *Limits to Growth* have been widely repeated, from the think tank extremes to the economic mainstream. All ignore the fact that precise prediction was never the aim of the *Limits* project, even though the media and most economists focused on dates and figures. The idea of *Limits to Growth* was to alert an apparently unconscious world to the longer-term consequences of exponential growth as the scale of the human enterprise ballooned. The Meadows team avoided exact dates and quantities and kept many axes on their graphs nonspecific as to timing. "In terms of exact predictions, the output is not meaningful" the original text stated,[31] and Dennis Meadows later described the purpose in this way: "We used it [World3] to determine the main behavioral tendencies of the global system."[32] This was a study of process and trend. The object was a broad understanding of the way the global economic system unfolds, not a set of exact forecasts. Though the MIT team reiterated this throughout their work and both the subsequent updates, critics rarely acknowledged these objectives, or perhaps did not even understand them. The critiques arising from or allied to the Bailey attack are all based on a misunderstanding of the purpose and research strategy of the Meadows team and an incorrect reading of one table.

How Close to the Mark?

Into the Future: Testing the Model

It was always my purpose to ask, as I drew to the end of this book, whether the projections made by the Meadows team in 1972 approximated actual conditions observable decades later. In fact, this question has been pursued by several investigators.

Graham Turner of the Sustainable Ecosystems Unit at Australia's Commonwealth Scientific and Industrial Research Organisation (CSIRO) compared the output generated by various runs of the World3 model in the *Limits to Growth* work with readily available public statistics for the period 1970–2000. Turner stresses the way the Meadows team was interested in process and trend rather than precise prediction, as noted above. Turner also emphasizes the crucial role of feedback loops and delayed effects in the behavior of the planetary system, and reflects on the distortions circulated by Bailey.[33]

But Turner's main purpose was to compare actual data from the real world for the period 1970–2000 with the data generated in three of the key scenarios run by the *Limits* team. The team ran the World3 program many times using differing assumptions about several key variables: population, services per capita (health and education), food per capita, industrial output per capita, consumption of nonrenewable natural resources, and persistent pollution. Turner examined the output from three of these runs: the "standard run," reflecting business as usual, where rates of growth, industrialization, and resource use continued on their 1970 trajectory; the "comprehensive technology" run, modeling what would occur if huge improvements in agricultural productivity, resource extraction, pollution mitigation, and birth control were achieved; and the "stabilized world" run, in which population and capital investment were stabilized alongside significant technological advance.[34]

The standard run of the World3 model resembled the world that actually transpired. For net population, Turner found that actual data for 1970–2000 agree closely with the standard run.[35] When he used electricity supply as his proxy for services per capita, the real-world data were again very close to the standard run; when he used literacy rates as the proxy for services, the standard run was more optimistic than reality. Food per capita has been slightly better than the standard run, though broadly similar, while industrial output per capita is virtually identical to the standard run.

In dealing with nonrenewable resources, Turner examined a range of data sets to reflect the ongoing uncertainty surrounding the future availability of energy resources in particular. If more speculative technologies such as nuclear fusion and effective carbon capture do not come into play and coal burning is limited owing to greenhouse concerns, the real world tracks World3's standard run fairly closely, though depletion in the real world is slightly slower than in the model. If much of the remaining coal is burned, the data approximate the resources output for the "comprehensive technology" run. For persistent pollution data, Turner used CO_2 emissions as a proxy since data on such pollutants as heavy metals, radioactive wastes, and organic pollutants were inadequate and insufficiently available as aggregated global totals. The actual rise in CO_2 emissions was somewhat lower than persistent pollution in the standard run.

In general, the real-world data do not match the "stabilized world" run at all, which is hardly surprising, since ideas about biophysical limits have been ignored in most policymaking since 1980. The "comprehensive technology" run is also shown to be wide of the mark, being much more optimistic than the observed data in most cases. But the standard, or business as usual, model (which trends toward collapse in the middle of the twenty-first century) is a close match, yielding outcomes that tally well with what has actually occurred. This result is compelling, in light of the complexity of the feedbacks between sectors that are incorporated into the model. Turner believes that the close correlation between the real world and the standard run supports the conclusion "that the global system is on an unsustainable trajectory unless there is substantial and rapid reduction in consumptive behaviour, in combination with technological progress."[36]

Turner also points out that the *Jevons paradox* or "rebound effect" is at the root of the ambiguous benefits brought to the system by technical efficiencies: significant gains in efficiency do not moderate consumption but rather facilitate expansion. Although such gains should theoretically reduce impacts, this does not happen in practice, something William Jevons, one of the pioneers of neoclassical economics, observed in the middle of the nineteenth century. For example, the carbon intensity of industrial production has declined for almost a century, while the rate of carbon emissions has continued to grow exponentially. Indeed, it is arguable that there is no real paradox here. As engineer Michael Huesemann notes, "technological innovation has never been used to stabilize the size of the economy; [its] main role has always been exactly the opposite, namely the enhancement of productivity, consumption, and economic growth."[37]

The systems ecologists Charles Hall and John Day have also compared the standard run with actual 2008 data and, like Turner, have found that the model's output matches population and industrial output per capita in 2008; for resources, they looked at specific resources such as copper, oil, soil, and fish and found that the actual data in 2008 were very close to the model's predicted values in the early twenty-first century. For pollution, they looked at CO_2 and nitrogen and, again like Turner, found levels that were close to, though somewhat less than, the standard run. Despite the prevailing perception of the abject failure of the *Limits* work, Hall and Day point out that, whatever occurs later, the model's performance has not been invalidated so far. "We are not aware," they write, "of any model made by economists that is accurate over such a long time span."[38]

Hall and Day also explore reasons for the decline in resource prices through the twentieth century. Like McNeill and Cleveland, they attribute the long run of cheap resources to the availability of cheap energy rather than to "technology," as economists tend to believe. In any case, "technology does not work for free," they argue, giving the example of US agriculture, where up to ten calories of petroleum are used to generate every one calorie of food. Since energy is the master resource on which everything depends, wealth is produced with huge amounts of oil and other fuels—to the extent that everyone in the developed world has the energy equivalent of somewhere between thirty and sixty "slaves" working all day, every day. In these circumstances, cheap fuel is of the essence. Hall and Day argue that energy return on investment (EROI) is more critical than price. EROI is a measure of how much energy is produced by the application of a given unit of energy invested. While the ratio between energy produced and energy expended for new oil in the United States stood at 100:1 in 1930, 40:1 in 1970, and 14:1 in 2000, they suggest that this ratio for oil production will approach 1:1 within a few decades. Hall and Day end their paper with a plea for the teaching of economics from a biophysical as well as a social perspective: "The concept of the possibility of a huge multifaceted failure of some substantial part of industrial civilization is so completely outside the understanding of our leaders that we are almost totally unprepared for it."[39]

Tim Jackson and the Myth of Decoupling

Ecological economists worldwide have continued to warn that indefinite economic growth is neither feasible in geophysical terms nor automatically beneficial to human beings. (For a brief account of some of the ecological economists and allied critics of economic growth working today, see the appendix). Tim Jackson's work for the UK's Sustainable Development Commission is one of many such challenges to the endemic valorization of growth and is particularly noteworthy for its critique of "decoupling," in which technical efficiencies are thought to solve resource and ecological constraints by enabling growth to continue while decreasing the level of resource inputs and pollution outputs. Ambiguities in the role of technical improvement are already involved in the Jevons paradox, however, where efficiency breeds expansion rather than reduced impacts.

Jackson asks whether a strategy of "growth with decoupling" could, in reality, deliver ever-increasing incomes for a world of nine billion people.[40] He notes that the amount of primary energy used to produce every unit of economic output (its energy intensity) has indeed declined quite steeply—by around a third—in the past thirty years or so, and even more sharply in the United States and the UK. Material intensities more generally have also been reduced in the advanced economies, and emissions intensities have followed suit in most cases. Global carbon intensity, for example, fell almost a quarter, from about one kilogram per dollar to just under 770 grams. This "relative decoupling" has been a long-term feature of advancing industrial economies and perhaps underlies the technological optimism so prevalent in mainstream economics. For about thirty years we have, in fact, been doing more with the same amount of inputs. Gross material throughputs, however, continued to increase, even in advanced economies, neatly demonstrating the Jevons paradox. More troubling, global downward trends in energy intensity have reversed since 2009 and begun to rise again. For some materials (iron ore, bauxite, copper, nickel), extraction is now rising faster than GDP, and cement production is growing some 70 percent faster than GDP.[41]

Jackson uses Ehrlich and Holdren's equation, I = PAT (Impact = Population × Affluence × Technology; see box 3.1), to explore the elements that constitute the human impact on nature. Jackson points out that it is the overall impact (I) that must ultimately be stabilized or reduced. He goes on to ask what amount of technological innovation (T) will be necessary to counter the continuing growth of populations (P) combined with the accelerating growth of per capita income, or affluence (A). While the standard argument in neoclassical economics holds that technology will create the space needed, Jackson, focusing on carbon emissions, observes that increasing population and affluence will, under business as usual, increase CO_2 emissions by 80 percent by 2050, leading to atmospheric concentrations far beyond those considered tolerable, let alone safe. To neutralize this trend, T would have to improve by approximately 7 percent per year—ten times faster than at present. Even that unlikely level of innovation does not address global equity problems, so performance would need to be better than this, and a further caveat is involved in the recalcitrant population curve—the UN's latest projected peak has risen by nearly two billion, to 10.9 billion.[42] For Jackson, this is the crux of the problem:

> The scale of improvement required is daunting. In a world of nine billion people, all aspiring to a level of income commensurate with 2% growth on the average EU income today, carbon intensit[y] ... would have to fall ... 16 times faster than it has done since 1990. By 2050, the global carbon intensity would need to be only six grams per dollar of output, almost 130 times lower than it is today.[43]

While we cannot rule out an unexpected and truly massive technological breakthrough, our current progress is hardly encouraging. In the light of the global inequities that make some degree of growth essential for large areas of the world, Jackson concludes that there can be "no credible, socially just, ecologically sustainable scenario of continually growing incomes for a world of nine billion people."

Michael Huesemann addresses these same issues and argues that it is the root causes of unsustainable behavior that must be addressed, namely, "our society's obsession with economic growth … driven by an excessive desire for affluence (A) and a lack of limits on population (P)." These, he points out, are not technological but social and ethical issues and will not be affected by ecoefficiency, which, he believes, will only postpone a "socially and economically disruptive day of reckoning." Pursuing technological solutions is, in his view, a confusion of means and ends and, while the goal remains economic growth, improvements in ecoefficiency merely promote this end and cannot generate its opposite.[44]

As far as its ability to indicate likely outcomes goes, the *Limits to Growth* standard run—where the business as usual scenario was modeled—closely matches the situation in the real world today. But though the performance of the World3 model far exceeds that of any known economic modeling over such periods of time, the message remains marginal to real-world policymaking.

15

Conclusion: The Planet and the Pie

As the historian Dipesh Chakrabarty argues, what is new about the pursuit of the study of history in the twenty-first century is the need to address the intersection between natural history and human history, something we have never really faced on a global scale before.[1] The key to this need for alignment and mutual enlightenment between natural and human history is the concept of scale, an insight brought to prominence by the ecological economists. Herman Daly and his colleagues perceived that the scale of the human project in relation to the scale of the planet had reached a new ratio where humans are no longer inconsequential.

Especially since World War II, the human project has altered—and continues to alter—the actual physical condition of the earth. Phosphorus and nitrogen cycles are being transformed, nitrogen very radically.[2] Emissions of CO_2 are not just warming the planet but acidifying the oceans, which jeopardizes corals and animals with carbonate-based shells, including much of the plankton at the base of the marine food chain. By 1986, human beings were already appropriating around 40 percent of the planet's photosynthetic product, leaving relatively little for all other species combined.[3] Rates of species extinction are estimated at 100 to 1,000 times preindustrial levels. Extraction of the earth's water for human use is close to the maximum in many places, with one quarter of rivers worldwide no longer reaching the sea; groundwater has been exhausted or grossly depleted in many others.[4]

While deniers of ecological crisis like to argue that notions of human impacts on the geophysical scale are laughable, this attitude reveals an ignorance of natural history. It is scientifically uncontested that humble cyanobacteria microscopically producing oxygen over two or three eons created an oxygen-rich atmosphere suitable for complex life, including ours. If algae can have planetary impacts—expressed very slowly, but unquestionably a geophysical force—big animals such as humans are obviously in a position to change the planet rather faster.

Conspicuous environmental problems accompanied the immense economic growth of the postwar world, a growth that depended on an unprecedented escalation in the scale of industrial production. For industry to remain profitable and to continue its accumulation of capital, this increase in scale was unavoidable and remains so. It is not surprising that large sectors of business embrace the denial or minimization of the problems it generates, arguing that growth is the solution to its own problems.

Underlying the debate about the limits to this relentless economic expansion lie the assumptions of the participants. Economists view the human economy as the primary system and the natural world as a sector of it. Economists who are well respected in their own profession have argued hyperbolic versions of the "no limits" case—that economic growth can easily continue for another 2,500 years, that a population of 3.5 trillion can be supported, or that "copper and oil come out of our minds."[5] Faith in the magic of prices and technology underpins these claims. For physical scientists and ecological economists, the planet is the primary system and self-evidently finite. Whatever pie we might be baking, they say, physical parameters are essential to understanding longer-term economic processes.

The emergence of environmental degradation on a hitherto unknown scale after World War II led not only to the public concern and legislative changes described in chapter 3 but also to renewed attacks from business on regulatory regimes that interfered with profit and accumulation. Any policy that enforces the internalization of environmental costs constitutes a threat to profit, the core corporate interest, and is depicted as undermining civilization. In the absence of a comprehensive strategy for reform, the collision of economic expansion with the finite planet, though moderated to varying degrees in particular instances, has continued to intensify.

Such a wide-ranging strategy is unlikely to be adopted by a corporate network for which growth is a basic precondition, intrinsic to its functioning. When operating smoothly, growth will inevitably stimulate more growth, and to be viable, corporate capitalism must keep expanding.[6] In Australia today, business organizations routinely insist that we need more people to make the economy grow and generate ever-increasing personal wealth, and that growth is imperative if we wish to repair environmental damage—and employ the increasing population they call for.[7] They do not engage with the proposition that this approach is a kind of pyramid scheme and cannot go on indefinitely; they merely assert that human ingenuity will prevail. This ignores the underlying contradiction—that an increasing population will inevitably demand further infrastructure and more growth, and that faster growth will compound the water scarcities and greenhouse gas emissions it is expected to ameliorate. In the absence of an infinite planet, growth cannot solve its own problems.

Quest for a Bigger Pie

Growth was not promoted by governments as a policy objective during the first half of the twentieth century,[8] though it was always prized and pursued by business. Yet business had grown nervous about overproduction by the early 1920s and had taken steps to ratchet up the level of consumption in response. The newly consolidated national US corporations were building a massive productive apparatus, and their products needed outlets. If basic needs were successfully met, it was urged that "old needs" must be replaced with new. As the radio extended the reach of advertisers into the daily life of people, new needs could be created more or less at will. The tactic of increasing consumption—to soak up production and drive further economic growth—rested on the harnessing of desire with its complex individual peculiarities rather than on the meeting of basic material needs, though first world working people did reap real benefits. The incitement to consume resurfaced after World War II, when it became an intrinsic element of the *bigger pie* approach to social ills.

After World War II, growth began to be advanced as the preferred solution to the many problems of poverty and human suffering recognized throughout the postwar world, and gradually became an avowed objective of governments. In first world countries, labor organizations were increasingly convinced that economic growth would provide what workers wanted most—secure and comfortable material conditions rather than fundamental reform of the capitalist system. With the "prosperity and welfare" agenda, the working class rapidly improved its material situation and the welfare state provided support for those who could not compete successfully in the market system; much of the first world's population has become the global rich. Conditions were especially propitious in the developed world before the crisis of the mid-1970s precipitated the neoliberal "revolution." Yet, though beneficial for its workers, the postwar welfare state was a growth state, the most spectacular ever seen up to that point.

Growth was also preferred to redistribution of land and resources in the third world. The *bigger pie* was advanced as the answer to everything and remains the dominant view for business, governments, and the international bodies allied to business.[9] Although Arndt found the limits critique to be ubiquitous in the 1970s and suggested it would probably put an end to the headlong pursuit of economic growth, this did not turn out to be the case; rather, its influence waned as neoliberalism transformed the economic discourse.

During the explosive growth of the first thirty years after World War II, there was little evidence that the swelling pie, though undoubtedly beneficial for first world populations, was enhancing the lives of people in the third world. This awareness troubled the men who founded the Club of Rome and was one of the five crucial problems the Meadows team set out to address. Contrary to some critical characterizations, neither the Club of Rome's founders nor the Meadows team were "middle-class greenies" who were happy to abandon the poor. Rather, the welfare of the poor was a major focus of their thinking.

"Development" was conducted in the belief that there was only one path to prosperity, and it required adoption of the Western template and the fostering of a "saving" class. No contradiction was perceived between extending the wealth of the first world and accommodating the needs of the third. The "development discourse" imposed Western norms of progress on third world countries, aiming to industrialize economies, "modernize" peasantries, abolish feudal relations under a marketplace template, and sweep "backward" cultures aside, supposedly in their own interest. The reality, though, was that even where GDP increased, income polarization often widened, poverty was not ameliorated, and low-income people sometimes ended up worse off.[10]

The development economists conceived of "progress" as a transhistorical process available to all who were willing to modernize. The concept of quasi-inevitable stages of economic growth drew on the economistic assumption that the availability of physical sources and sinks is irrelevant, given the cleverness of human technologies. They assumed that the Western pathway to wealth could be pursued in any context and disregarded the specifics of the so-called empty world, ripe for exploitation, that furnished Europe with its means.

By the middle of the 1970s, it was clear that the postwar boom was over, a fact complicated—and possibly partly precipitated—by the oil crises of that decade. All of this occurred soon after the dawning of awareness of environmental crisis. Pressed by scientists pointing to environmental decline, UN bodies began attempting to develop what became known as sustainability. However, even as they struggled somewhat ineptly to incorporate the concept, the onus for development was being handed over to the free market and the TNCs that controlled it. There is little evidence that transnational capitalism has had any interest in designing operations that can continue indefinitely. The prevailing faith in technology and substitution involves the notion that it is *rational* and *efficient* to liquidate resources before moving on to the next option; in this scenario, long-term sustainability is neither necessary nor desirable.

In the free market era, when the user pays principle replaces the concept of the common good, private investment is regarded as the appropriate source of aid. Direct foreign investment quintupled in the 1990s and has become the dominant source of financial assistance to developing countries. Profit-based, its primary aim is not to serve the needs of the poor so much as to expand markets for people who can pay. Growth has generated a larger class of such people, but at the same time, hundreds of millions are excluded. The poorest countries are not favored for direct investment, and the poorest people in emerging countries are in much the same situation.[11] John Milios argues that the intent is a "class offensive," one aimed at "reshuffling the relation of forces between capital and labour on all social levels to the benefit of capital."[12]

If this is not the case, it certainly looks like it. The UNDP's 2005 report found that "the world's richest 500 individuals have a combined income greater than that of the poorest 416 million," while Oxfam calculated in

2014 that the richest 85 *individuals* own wealth equal to that of the poorer half of the world's population, more than 3.5 billion people.[13] Citigroup's second Plutonomy Report is also instructive; it points to the boom in luxury consumption by the rich and ultra-rich and the superiority of an investment strategy aimed at "businesses selling to or servicing the rich, be it for example luxury goods, stocks or private banks." The Citigroup analysts are clear that "global capitalists are going to be getting an even greater share of the wealth pie over the next few years, as capitalists benefit disproportionately from globalization and the productivity boom, at the relative expense of labor."[14]

Worldwide inequality has not been remedied, despite the claims of CEOs and IMF bureaucrats. As of 2004, 95 percent of the first world's population fell within the top quintile of world income, making our middle classes and employed working classes part of the global rich.[15] UNICEF's 2011 report suggests slightly lower estimates but still places some 83 percent of the first world population in the top income quintile in 2007.[16] Claims for the globalization era, which is supposed to have lifted millions out of poverty, are overblown. Some sections of the Chinese population and smaller numbers elsewhere have benefited, but the wealth generated during the neoliberal era has overwhelmingly gone to those who were already rich, most of it to the top decile, with only 10 percent going to people in the poorer half of the world's population. The Brundtland Commission's modest proposal that "part of the increases in the income of the rich should be diverted to the very poor"[17] has been ignored, and the opposite has occurred. Internationally, despite a world pie that has grown eight to ten times larger than in 1950, the slices available to the poorest people have certainly not; immense inequality persists within and between nations. Even if higher percentages of people are better off—which may be the case—the gross numbers of people in difficulty (with incomes less than $2 a day) have been similar for the last three decades, and the numbers of those without discretionary income (the "non-middle class") are also static and expected to remain so.[18]

World Council of Churches economic justice executive Rogate Mshana points to the contrast between discretionary spending in the first world and unmet needs in the third:

Today, tremendous amounts of money are being expended on superfluous and even life-destroying goods. For example, in 2005 alone, global military spending reached well over US$1 trillion (about US$3.1 billion daily) with the United States accounting for nearly half that amount. Europeans spend around US$105 billion on alcohol and US$59 billion on cigarettes per year. The UN estimates that a budget of US$150 billion is required to reach the Millennium Development Goal (MDG) of halving poverty by 2015; the annual amount of US$25 billion would be sufficient to eliminate hunger and malnutrition as well as provide clean drinking water for everyone in the world.[19]

The neoliberal version of the bigger pie expects to address poverty by integrating everyone into the world market and building an expanding global consumer class. In this version, the entire world population of seven billion ascends the stairway to imitative affluence, as described in chapter 5. No account of any physical parameters is thought necessary. The Chinese political economist Minqi Li, on the other hand, recognizes that the ecological limits of world resources and sinks are likely to preclude what might otherwise have been the ascent of China to world dominance over the next century—just as the United States rose to dominance as two world wars weakened the British Empire. The sheer quantity of resources needed to render everyone—or even most people—middle class as we now understand the term seems absolutely beyond the capacity of planet Earth.[20] In the words of Wolfgang Sachs, "The style of affluence in the North cannot be generalised around the globe, it is oligarchic in its very structure."[21]

The history of development shows that, after nearly seventy years, a much-multiplied pie has utterly failed to yield sufficiently large slices to afford everyone even modest security. While the middle class in a few selected countries has expanded, the billions at the bottom of the scale remain poor. The plan to generalize first world affluence to the rest of the world is shown to be a dangerous folly. As Sachs insists, what is required is "the revision of goals rather than the revision of means."

Again and again, it emerges that the much-vaunted efficiency of neoliberal economics means efficient profit-making only and serves as an ideological excuse for excluding those who cannot pay. What is described as "optimal allocation" is the circulation of goods according to the payment of the highest price. While economists like to think this will equate to the most productive use, it seems clear that cigarettes and whisky, however profitable they may be, do not constitute a more productive use of resources than bowls of rice and clean water.

Propaganda and Politics

Given such grave doubts about the availability of sources and sinks and the comprehensive failure of Truman's stated project for the oppressed, how can ongoing growth remain the accepted wisdom? In the preface to their thirty-year update to the *Limits to Growth*, Meadows and colleagues noted that perception often carries more weight than material reality. They point to

> those groups (largely comprised of economists) who have spent the past 30 years pushing the concept of free trade. Unlike us, they have been able to make their concept a household word. Unlike us, they have convinced numerous politicians to fight for free trade. . . . Ecological overshoot seems to us to be a much more important concept in the twenty-first century than free trade. But it is far behind in the fight for public attention and respect.[22]

Here, the Meadows team is wrestling with the same puzzle that launched this study—the outrageously successful discourse of mainstream economists, in contrast to the concurrent eclipse of scientists' warnings of ecological peril. The scientists, of course, issued very little literature for public consumption and were engaged in research rather than shaping public opinion, while economists, on the other hand, were embedded within the ruling ideology of material progress and the burgeoning propaganda institutions of big business. The huge power commanded by the new corporations that consolidated at the beginning of the twentieth century was channeled through a diverse web of propaganda arrangements, utilizing the ever-changing technologies of the new era.

The machinery assembled over the first sixty years of the century, described in chapter 10, was ready to respond when environmental concerns emerged in the 1960s. From 1970, a renewed burst of advocacy for private enterprise, disguised as an educational service to the community, reached "almost everyone," according to *Fortune* magazine.[23] Business's propaganda techniques were further institutionalized with the establishment of a massive and well-paid in-house intelligentsia capable of defining both government policy and media debate.

Concurrently, old political alignments began to shift. Despite a strong bipartisan consensus on environmental matters before 1972 in the United States,[24] the conservative side of politics moved toward categorical opposition to environmental regulation, a trend that coincided with the rise of neoliberalism (itself ably propagated by the "free market" think tanks). Indeed, the antiregulatory strand of neoliberal doctrine was a perfect weapon against environmentalism. With its roots in the business struggle against the control of toxic products for more than a century, the new mantra of deregulation appealed to criteria of supposed cost-effectiveness.

Free market fundamentalism emerged as the dominant ideology of policymakers at the same time that evidence of anthropogenic global warming began to coalesce and become accepted by the scientific establishment. Although it has been suggested that the two developments might have been coincidental,[25] this is probably not entirely so. Environmental decline and its attempted remedies were part of the complex of problems that neoliberalism was responding to. The attack on environmental amelioration was an intrinsic element of free market ideology, especially in the United States, and extended to attacks on sectors of the scientific community. Once scientists began investigating the damage caused by postwar industrialization, business interests and their mouthpieces began a campaign to undermine science. Regarded as an indispensable foundation of progress and an engine of invention for a century or more, the science mainstream came to be seen as a potential enemy of business in the course of two or three decades. Professionals subject to peer review were attacked as proponents of "junk science" while industry-funded front organizations professed a commitment to "sound science." This attack on science emerged during the tobacco industry's campaign to create doubt about the link between smoking and cancer and has been used ever since to foster uncertainty where uncertainty does not exist.

The elites of the early twentieth century were forthright. They were open about their intention to shape opinion in the interests of the business community they were allied to, and to maintain the power it had exercised through the nineteenth century. Today's elites, despite railing against what they call "political correctness" or "the invisible muzzle," do not admit openly, as Bernays did, that their central objective is the protection of corporate interests from meddling citizens. Nonetheless, today's think tank networks have quietly admitted their intention to transform the policy agenda of government along lines congenial to business; they have accordingly recruited bright young intellectuals who would be able to influence public opinion and the policy of governments, and have marshalled an intellectual artillery to target "enemy strong points."

There is a plethora of evidence that the "free enterprise" think tanks of the late twentieth century were funded—and continue to be funded—by industries intent on curbing democratic regulation. Similarly, there is evidence that the same polluting industries buy direct influence in governments and bureaucracies, exert direct influence over media conglomerates that belong to the same business class, and campaign relentlessly to depict free market capitalism as a preordained aspect of the natural order. The think tank operatives of the last thirty to forty years have indeed softened up the positions of their opponents and, despite their dense connections with the conservative wing of politics in their respective countries, have presented themselves as "nonpartisan" or "independent," and enjoyed the same tax deductions as charities. Yet, on closer inspection, the idea that the free market is a spontaneous outgrowth of human nature, inexorable and natural, is contradicted by the immense effort expended in advertising it.

A Massively Inconvenient Truth

Even at the outset, when *Limits* was first published and some institutions in the developed world seemed open to its ideas, economists, both academic and popular, went on the attack. For nearly forty years, questions about physical limits have met systematic derision from business interests and mainstream economists. Most of the vocal critics of *Limits* have ruled out any notion of slowing or restraining growth. Some simply ignored the MIT scenarios that moderated growth and led to stability.[26] "Club of Rome" became easy shorthand for dismissive comments about prophets of doom.

In part because of this barrage of scorn, questions about the continuation of economic growth in a finite world remain, largely, excluded from public discourse and policymaking. Galbraith foresaw in 1958 that the "gargantuan and growing appetite" he perceived in 1950s America would, in the long run, need to be curtailed. Instead, the appetite for ever more consumer goods has deepened in the first world, and its extension to sections of the new middle classes in China, India, and Latin America is celebrated as a boon for all. We continue, however, to face extreme versions of the problems set out forty years ago by the Meadows team: accelerating industrialization, continuing population growth, extensive malnutrition, the depletion of nonrenewable resources, and environmental decline.

If ongoing growth had provided security and a decent living for all, a more compelling case for the "progress" route could, perhaps, be made. As it is, prosperity is concentrated among a privileged minority and material security is confined to no more than half the world's many people. In basing solutions to third world poverty on economic growth and the stimulation of a swelling consumer class, we are provoking a huge and ongoing increase in energy and resource consumption and a concomitant increase in all forms of pollution, including greenhouse gas emissions. Despite some attempts from some governments—and a global recession—these are increasing at an accelerating rate, and the human project pushes ever closer to what Johan Rockström and colleagues describe as the "planetary boundaries" and the danger of crossing tipping points, with unknowable consequences.[27] Although climate often occupies center stage in the discussion of environmental crisis, it should not be thought that a transition to plentiful low-carbon energy alone, even if we could manage it, is all that is needed to redress the multiple crises.

Ecological economists continue with their work on the steady-state economy and are developing theories of what might be necessary to implement such an economic system (see the appendix). They are concerned about both ecological destruction and social injustice, and want to ameliorate both. Their contribution to the understanding of the perils of growth has been immense. Some socialists are also focused on these issues and, critical of the idea that a steady-state economy could come into being within capitalism, argue that the slowing, arrest, or reversal of growth will require a transition to some form of socialism. They criticize the ecological economists for believing that the current economic system could accommodate the drastic measures required to curtail growth. Both proposals—a transition to socialism or the taming of the capitalist economy—seem equally hard to imagine in the neoliberal era, but, to hijack Margaret Thatcher's famous expression, "there is no alternative."

Herman Daly's 2008 ten-point program is an excellent example of the sweeping changes ecological economists consider necessary.[28] Daly is no socialist, but most of the items on his agenda are totally unacceptable to corporate capitalism in the neoliberal world. His program includes ecological tax reform; limitations on unequal income distribution; the re-regulation of international commerce; the downgrading of the IMF, World Bank, and WTO; the abolition of fractional reserve banking;[29] stabilization of the population; and the transfer of the remaining commons to public trust. Under the current economic system, there seems little to no chance that any of these measures would be adopted by governments that exist at the pleasure of market forces. As I have argued in detail, many of these precise measures have been resisted and blocked in the past, especially since the 1970s, while the conduct of the IMF, the development of the WTO, and the continuing pursuit of "free trade partnerships" such as the Trans-Pacific Partnership and the Transatlantic Trade and Investment Partnership constitute conscious and ongoing boosts to the power of transnational corporations.

Daly also suggests that "instead of treating advertising as a tax-deductible cost of production we should tax it heavily as a public nuisance." The removal of tax deductibility for advertising could be very beneficial in reducing consumer-driven growth; again, it is difficult to imagine actual legislation to this end, let alone punitive measures. The cultural change that Daly advocates is incompatible with the system of consumer capitalism in which we are enmeshed. Even if we argue that capitalism can survive without growth in material production, which I doubt, Daly's reforms would not be welcomed. And yet they represent the barest of minimums that we need.

We confront a massively inconvenient truth, one that will not respond to changing light bulbs and other gestures of individual responsibility. Such actions won't hurt, of course, but they won't on their own alleviate the problems we face. Structural change is indispensable. We need a different kind of economy, one designed to meet needs rather than create them. To achieve such outcomes, we will need to disturb the business universe and the tales of progress and prosperity it feeds us, and substitute alternative visions and ideals.

We must also restore democratic norms to the conduct of elections in our plutocratic democracies, and reestablish the preeminence of our elected institutions, liberating them from the market's "golden straitjacket." Above all, we need to abandon the consumer path to human advancement and the reduction of our choices to monetary terms. The consumer template for the human future has outworn its usefulness. Stimulating consumption in the interests of growth and chasing economies of scale was, perhaps, suitable for the "empty world." In the "full world" (and getting fuller) we need redistributive justice within and between countries and a plan for the first world to reduce its material demands to allow space for the rest of the world to reach material security.

It remains for others to invent pathways to solutions for these difficult problems. My object has been to illuminate the reasons for the ideological dominance of growth, and to foster an awareness of the actual realities—human and ecological—that contradict its confident discourse. Challenging the manufactured truths of think tanks and advancing a sense of reality in the public arena are the critical next steps.

Appendix: Selected Critics of Growth, 2013

Critics of growth, though far fewer than its admirers, are found throughout the world. The following list is by no means exhaustive, but aims to give the interested reader a guide to the scope of the field. Many mentioned here are allied with ecological economics or the geophysical sciences, and some, like myself, are people from a range of other disciplines who have found their arguments persuasive.

CASSE (US), http://steadystate.org

About

The Center for the Advancement of the Steady State Economy (CASSE) is based in the United States but has chapters worldwide. It's first policy aim is stated as:

> First and foremost, adopt the right macro-economic policy goal—a steady state economy that features sustainable scale, fair distribution of wealth, and efficient allocation of resources. A prerequisite to adopting this macro-economic policy goal is a cultural shift from the pursuit of lifestyles driven by endless economic expansion and unsustainable consumerism to lifestyles driven by the search for long-term prosperity and sustainable consumption that fulfils people's needs.

Suggested Reading

CASSE provides an excellent reading list that encompasses classic ecological economics texts from the 1960s on and much of the current work worldwide, including that of Brian Czech, Rob Dietz, Richard Heinberg, and Tim Jackson. It is available at http://steadystate.org/discover/reading-list.

CASSE also cites Herman Daly's 2008 paper for the Sustainable Development Commission, UK (April 24, 2008), "A Steady-State Economy." This paper is an excellent summary of Daly's steady-state economy and appears in the main reference list.

The first chapter of the following book is available online:

Dietz, Rob, and Dan O'Neill 2013. *Enough is Enough: Building a Sustainable Economy in a World of Finite Resources*. San Francisco: Berrett-Koehler. http://steadystate.org/discover/enough-is-enough.

Two US contributions not listed by CASSE would be added to my "must-read" list:

Costanza, Robert, Gar Alperovitz, Herman Daly, et al. 2013. Building a sustainable and desirable economy-in-society-in-nature. In *State of the World 2013: Is Sustainability Still Possible?* Washington, DC: Island Press.

Extended version available at United Nations Division for Sustainable Development: http://sustainabledevelopment.un.org/content/documents/Building_a_Sustainable_and_Desirable_Economy-in-Society-in-Nature.pdf.

Huesemann, Michael, and Joyce Huesemann. 2011. *Techno-fix: Why Technology Won't Save Us or the Environment*. Gabriola Island, BC: New Society Publishers.

Peter Victor (Canada)

About

The Canadian economist Peter Victor is engaged in analysis of the issues involved in growth, slowing growth, and, possibly, reversing growth. Victor's major book attempts a model of slowing growth in the Canadian economy. He is agnostic as to whether capitalism could accommodate a no-growth or degrowth economy, but he argues that "green growth" is not a feasible alternative.

Suggested Reading

Victor, Peter. 2008. *Managing without Growth: Slower by Design, Not Disaster.* Cheltenham, UK: Edward Elgar.

Victor, Peter. 2010. Questioning economic growth. *Nature* 468:370–371.

Victor, Peter, and Gideon Rosenbluth. 2006. Managing without growth. http://www.greenparty.ca/files/Peter_Victor-No_growth.pdf.

***Décroissance (Europe)*, http://www.degrowth.org**

About

In Europe, the *décroissance* (degrowth) movement has emerged in the past decade and has held conferences in Paris (2008), Barcelona (2010), and Venice (2012), as well as an American meeting in Montreal (2012). Serge Latouche, emeritus professor of political economy at the University of Paris-Sud, is prominent in the movement, as is journalist Hervé Kempf.

Suggested Reading

List of publications: http://www.degrowth.org/publications.

Kempf, Hervé. 2008. *How the Rich Are Destroying the Earth*. White River Junction, VT: Chelsea Green.

Latouche, Serge. 2006. The globe downshifted. *Le Monde Diplomatique*, January 13. http://mondediplo.com/2006/01/13degrowth?var_recherche=Serge+Latouche.

Latouche, Serge. 2007. De-growth: An electoral stake? *International Journal of Inclusive Democracy* 3 (1) . http://www.inclusivedemocracy.org/journal/vol3/vol3_no1_Latouche_degrowth.htm.

Latouche, Serge. 2010. *Farewell to Growth*. Cambridge: Polity Press.

Montreal conference papers: http://www.montreal.degrowth.org/papers.html.

Proceedings of the Paris conference: http://events.it-sudparis.eu/degrowthconference/en/appel/Degrowth%20Conference%20-%20Proceedings.pdf.

Proceedings of the Barcelona conference: http://www.barcelona.degrowth.org/Proceedings-new.122.0.html.

Attac

About

In Germany, Attac (originally founded in 1998 as the Association for the Taxation of Financial Transactions and Aid to Citizens) is the organizational hub of the post-growth movement. Attac held a "Beyond Growth" conference in Berlin in 2011, opened by Vandana Shiva.

Suggested Reading

An account of the key themes of the Berlin Conference was written for FEASTA (Foundation for the Economics of Sustainability) by the English economist and community campaigner Brian Davey. It is available at http://www.feasta.org/2011/06/10/what-could-a-post-growth-society-look-like-and-how-should-we-prepare-for-it.

Degrowth (UK)

About

In the UK, the new economics foundation (nef) has been at work on similar issues since the 1980s—"a new model of wealth creation, based on equality, diversity and economic stability," and on "economics as if people and the planet mattered."

Suggested Reading

Jackson, Tim. 2009. Beyond the growth economy. *Journal of Industrial Ecology* 13 (1): 487–490. http://steadystate.org/wp-content/uploads/Jackson_2009_Beyond_the_Growth_Economy.pdf.

See the main reference list for Jackson's book, *Prosperity without Growth*.

Simms, Andrew, and Victoria Johnson. 2010. *Growth Isn't Possible*. London: nef. http://www.neweconomics.org/publications/entry/growth-isnt-possible.

Woodward, David, and Andrew Simms. 2006. *Growth Isn't Working*. London: nef. http://www.neweconomics.org/publications/entry/growth-isnt-working.

Australian Work

About

Richard Sanders, who was commissioned to write an appraisal of the growth problem for the Australian Conservation Foundation (ACF), is cautious about the feasibility of addressing headlong growth within a market framework: he holds that credit creation should be a power of the public sector, not private banks, and expresses doubts about the much-vaunted efficiency with which markets achieve resource allocation. He is equally adamant that "green growth" is a fable disconnected from reality. A very radical change is needed, in Sanders's view.

Like other ecological economists, he stresses the destructive role of fractional reserve banking, where banks simply "create" money by lending it out and are unconstrained by the need to hold reserves that match the loans—or even come close to it. Money created in this way in the *virtual finance* sector is nevertheless a claim of real wealth, since it can be exchanged for real things. This system (which underpins the financialization described in chapter 6) and the debt it fosters are aspects of the pyramid scheme that relies on the assumption of growth continually compounding at the rate of interest.

Clive Hamilton's early work focused on the addictive and disease-like manifestations of rampant consumption in first world societies and the ways in which it corrupts our social and political processes. More recent books venture into the politics of climate denial.

Paul Gilding takes the view that growth can be curtailed within capitalism. This will happen, he argues, because it must, rather than by choice or at the behest of business. His book appears in the main reference list.

Simon Michaux, a mining engineer with many years experience in the industry, proposes that a resource crisis is almost upon us, driven by the exhaustion of high grade, accessible ores and the ever-increasing energy, water, waste and environmental impact involved in extraction. His lecture on this subject is in the main reference list.

Suggested Reading

Hamilton, Clive. A full list of his numerous publications can be found at http://www.cappe.edu.au/docs/staff-cvs/hamilton.pdf.

Sanders, Richard. 2009. Discussion paper 2. In *Future Economic Thought*. ACF. http://www.acfonline.org.au/future-economic-thought.

Australian Organizations

The ACF released a paper analyzing growth in 2010. Titled *Better Than Growth: The New Economics of Genuine Progress and Quality of Life*, it is a summary of the issues and some possible remedies, written for popular consumption: http://www.acfonline.org.au/sites/default/files/resources/ACF_BetterThanGrowth.pdf.

The CSIRO's Sustainable Ecosystems Unit has produced two comprehensive reports on the physical impact of different immigration levels, the second in collaboration with Flinders University. These reports analyze the physical parameters that might limit expanding Australian populations.

Foran, Barney, and Franzi Poldy. 2002. *Future Dilemmas: Options to 2050 for Australia's Population, Technology, Resources and Environment*. Canberra: CSIRO Sustainable Ecosystems. http://www.cse.csiro.au/publications/2002/fulldilemmasreport02-01.pdf.

Foran, Barney, and Franzi Poldy. 2002. *Dilemmas Distilled: A Summary and Guide to the CSIRO Technical Report*. Canberra: CSIRO Sustainable Ecosystems. http://www.cse.csiro.au/publications/2002/dilemmasdistilled.pdf.

Sobels, Jonathan, Sue Richardson, Graham Turner, et al. 2010. *Long-Term Physical Implications of Net Overseas Migration: Australia in 2050*. National Institute of Labour Studies, Flinders University, Adelaide. http://www.immi.gov.au/media/publications/research/_pdf/physical-implications-migration-fullreport.pdf.

The team at Monash University's Centre for Population and Urban Research also looks critically at ongoing growth in Australia; it focuses on the impact of the high levels of net immigration required to achieve "pre-set targets for economic growth." Its website is http://artsonline.monash.edu.au/cpur.

One of Bob Birrell's papers, "Population, Growth and Sustainability," is available here: http://parlinfo.aph.gov.au/parlInfo/search/display/display.w3p;query=Id%3A%22library%2Fprspub%2F417066%22.

Other Australian Critics of Growth

Michael Lardelli is a geneticist at the University of Adelaide who is connected with the Association for the Study of Peak Oil and Gas (ASPO). Links to his numerous articles on *Limits* issues are listed on his university website under "Popular Media and Other Presentations/Publications": http://www.adelaide.edu.au/directory/michael.lardelli.

Mark O'Connor is a poet and environmentalist who, in 2008, published *Overloading Australia: How Governments and Media Dither and Deny on Population* with William J. Lines (Canterbury NSW: Envirobooks). It is now in its fourth edition. Extracts are available at http://www.australianpoet.com/overloading.html.

See below for comment on Ted Trainer.

Socialist Critics

About

Richard Smith's socialist critique is concise and telling. Smith acknowledges that Daly and the ecological economists are on the right track as far as the need to put an end to growth and redistribute wealth equitably, but Smith does not agree that the market allocates resources efficiently.

Like Jackson (2009), Smith argues that "decoupling" energy and material flows from GDP growth is a mathematical impossibility. Unlike Jackson and many of the ecological economists, he does not believe that growth can be tackled within a capitalist economic system. He points to the need for very significant regulation, something anathema to business throughout the century since consolidation into gigantic corporations commenced—and no doubt earlier still. He also argues that growth is an inbuilt aspect of the system, not an optional extra. This view accords with my own analysis throughout the present work. The cultivation of consumption for its own sake was an intentional strategy of business since the eighteenth century and was adopted as a key bulwark of the system from the 1920s.

Australian Ted Trainer has been working on these same issues for many years and is in broad agreement with Smith. Links to many of Trainer's articles and essays are available at https://socialsciences.arts.unsw.edu.au/tsw/.

David Harvey, one of the world's foremost Marx scholars, argues that one of the fundamental contradictions of capital today is its dependence on continuing economic growth. A short statement that echoes Daly's distinction between growth and development is found here: http://www.youtube.com/watch?v=uOsKuyh5ps0.

Harvey's argument is more fully elaborated in "The contradictions of capital," a lecture he gave at Warwick University: http://www.youtube.com/watch?v=8UD-QqYFJqY.

Suggested Reading

Smith, Richard. 2010. Beyond growth or beyond capitalism? *real-world economics review* 53:28–42. http://www.paecon.net/PAEReview/issue53/Smith53.pdf.

Smith, Richard. 2011. Green capitalism: The god that failed. *real-world economics review* 56:112–144. http://www.paecon.net/PAEReview/issue56/Smith56.pdf.

Smith, Richard. 2013. Capitalism and the destruction of life on Earth: Six theses on saving the humans. *real-world economics review* 64: 125–151.

Trainer, Ted. 2011. The radical implications of a zero growth economy. *real-world economics review* 57:71–82. http://www.paecon.net/PAEReview/issue57/Trainer57.pdf.

Notes

Introduction

1. Meadows et al. 1972.
2. Meadows et al. 1972, 27.
3. Buell 2003.
4. Oreskes and Conway 2008, 66.
5. Wallerstein 1974, 347–357.
6. Cited in Fones-Wolf 1994, 16.
7. CNN *Money* 2012.
8. Broad 2006. Development economist Robin Broad's study of the World Bank found bias in its hiring and promotion practices, the selective application of peer review, the exclusion of dissenting voices, and manipulation of data. She challenges the objectivity of its work.
9. Mooney 2010.
10. This is not intended to exclude acceptance of the notion, in physics, of the influence of the observer on quantum reality or more recent developments in science such as chaos and complexity theory.

1 Economic Growth: Origins

1. Crosby 1986, 71–103.
2. Ponting 1993, chaps. 7, 9.
3. Clark 2008.
4. Berman 1981, 53.
5. Nef 1969, 24–26.
6. Berman 1981, 58.
7. Biggins 1978, 53; Berman 1981, 49.
8. Bowden 1965, 65.

9. Wolf 1997, 274–78.

10. Campbell and Laherrère 1998, 81; BP 2012, 8.

11. McNeill 2001, 298–9.

12. McNeill 2001, 6.

13. Marx 1954a [1848], 10.

14. Marx was also a close follower of contemporary scientific research, so it is not possible to predict what attitude he might have taken to the "limits" theorists of the late twentieth century, even though it is usually assumed that he would dismiss them because of their perceived similarity to Malthus.

15. Marx 1954b [1887], 229, 257.

16. Marx 1959 [1894], 813.

17. Marx 1954b [1887], 474.

18. Dead zones occur when fertilizer-rich runoff leads to algal blooms that exhaust the available oxygen.

19. Cordell and White 2008, 1, 6.

20. Smith 2002, 1654.

21. Rockström et al. 2009, 32.

22. Den Biggelaar et al. 2003, 3; Arnalds 2009, 43; Montgomery 2007, 4.

23. Marx 1959 [1894], 820; 1954a [1848].

24. Farrelly 2006.

25. Clausen 2007, 48–49.

2 Economic Growth: Perceptions

1. Larry Summers was the World Bank's chief economist at this time. Summers is notorious for signing a 1991 memo drafted by his staff which held that "the economic logic behind dumping a load of toxic waste in the lowest wage country is impeccable" and "under-populated countries in Africa are vastly UNDER-polluted, their air quality is probably vastly inefficiently low compared to Los Angeles or Mexico City" (cited in Pellow 2007, 9). Summers was President Clinton's treasury secretary and has also served with the Obama administration.

2. Daly 1999, 16–17.

3. Daly 1999, 15.

4. Boulding 1966, 303.

5. Sachs 1999, 26–27.

6. Solow 1974, 1.

7. Beckerman 1972, 332–333.

8. Ponting 1993, 18.

9. Ponting 1993, chaps. 4, 13.

10. Ponting 1993, 41.

11. Montgomery 2007; Smith 1995.
12. Ponting 2002.
13. Ruddiman 2005, 88–93. Ruddiman suggests that rice-paddy agriculture may have magnified the effect.
14. McNeill 2001, 8.
15. Durning 1994, 22–23.
16. McNeill 2003, 264.
17. McNeill 2003, 268.
18. Clark 2008.
19. Georgescu-Røegen 1971, 278; 1975, 353.
20. Perlez and Johnson 2005; Michaux 2013.
21. Georgescu-Røegen 1971, 19. "The Great Migration" is the westward movement of the Central Asian herders (known popularly as barbarians) into Europe, a migration sometimes argued to have contributed to the fall of the Roman Empire.
22. Osnos 2006.
23. International Energy Agency (IEA) 2010, 23.
24. Ehrlich, Ehrlich, and Holdren 1993 [1977], 73.
25. Campbell 1999; Hirsch 2005; Klare 2005; Simmons 2005; IEA 2010.
26. Also known as "shale oil," this is liquid petroleum still locked in rock (see chapter 14). It should not be confused with "oil shale" (kerogen), a solid bituminous substance that has yet to be successfully extracted.
27. Carey and Carter 2007; Pimentel 2003.
28. Bleizeffer 2006.
29. Solo 1974, 515–517.
30. Daly 1991a, 226.
31. Daly 1991b, 37.
32. Chew 2001, 9.
33. Ruddiman 2005.
34. Meadows, Meadows, and Randers 2004, 17–49, esp. 21.
35. International Monetary Fund (IMF) 2013b, 149.
36. Daly 1991b, 39.
37. Daly 1991b, 40.
38. Daly 1978, 14–48.
39. Daly 1991b, 35–36.
40. Mill 1848, 306–311.
41. Norgaard 1994, 32.
42. Norgaard 1994, 7, 33.

43. Truman 1949.
44. Norgaard 1994, 79.
45. Norgaard 1994, 126–127.
46. Norgaard 1994, 122.
47. Norgaard 1994, 124.
48. Public Citizen 2003.
49. Norgaard 1994, 131–132.
50. Kaysen 1972, 663.
51. E.g., Diamond 2005; Chew 2001; Ruddiman 2005.
52. Solow 1974, 11.

3 The Limits to Growth Debate

1. DeLong 1997.
2. DeLong (1998) produces three columns of estimates of comparative GDP over time, according to differing criteria. His preferred model amplifies the estimate to account for the value of new commodities not available in earlier periods. Even if this weighting is excluded, the annual increase in world GDP in the last decade of the twentieth century equates to half the total world economy in 1900.
3. IMF 2000, 150–151.
4. Kraft 2000, 21.
5. Beck and Kolankiewicz 2000, 123–124.
6. Sinding 2007, 8.
7. Foster 1998; Ross 1998.
8. Kraft 2000, 23.
9. Darling 1970.
10. Macinko 1973.
11. Goldsmith and Allen (with Davoll, Lawrence, and Allaby) 1972, 30.
12. Meadows et al. 1972, 27.
13. UN Environment Programme (UNEP) 2007, 2012.
14. Verney 1972, 418.
15. Strong 1972, 414.
16. Golub and Townsend 1977.
17. Strong 1972, 414–417.
18. Barney 1982, preface to vol. 1.
19. Barney, Freeman, and Ulinsky 1981; Voyer and Murphy 1984.
20. Ponting 1993.
21. Braudel 1981, 73–78.

22. Benedictine sister Joan Chittister, cochair of the Global Peace Initiative of Women, speaking to Rachael Kohn on January 31, 2010 (Kohn 2010).

23. Malthus 1976 [1798].

24. Polanyi 1944, 82.

25. Malthus 1989 [1803], 127.

26. Malthus 1976 [1798], 135.

27. Ricardo 1952 [1817], 175.

28. Osborn 1948, 65.

29. Osborn 1948, 179–192.

30. Vogt 1948, 110.

31. Vogt 1948, 34–36, 63–67.

32. *Time* 1948; Brandt 1950; Fisher 1949.

33. Foster 1998.

34. Ross 2000, 8.

35. Hardin 1968, 132.

36. Kennedy 2003.

37. Ciriacy-Wantrup and Bishop 1975.

38. Shiva 2002, 28–31.

39. Ostrom 1990.

40. Hardin 1974a, 1974b.

41. Speaking to Robyn Williams on October 29, 2011 (Williams 2011b).

42. Ehrlich 1968, 100.

43. Hay 2002, 186.

4 *The Limits to Growth* and Its Critics

1. Cited in Barney 1982, 607.

2. Goldsmith and Allen (with Davoll, Lawrence, and Allaby) 1972.

3. Beckerman 1972, 328.

4. Gillette 1972.

5. Beckerman 1972.

6. Beckerman 1974.

7. Koehler 1973; McGinnis 1973.

8. *Economist* 1972; Gillette 1972; Gordon 1973; Koehler 1973; Pavitt 1973; Solow 1973; Walls 1973; Ohlin 1974; Boserup 1978; McGinnis 1973.

9. Maddox 1972; Cole et al. 1973.

10. Dryzek 1997, 41; Hay 2002, 185–187, 205.

11. Freeman 1973; Nordhaus 1973; Pavitt 1973.

12. McGinnis 1973, 296.

13. Freeman 1973; Kiernan 1972; Passell, Roberts, and Ross 1972, 1.

14. Meadows et al. 1973, 218, 235.

15. Meadows et al. 1973, 222–223, 231–235, 237.

16. Kneese and Ridker 1972.

17. Nordhaus 1973; Forrester 1971.

18. Nordhaus 1973, 1158, 1165, 1182–1183.

19. Forrester, Low, and Mass 1974, 170–178, 171.

20. Forrester, Low, and Mass 1974, 180–181.

21. Forrester, Low, and Mass 1974, 184.

22. Journalist Ticky Fullerton (2002) asked economist Chris Murphy about geophysical parameters in economic models: "Why don't mainstream models take account of things like this?" Murphy: "Well they do when environmental outcomes are a major part of the policy issue that's being considered."

23. Kaysen 1972, 664.

24. Beckerman 1972, 338.

25. Solo 1974, 517.

26. Solow 1973.

27. Meadows et al. 1973, 225.

28. Landers 2007.

29. Silk 1972.

30. Solow 1973, 41.

31. Beckerman 1974, x.

32. Trembath 2007; Department of Foreign Affairs and Trade (Australia) 2007. See chapter 6 for a full analysis of this paper.

33. Meadows et al. 1973, 236–237.

34. Krauss 2008.

35. Golub and Townsend 1977.

5 Growth and Consumerism

1. This status is sometimes claimed for the British East India Company which was founded a year or so earlier, chartered by the English government, and set up along similar lines. The East India Company began by funding specific voyages, however, and did not have tradable shares until 1661.

2. Sayle 2007, 25.

3. Smith 1852 [1776], 6. Though Smith is often used to support the exculpation of greed, some scholars, such as John Ralston Saul (1997), argue that Adam Smith's *Wealth of Nations* cannot be read without reference to his other major

work, *The Theory of Moral Sentiments*. Though Smith argued that self-interest *can frequently* lead to the common good of society as a whole, he did not hold that it always and necessarily does so.

4. McKendrick, Brewer, and Plum 1982, 19.

5. PBS 1980, vol. 7; Edney 2005.

6. Friedman 1962, 133.

7. Jones et al. 2007, 32–36.

8. Dowie 1977.

9. Maruyama 1996.

10. Similar conditions also apply in first world countries where "guest workers"—and sometimes forced labor (Badkar 2011)—have been introduced, for example farm laborers in Florida (Crick 2007) or southern Italy (Wasley 2011).

11. Hobsbawm 1989, 316–319.

12. The Australian government report released for the Asia-Pacific Economic Cooperation forum in late 2007 reflects the depth of this concept in modern policy thinking and is discussed in chapter 6.

13. McKendrick, Brewer, and Plum 1982.

14. Fraser 1981; Leach 1993, 20; Whitwell 1989, 7–11.

15. Leach 1993, 3.

16. Somewhat modified, in Germany's case, by crippling reparations.

17. Cherrington 1980, cited in Beder 2004a, 1.

18. Mill 1848, 308.

19. Hunnicutt 1996, 1–6.

20. Hunnicutt 1988, 37–65.

21. Cited in Hunnicutt 1988, 38.

22. Allen 1931, 128.

23. Bernays 2005 [1928], 84.

24. Cowdrick 1927, 208, cited in Hunnicutt 1988, 42.

25. Cited in Hunnicutt 1988, 57.

26. Committee on Recent Economic Changes 1929, xv, xviii.

27. Cited in Hunnicutt 1988, 42.

28. Schumpeter 1975 [1942], 82.

29. Whitwell 1989, 11–15.

30. Ewen 1996, 228.

31. Speaking to Phillip Adams on March 25, 2008 (Adams 2008).

32. Ewen 1996, 179.

33. Packard 1959, 308.

34. Galbraith 1958, 92.

35. Galbraith 1958, 93, 95–96.
36. Galbraith 1958, 97.
37. Cited in Kettles 2008, 47.
38. Packard 1963 [1960], chaps. 6, 7.
39. Whitwell 1989, 34.
40. Ewen 1996, 10–11.
41. Bernays 1961 [1923], 34–35.
42. Bernays 2005 [1928], 37–38.
43. Bernays 2005 [1928], 63, 75.
44. Le Bon 1960 [1895], 23, 112, 114–115.
45. Bernays 2005 [1928], 73, 38, 71.
46. Marcuse 1968, 9.
47. Marcuse 1970, 62, 67.
48. Schor 1991, 2.
49. Marcuse 1969, 4–6.
50. Harmer 2008.
51. IMF 2009, 1.

6 The Rise of Free Market Fundamentalism

1. A term revived after World War II by dissenting scholars who sought to situate economies in their historical, political, and cultural settings.
2. Smith 1852 [1776], 184.
3. E.g., Ormerod 1994; Saul 1997.
4. Rand's novels were first published in the 1940s and 1950s.
5. Keay 1987.
6. Friedman and Friedman 1998, 605.
7. Hayek 1976 [1944], 10, 78.
8. Tribe 2009, 75–6.
9. UN 1948.
10. Hayek 1976 [1944], 19.
11. Stretton 2000, 219.
12. Nadeau 2008; Mirowski 1984.
13. Ormerod 1994, 28.
14. Stretton 2000, 637.
15. Stretton 2000, 80.
16. Roosevelt 1941.

17. Such as Dr. Brian Walker of Australia's CSIRO.

18. Walker and Salt 2006.

19. Sheeran 2008.

20. Woolley 2008; Phillips 2006, 266.

21. Nor does "high-frequency trading," which is conducted by computers according to algorithms, meet any needs other than the generation of speculative profit through arbitrage. It now constitutes somewhere between 20 and 50 percent of trading on the Australian Stock Exchange and, like Woolley's momentum trading, does not reflect any aspect of the real underlying value of the stock being traded (Mares 2011).

22. Woolley 2008.

23. The assembly of these practices is examined in detail in part III.

24. Cited in Cochran and Miller 1942, 343.

25. Fraser 2008.

26. He did not subscribe to the theory of automatic tendency toward equilibrium (Keen 2001, 300), for example, though this did not preclude a faith in the superiority of market processes.

27. Hayek 1976 [1944], 68, 15, 42.

28. Harris 1997.

29. Cockett 1995, 307; Burton 2004. See also part III.

30. Thorsen and Lie 2006, cited in Mirowski 2009, 449; Thorsen 2009, 16.

31. Lapham 2004. See part III for the role of family foundations.

32. Klein 2007, 59–62, 80–81.

33. These arrangements were agreed to at the Bretton Woods Conference in 1944 and aimed to avoid the monetary chaos and competitive devaluations that had characterized the 1930s. The US dollar became the world's reserve currency. It was pegged to gold at $35 an ounce, and other currencies were given fixed exchange rates.

34. Frieden 2006, 339–342.

35. Helleiner 2008, 222.

36. Hammes and Wills 2005, 502, 504–509.

37. Transnational corporations based in Western Europe and Japan had, however, increased in number from about 1950.

38. Hayek 1976 [1944], 1945.

39. Duménil and Lévy 2005, 11.

40. Frieden 2006; Campbell 2005, 188–190.

41. Frieden 2006, 343.

42. Frieden 2006, 339–342.

43. The winter of 1978–1979 preceded the 1979 election won by Thatcher's Conservatives. The Callaghan Labour government had succeeded in reducing

inflation by securing union cooperation in limiting wage increases. Attempts to retain a cap on wages at the end of the accord period led to conflict between government and unions, and many strikes affected essential services during an extremely cold winter.

44. PBS 2002, chap. 13.

45. Ranelagh 1991, ix.

46. PBS 2002, chap. 13.

47. Frieden 2006, 372–373.

48. Ebenstein 2003, 52.

49. "The Brick" was a 500-page neoliberal handbook drafted by a ten-man group assembled by Orlando Saenz, president of the National Association of Manufacturers in Chile. Eight of them had studied at the University of Chicago. Saenz took this step following a meeting between the leaders of Chile's top businesses that had decided that Allende's government was "incompatible with freedom ... and the existence of private enterprise" and had to be overthrown (Klein 2007, 70–71).

50. Broder 2002.

51. Blumenthal 1986, 33; Heritage Foundation 2010.

52. Beder 2006b, 29.

53. Harris 1997.

54. Friedman 1999, xvii.

55. Friedman 1999, 9.

56. Klein 2002, 200.

57. Solomon 2003.

58. Friedman 1999, 9–11.

59. Lynch 2008.

60. Dobbs 2008.

61. Department of Foreign Affairs and Trade (Australia) (DFAT) 2007, 21.

62. DFAT 2007, 1–3, 20–21, 23.

63. Renner 2008.

64. World Bank 2007; DFAT 2007, 6, 8–15, 18–19.

65. Garnaut 2008, 9.

66. Wackernagel et al. 1997.

67. The 2011 data released in the BP annual statistical review for 2012 estimated global consumption at almost 89 million barrels per day (mb/d) and global production at just over 84 mb/d. US per capita consumption was calculated from the BP statistics to be 0.061 mb/d, which was 8.5 times greater than China's per capita figure. At US levels, China would require 82 mb/d.

68. The World Resources Institute (2005) *Earth Trends* website estimated global paper consumption at 352 million metric tons per annum for 2005. US annual

per capita consumption was estimated at 297 kilograms, six times greater than Chinese per capita consumption of 44.7 kilograms. At US levels, China would require 353 million metric tons.

7 "Development" and Globalization

1. Smith 1852 [1776], 258–259.

2. Galeano 2008, 39–40; Prakash 1991, 115; Chang 2009, 7; Blom 2008, 93–94, 99–100.

3. UNICEF 2009, 2.

4. Food and Agriculture Organization of the United Nations 2009; 2010, 4; 2013.

5. Arndt 1981.

6. Truman 1949.

7. See Agarwala and Singh 1958.

8. Rostow 1960.

9. Lewis 1954, 416–417.

10. UN Department of Economic and Social Affairs 1951.

11. The controversial economist Greg Clark (2008) has even hypothesized that the dispossessed and overworked laborers were so downtrodden that they had few surviving offspring and literally died out. Clark suggests that the descendants of the new entrepreneurial classes, battling each other for a stake in the new order, represent the major genetic heritage of the modern English.

12. Arrighi, Silver, and Brewer 2003.

13. Escobar 1995, 5.

14. Escobar 1995, 80.

15. Li 2009, 74.

16. Frank 1978, 154.

17. Arrighi, Silver, and Brewer 2003, 8–11.

18. Harris 1986, 202.

19. Ruggiero 1997.

20. Arrighi, Silver, and Brewer 2003, 14–15.

21. Easterly 2001, 135–136, 154.

22. When Adam Smith celebrated the virtues of free trade, he was largely referring to trade within Britain, which was impeded between cities and counties. Ricardo's notion of "comparative advantage," still trotted out by neoliberals, assumed that capital would *not* be mobile between nations, which is far from the case today, when capital is so mobile that it can be redirected with a few taps on a keyboard.

23. Birdsall 2007.

24. Department of Foreign Affairs and Trade (Australia) (DFAT) 2012.

25. Griffiths 2002; Ben-Ari 2002, 9; Soares 2005.

26. Chang 2002, 2.

27. Chang 2002, 22, 30.

28. Wade 2004, 581.

29. Vidal 2008; 2007.

30. International Air Transport Association 2013.

31. Payer 1991, 39–41. In 1981, "Latin America was lent an incredible $61 billion, of which $45 billion immediately went back to the banks as repayments" (Hanlon 2002, 28).

32. Brzoska 1983, 274–275; Perlo-Freeman and Perdomo 2008; Jubilee Research 1998.

33. Mandel 2006, 16–20.

34. Mandel 2006, 31; Hanlon 2006, 211. Precedents for cancellation of debts date from the refusal of the United States to honor Cuba's debt to Spain in 1898 and extend to the 2003 write-off of part of the Iraqi debt contracted by Saddam Hussein.

35. Hanlon 2006, 215–216.

36. Karliner 1997.

37. Buckley 2002/2003.

38. Speaking to Stephen Long on September 22, 2009 (Long 2009).

39. UN Conference on Trade and Development 2004, 10.

40. Hanlon 2002, 27–28.

41. George 1992, xv–xvi.

42. Edward 2006, 1667–1668.

43. Reddy and Pogge 2005.

44. Reddy and Pogge 2005.

45. Edward 2006, 1680.

46. Wade 2004, 570–571. More recently (2013), Gallup estimated median per capita household income at $2920 (PPP adjusted). Although significantly greater than Wade's figure for 1999, it remains low—an indication of the persistent gulf between the richer and poorer halves of the world's population (Phelps and Crabtree 2013).

47. Woodward and Simms 2006, 1.

48. Reddy 2008.

49. Wade 2007.

50. Johnston 2005.

51. UN Development Programme 2005, 17–18.

52. Henry 2012; Shaxson Christensen, and Mathiason. 2012.

53. Davies et al. 2008, 7.

54. Edward 2006, 1673.

55. Roy 2002, 2–3.

56. Edward 2006, 1673–1680. In Edward's study period (1993–2001), the poorer half of the world's population received less than ten percent of the increased income between them; by 2014, these 3.5 billion people were calculated to own about the same as the 85 richest individuals on earth (Fuentes-Nieva and Galasso 2014).

57. Krueger 2004.

58. Woodward and Simms 2006.

59. Murdoch 2008.

60. World Bank 2007, 73; DFAT's APEC report (2007, 6) was based on these figures.

61. UN Millennium Development Goals 2013, 42–46; World Bank 2012, 72.

62. Buckley (2009). In 1999, these five were Korea, Taiwan, Singapore, Hong Kong, and Mexico, though Brazil might have to be added as a sixth by 2011, and coastal China would also be a candidate (Buckley, pers. comm.). In several of these examples massive inequality persists. Such polarization makes averages suspect.

8 Growth and "Sustainable Development"

1. World Commission on Environment and Development (WCED) 1987, 40–41.

2. WCED 1987, 43–52, 59.

3. WCED 1987, 67; see chap. 7.

4. WCED 1987, 85.

5. WCED 1987, 5–6.

6. WCED 1987, 1.

7. Sachs 1999; *Ecologist* 1993.

8. Middleton, O'Keefe, and Moyo 1993.

9. UN Conference on Environment and Development (UNCED) 1992a, 1992b.

10. Grubb et al. 1993, 98–103.

11. Khor 1997; Middleton, O'Keefe, and Moyo 1993, 2; Grubb et al. 1993, 106, 125.

12. E.g., *Economist* 2003; Goodman 2011. The only approximation of the phrase I found in contemporary records was written by Elmer-DeWitt (1992) for *Time* magazine, who attributes it to "US negotiators." The phrase might also have arisen when the *Guardian* reported the president saying that "we cannot shut down the lives of many Americans by going extreme on the environment" (Walker et al. 1992, 1).

13. Middleton, O'Keefe, and Moyo 1993, 4, 11.

14. Grubb et al. 1993, 86.

15. Andrews 2001. This attitude is mirrored in Australia, though rarely stated quite so explicitly.

16. UN Centre on Transnational Corporations (UNCTC) 2013. Plehwe (2009, 32–33) notes the role of the Heritage Foundation (see part III) in the demise of the UNCTC. The Heritage Foundation's "applied neoliberal policy knowledge" played a key role in undermining the UNCTC and shifting development emphasis from notions of economic independence and sovereignty to those of "good governance and corporate citizenship expressed by the amicable relations between corporate and political leaders in the UN Global Compact frame."

17. Sachs 1999; Redclift 2005; Pandit speaking to Latha Jishnu, October 19, 2008 (Jishnu 2008); Daly 1993, 267–273; 1996, 166–7. Growth here refers to growth in *material* flows—extraction of resources for physical production and the resulting waste. See also chapter 2.

18. Sachs 1999, 29.

19. Seabrook 2002.

9 Growth and Its Outcomes for the Poor

1. More 1999 [1516], 101.

2. Lynch 2004.

3. Landesa Rural Development Institute 2011, 2. Landesa, with headquarters in Seattle, works to secure land rights for the world's poorest 2.47 billion people, living on less than $2 a day.

4. Taylor 2005a, 2005b, 2005c.

5. Elkington and Lee 2005.

6. Balazovic 2011.

7. McLaughlin 2012.

8. Padel and Das 2007, 25.

9. Wasley 2009.

10. Fernandes 2006, 113–115.

11. Lewis 2006.

12. Guha 2005.

13. *Business Standard* 2006.

14. Ray and Chaudhury 2008.

15. Goldman 2005, viii–ix.

16. Speaking to Norman Swan on January 25, 2010 (Swan 2010). Sen was convicted and sentenced to life imprisonment under antiterrorist laws on what appear to have been trumped-up charges relating to his visits to prisoners. After

several periods of imprisonment, he was granted bail by the Indian Supreme Court in April 2011, pending the hearing of his appeal.

17. Patnaik 2007, 3132–3135.
18. Patnaik 2009, 69.
19. International Institute for Population Sciences and Macro International 2007, 273.
20. Page 2007.
21. Magnier 2013.
22. Friedman 1980.
23. Davis 2004, 5.
24. UN Department of Economic and Social Affairs (UN/DESA) 2006, 19.
25. Yeung and Lo 1996, 41.
26. UN/DESA 2006, 2.
27. World Bank 2008, ii.
28. Davis and Belkin 2008.
29. UN Human Settlements Programme (UN-Habitat) 2003a, xxv–xxvi; 2003b, 12.
30. UN-Habitat 2003a, xxvii, 46.
31. Mulama 2006; UN-Habitat 2003b.
32. Brugmann 2009.
33. Chu 2008; Apte 2008.
34. De Launey 2006; Amnesty International 2008.
35. Carmichael and Nara 2001.
36. Rostow 1960, 10.
37. Murdoch 2008.
38. Edward 2006, 1677.
39. Reddy 2008.
40. Edward 2006, 1673–1680.
41. Lappé, Clapp, Anderson, et al. 2013a.
42. Food and Agriculture Organization (FAO) 2009, 2010.
43. UN Millennium Development Goals (UNMDG) 2009, 11.
44. Lappé, Clapp, Anderson, et al. 2013a.
45. UNMDG 2013, 6. Extreme poverty is defined by the World Bank's $1 a day metric.
46. FAO 2012, 55.
47. FAO 2012, 8, 4–5, 15–27.
48. Lappé, Clapp, Anderson, et al. 2013b, 9–10.
49. FAO 2012, 27.

50. Lappé, Clapp, Anderson, et al. 2013b, 13–14; Guereña 2010, 25–28.

51. ILO 2006, 13; UNMDG 2013, 6, 8.

52. Ekins 1991, 250–252.

53. Goodman 2009a.

54. Goodman 2009b.

55. Kempf 2008, 47–8; UN Development Programme (UNDP) 2005, 18.

56. Hervé Kempf describes this global oligarchy as a "predatory and rapacious ruling class ... blind to the explosive power of manifest injustice. And blind to the poisoning of the biosphere" (Kempf 2008, 59)

57. Vitali, Glattfelder, and Battison 2011, 3–4, appendix S1. The top fifty holders of control are listed in appendix S1, 17.

58. Cited in Matthews 2008.

59. Musgrove 2006; Connor and Dent 2006, 4, 8; Adams and McLaughlin 2009; *Sydney Morning Herald* 2008a.

60. Bengali 2006; Ackerman 2008; Egan 2007; Hamer 2007. Corporate miners have also operated without precaution, fouling rivers and destroying food gardens and fishing grounds (Burton 1999; Forero 2009; Gumbel 2007; Perlez and Bonner 2005); in 2006, a gas drilling rig near Surabaya in Indonesia pierced a toxic mud reservoir that continued to inundate fields and villages for years (Mydans 2008).

61. Kahn and Landler 2007.

62. Auffhammer and Carson 2008.

63. Environment News Service 2008.

64. Shiva 2008, 45.

65. Shiva 2008, 45.

66. Waldman 2005.

67. Shiva 2008, 49–52.

68. World Bank 2012, 72.

10 Propaganda

1. Arndt 1978, 71, 73.

2. Arndt 1978, 18, 27–37.

3. Leach 1993, 8.

4. Arrighi 1994, 241, 281; Mitchell 1989, 82.

5. Chandler 1977, 321.

6. Chandler 1977, 334, 328.

7. Chandler 1977, 317.

8. Arrighi 1994, 287; Marchand 1998, 10; Chandler 1977, 368–372.

9. Chandler 1977, 241; Arrighi 1994, 270–71.

10. Arrighi 1994, 281, 72. These internal trades facilitate the (perfectly legal) tax avoidance practised by many TNCs, in which profits are shifted to jurisdictions that levy little or no tax.

11. CNN *Money* 2012.

12. Leach 1993, 6–8; Lamoreaux 1985, 159.

13. Marx 1954b [1887], 714–715.

14. Griffin, Wallace, and Rubin 1986, 155.

15. Cited in Marchand 1998, 42.

16. Le Bon 1960 [1895], 9.

17. Ewen 1996, 41–43.

18. *New York Call*, January 28, 1915, cited in Ewen 1996, 79–80.

19. Marchand 1998, 43.

20. Creel 1920.

21. Creel 1918, 188.

22. Ewen 1996, 102–104, 121, 423.

23. Lippmann 1920, 5, 12–13.

24. Cited in Cochran and Miller 1961 [1942]), 333.

25. Bernays 2005 [1928], 54, 48.

26. Ewen 1996, 219.

27. Ewen 1996, 223.

28. AT&T president, cited in Cochran and Miller 1961 [1942], 340.

29. Cochran and Miller 1961 [1942], 342.

30. Soule 1947, 142.

31. Cochran and Miller 1961 [1942], 338.

32. Cited in Lee 1929, 37.

33. Ewen 1996, 180.

34. Soule 1947, 142.

35. Cited in Marchand 1998, 4.

36. Bernays 2005 [1928], 94.

37. Lane 1950, 18–19. This was the first congressional investigation of NAM's propaganda activities. Although numerous bills to regulate lobbying were subsequently introduced, none was passed (Carey 1997, 21).

38. Walker and Sklar 1938, 428.

39. Cited in Marchand 1998, 202–203.

40. Fones-Wolf 1994, 25.

41. NAM 1936, "What is your American system all about?," cited in Ewen 1996, 303–305.

42. *Fortune* 1938, 51.
43. Roosevelt 1944, 111–116.
44. *Fortune* 1938, 51.
45. NAM action chart reproduced in US Congress 1939, 283.
46. Ewen 1996, 314.
47. Walker and Sklar 1938, 433.
48. Lippmann 1927, 37–38.
49. Ewen 1996, 312.
50. I am indebted to Ewen's book (1996) for alerting me to the existence of this revealing photograph, showing a breadline under an American Way billboard during the 1937 Louisville Flood, Kentucky. It was taken by the celebrated photographer Margaret Bourke-White.
51. Walker and Sklar 1938, 123.
52. Ewen 1996, 312–320.
53. US Congress 1939, 218.
54. US Congress 1939, 178.
55. Ewen 1996, 327.
56. Ewen 1996, 322–335.
57. Ewen 1996, 234–235.
58. Marchand 1998, 214, 226, 207.
59. US Chamber of Commerce, cited in Marchand 1998, 137; Wall 2008, 48–49.
60. Marchand 1998, 235.
61. NAM executives conference, cited in Ewen 1996, 340.
62. Marchand 1998, 336; chairman of NAM's executive committee in 1942, cited in Fones-Wolf 1994, 26.
63. Marchand 1998, 342–349.
64. Chomsky 2001.
65. Cited in Marchand, 322.
66. Scott 1946, 14.
67. The Psychological Corporation was founded to support applied psychology in areas such as advertising and vocational testing and selection (Sokal 1981).
68. Link 1948, 14; Link cited in Carey 1987a, 8.
69. Carey 1987a, 8.
70. Fones-Wolf 1994, 3, 53.
71. Cited in Ewen 1996, 386.
72. *Fortune* 1949, 68.
73. Griffith 1983, 389; Carey 1987a, 9.

74. Bernays 2005 [1928], 37–38.
75. See Mark Crispin Miller's introduction in Bernays 2005 [1928], 15–18.
76. Walker and Sklar 1938, 121.
77. Brinkley 1989, 98–100.
78. Arndt 1978, 13.
79. Tom Yntema, CED board member, 1947, cited in Maier 1987, 65.
80. Maier 1987, 126.
81. Griffith 1983, 390–391; Ad Council president, cited in Griffith 1983, 393.
82. These merged to become the AFL-CIO in 1955. Union membership in the United States had tripled to nine million in the decade up to 1939 and jumped to 15 million in the course of the war; the union movement's postwar agenda embraced tax reform, unemployment insurance, price controls, an increase in the minimum wage, and above all full employment. These objectives reflected values that the Great Depression had fostered, such as social justice and a "fair share" for everyone, redistribution where required, and the right to organize.
83. Griffith 1983, 400–401.
84. Griffith 1983, 402; Fones-Wolf 1994, 51.
85. Carr 1949, 2.
86. Beder 2006a, 45–48.
87. Beder 2006a, 53.
88. Beder 2006a, 52, 59. Beder is a professor in the School of Humanities and Social Inquiry at the University of Wollongong.
89. Fones-Wolf 1994, 3.
90. Cited in Green and Buchsbaum 1980, 14.
91. Maier 1987, 130.
92. Murdoch 2008.
93. Maier 1987, 68.
94. Maier 1987, 123, 142–143.
95. Maier 1987, 146.
96. Vogt 1948, 63–67.

11 Sleight of the Invisible Hand

1. Beder 2006a, 58.
2. Walker and Sklar 1938, 113.
3. Sandbach 1978, 108.
4. Powell 1971.
5. Lapham 2004.
6. Powell 1971.

7. Edwards 1997, 8–9.

8. Kaiser and Chinoy 1999; Smith 1991, 200.

9. The Manhattan Institute was originally called the International Center for Economic Policy Studies when it was founded by Anthony Fisher and William J. Casey. It was renamed in 1981. Fisher was also instrumental in the founding of the Pacific Institute.

10. The Heartland Institute provided Australia's Senator Steve Fielding with his climate change denialist talking points in 2009. Fielding, who represented the Family First party, to the right of the conservative Coalition, attended Heartland's Second International Conference on Climate Change in New York, where he heard from the world's leading proponents of denial (Heartland Institute 2013).

11. Lapham 2004; Mayer 2010.

12. Jones 2006.

13. Rippa 1988, 270.

14. Hirsch 1975, 64–65; Parenti 1986, 72.

15. Hirsch 1975, 70–71; Parenti 1986, 73–74.

16. Cited in Parenti 1986, 74.

17. Beder 2006a, 65–67. Surveys designed to demonstrate "economic illiteracy" had already been deployed in the late 1940s by the Opinion Research Corporation (ORC). Drawing on original ORC documents, Beder (2006a, 39–42) reproduces a series of the self-serving questions used at that time and notes that "executives scored much better than teachers or high school students."

18. Beder 2002a, 18.

19. Lever 1978, 71; Weaver 1977, 188.

20. Beder 2006a, 73; 2002a, 19.

21. Stone 1991, 200–204.

22. Bruce Herd, unpublished PhD thesis, cited in Murray 2006, 257.

23. American Enterprise Institute 2010.

24. Carey 1997, 91.

25. AIMS 1972. AIMS began as Aims of Industry. It was known as Aims for Freedom and Enterprise in the late 1970s and is now called Aims of Industry again, or simply AIMS.

26. Cockett 1994, 306–308; PR Watch 2004.

27. Murray 2006, 156.

28. Liberal Party of Australia—Queensland Branch 2007. Australia's Liberal Party is the leading party of the Right, though it has governed since 1949 in coalition with the rural-based National Party.

29. *IPA Review* 1955; Scott 1950, 11, 429–459.

30. Moore and Carpenter 1987, 156–157.

31. Cahill 2004, 293.

32. Oliver Smedley, cited in Cockett 1994, 131.

33. Lindsay 1996.

34. Beder 2006a, 45.

35. Friedman 1999, 105–117.

36. Humphrys 2004.

37. Stone 1991, 200–201.

38. Cited in Murray 2006, 156.

39. Kristol 1977, 3–4.

40. Bagnall 2004.

41. Lindsay 1996, 17; Cahill 2005.

42. Lindsay 1996, 19–20.

43. Hugh Morgan secured funding from both local corporations and transnational corporations: CRA, BHP, Santos, Shell, and the Adelaide *Advertiser,* as well as Western Mining Corporation. All but the *Advertiser* are mining and energy companies (Cahill 2004, 211).

44. Bagnall 2004, 25.

45. Carey 1987a, 10.

46. Allen 1976.

47. Cahill 2010, 15.

48. Keavney 1978, 66.

49. Allen 1976.

50. Taibbi 2009.

51. Labor prime minister Gough Whitlam was dismissed by the governor-general on November 11, 1975, in a constitutional crisis precipitated by the conservative Coalition, which was elected to office in the subsequent poll.

52. Allen 1976.

53. Harris 1978, 11.

54. Bagnall 2004, 25.

55. Carey 1987a, 11.

56. Oreskes and Conway 2010, 1–35; Proctor 2012.

57. Ivens 1978, v.

58. Carey 1987a, 11; 1987b, 169.

59. Cockett 1995, 3–4.

60. Cahill 2004.

61. Sandbach 1978, 99–101, 105–107.

62. Robert Gottlieb, cited in Buell 2003, 10.

63. Beder 2002a, 16–21.

64. Green and Buchsbaum 1980, 81.

65. Carey 1997, 89.

66. MacDougall 1980, 14.

67. Beder 2002a, 18; Green and Buchsbaum 1980, 7.

68. Green and Buchsbaum 1980, 7; Cushman 2001.

69. Green and Buchsbaum 1980, 68, 73, 90.

70. Austin 2002, 85–86; Green and Buchsbaum 1980, 94.

71. The key question was phrased as follows: "Those in favor of setting up an additional consumer protection agency on top of all the other agencies," a question that falsely implied that consumer protection measures were already in place and a consumer protection agency would be redundant. This was not the case.

72. Green and Buchsbaum 1980, 113, 109.

73. Carey 1997, 94.

74. Vogel 1990, 193.

75. Beder 2002a, 17–18.

76. Rowell 1996, 22; Beder 2002a, 22.

77. Beder 2002b, 32.

78. Austin 2002, 93, 90; Revkin 2009.

79. Austin 2002, 85–86; Rowell 1996, 14–22; Beder 2002a, 47–62; Greenpeace 2007.

80. R. L. Barry, cited in Rowell 1996, 14. See also chapter 12.

81. Austin 2002, 85–86, 90.

82. Mayer 2010.

83. Austin 2002, 83, 89.

84. Dolny 1996.

85. Cushman 1998.

86. Walker 1998.

87. Walker 1998.

88. Sample 2007.

89. Gelbspan 2005.

90. Begley 2007.

91. Cubby 2012.

92. Mooney 2005.

93. *Mother Jones* 2005.

94. Greenpeace 2013; see Adam 2006 and Ward 2009 for Bob Ward's efforts to force Exxon to be accountable.

95. Snowe and Rockefeller 2006. See Brulle (2013) for recent analysis of how the opposition to climate science in the US has been funded since 2003. A feature is the increasing importance of the "black box" Donors Trust and Donors Capital Fund, in which contributions cannot be traced to contributors. When these

entities are added to the leading family foundations (Scaife, Koch, Bradley, Howard, Pope, Searle and Templeton) $200 million has been donated in the period from 2003 to 2010, primarily to the think tanks (AEI, the Heritage Foundation, the Hoover Institution, and the Cato Institute being prominent recipients). Opposition to climate science continues to be integrated into the overall US conservative movement, while the newly active "dark money" funds have joined foundations as the principal vehicles for corporate funding.

96. Associated Press, cited in Rowell 1996, 7; Green and Buchsbaum 1980, 15.

97. Canan and Pring 1988, 506.

98. Beder 2002a, 63–74; 2004b, 1–3; Walters 2003.

99. An exception was the suit brought by Tasmanian timber company Gunns against twenty environmentalists, including Greens senator Bob Brown and the Wilderness Society in 2004. It failed to silence anyone and left Gunns out of pocket (Ogle 2010), mainly because Brown had considerable support and a high public profile.

12 The Free Market Assault on Environmental Science

1. Dunlap and McCright 2010, 241–243.

2. Lahsen 2008, 213–214.

3. Dunlap and McCright 2010, 243; Jacques, Dunlap, and Freeman 2008, 353; Buell 2003.

4. Begley 2007.

5. Jacques, Dunlap, and Freeman 2008, 361.

6. Jacques, Dunlap, and Freeman 2008, 354.

7. Duncan 1985, 40.

8. Jacques, Dunlap, and Freeman 2008, 360; 363–4.

9. Jacques, Dunlap, and Freeman 2008, 361–363.

10. Riley Dunlap at the AAAS, broadcast on ABC Radio by Robyn Williams on April 3, 2010 (Williams 2010).

11. Bush 2003.

12. Cited in Austin 2002, 80.

13. Begley 2007.

14. Mooney 2005.

15. Bagla 2002.

16. Eilperin 2005.

17. Revkin 2005.

18. Michael E. Baroody, NAM executive vice president, May 16, 2001, cited in Edwards 2004.

19. US Chamber of Commerce, July 19, 2001, cited in Edwards 2004.

20. Dolny 1996.

21. American Council on Science and Health (ACSH) 2013.

22. Rampton and Stauber 2001, 228.

23. Consumer Reports 1994, 319; Markowitz and Rosner 2002a, 288.

24. Austin 2002, 81–82.

25. Lapham 2004; Austin 2002, 82.

26. These appointments were specified on the 2004 SEPP website as of March 22, 2011. The site no longer carries any link to such details.

27. Greenpeace 2004b. Greenpeace's Exxon Secrets website, ExxonSecrets.org (Greenpeace 2013), allows one to explore the multiple weblike connections between think tanks and front groups, their personnel, and major corporations.

28. Jacques, Dunlap, and Freeman 2008, 356.

29. Bailey 1993; 2002.

30. Shnayerson 2007.

31. Humphrys 2004; Paxman 2006.

32. Jacques, Dunlap, and Freeman 2008, 370.

33. Simon 1981; Barney 1982; Simon and Kahn 1984.

34. Oreskes 2004; Anderegg et al. 2010; Cook et al. 2013; IPCC 2001, 2007, 2013.

35. Karl 2007.

36. Newport 2010.

37. CPI Strategic 2009.

38. Jeffords was by then an independent. He had defected from the Republican Party some years earlier.

39. Cited in Oreskes and Conway 2008, 76.

40. Jackson 2010.

41. Legacy Tobacco Documents Library 1969, 4.

42. Oreskes and Conway 2010, 15, 142–143.

43. Oreskes and Conway 2010, 244–255; Burke 1993.

44. Nierenberg 1984; Golden 1982; Oreskes and Conway 2010, 103–106.

45. Oreskes and Conway 2010, 113, 118, 125–129.

46. US Senate 2007; Cohen and Dreier 2009; Krupp 2002.

47. Oreskes and Conway 2010, 169–174.

48. National Research Council (NRC) 1983, 151, 449, 474–482. The assessment of the status of climate modeling was conducted not by specialist climate modelers but by Nordhaus, while Schelling wrote the last chapter on policy implications. The judgments of the scientists, some much more alarmed than the economists, appeared between these chapters and were not much reflected in the chapters that framed the report.

49. NRC 1983, 1–4.

50. Shabecoff 1983.

51. Gelbspan 2005.

52. Pearse 2007, 22.

53. Pearse found that the "missing in action" sectors of the business community failed to pursue a climate policy in opposition to the AIGN for a range of reasons, some ideological and some structural. Australian insurers, for example, were lukewarm on climate policy, despite potential heavy losses from global warming; the industry harbored climate change opponents and was more focused on financial issues and its carbon-intensive investment portfolio than on its future possible losses. See Pearse 2007, 176–192.

54. Pearse 2007, 227–238.

55. Pearse 2007, 228–238.

56. Woodley 2011; AAP 2011.

57. Pearse 2009, 31, 37.

58. Pearse 2009, 25–27, 38–40, 48.

59. Warren 2008; Brewster 2008.

60. Pearse 2009, 69.

61. Smil 2006, 21. Vaclav Smil cannot be regarded as a biased environmentalist. He has published with the American Enterprise Institute and the free market website techcentral.com.

62. Australian Treasury 2008, 155. Calculations by Pearse 2009, 115.

63. Pearse 2009, 67–68.

64. World Coal Association 2013.

65. Jacobs 2011.

66. Lane 2011. At the Kyoto conference, the Howard government made its agreement to join the Kyoto Protocol conditional on the adoption of an ETS; this stipulation was included at the insistence of both Australia and the United States—along with that of other developed nations (Leggett 2001, 317; Mugliston 1998, 3).

67. The CEO of the Australian Chamber of Commerce and Industry, Peter Anderson, wrote: "We don't need to take a blow for the global team if other nations can't get their act together. And they can't" (Anderson 2011).

68. Productivity Commission (Australia) 2011. Though characterized, especially by its opposition, as a tax, the measure was to default to an ETS after three years. Although this ETS would limit overseas purchase of permits to 50 percent until 2020, the longer term future of outsourcing would not be determined until then. Strengths of the new policy were the addition of significant funding for clean energy—with "clean coal" deleted from this category—as well as the establishment of independent institutions to advise on the science, disperse the funds, monitor progress, and administer the pricing mechanism (Department of Climate Change and Energy Efficiency [Australia] 2011, 31–32, 63). The newly elected

Abbott government has pledged to abolish the carbon price and the Clean Energy Finance Corporation. It disbanded the advisory Climate Commission within days of being elected.

69. Goodman 2010b.

70. Luntz 2002. Luntz has claimed that the environmental section of his booklet was written in the mid-1990s (PBS 2006), suggesting that it did not represent his own views about global warming in 2002 when it was leaked. If this is true, however, Luntz is instructing his clients to lie when that is to their advantage. Whatever Luntz's actual personal position, his memo shaped Bush's environmental policy at the time.

71. Luntz 2002, 137–8.

72. Oreskes 2004.

73. Luntz 2002, 136–139.

74. Leggett 2001, 319. It was acknowledged at Kyoto that developing countries were responsible for very little of the CO_2 already in the atmosphere and were entitled to different targets. The US Senate, however, had already signaled its unwillingness to ratify a treaty that did not specify obligations for developing countries. Although the Senate's resolution (S. Res. 98) has been interpreted as being consistent with the principle of "common but differentiated responsibilities" (Harris 1999), nonetheless, it could also be interpreted as demanding action from developing countries. This approach was further emphasized as Bush took over the White House.

75. Luntz 2002, 136–137, 141.

76. Luntz 2002, 132.

77. Luntz 2002, 143.

78. Mooney 2004.

79. Ong and Glantz 2001, 1749–1750.

80. Markowitz and Rosner 2002a, 179.

81. Markowitz and Rosner 2002a, 290–292.

82. Markowitz and Rosner 2002b, 502.

83. Lead was prominent among early toxins, and reformers such as Alice Hamilton were already urging protection for workers in the unregulated lead industry and labeling of paint so that painters would know when lead was present.

84. Markowitz and Rosner 2002b, 502; 2002a, 7, 178.

85. Markowitz and Rosner 2002a, 9–11.

86. Monbiot 2006.

87. Greenpeace 2004a.

88. JunkScience.com.

89. ACSH 2013.

90. Marshall Institute 2009.

91. Marshall Institute 2013.

92. Singer 2000.

93. Brown 1996.

94. Herrick and Jamieson 2001.

95. Cited in Oreskes and Conway 2010, 59–62.

96. Oreskes and Conway 2010, 197–211.

97. Austin 2002, 83–84.

98. McCright and Dunlap 2000, 510–513.

99. Friedman 1999, 107.

100. McNeill 2001.

101. US Energy Information Administration 2013.

102. Cited in Rowell 1996, 71.

103. Long 1991.

104. Helvarg 1994, 7.

105. Egan 1991.

106. Krakauer 1991.

107. Arnold 1981, 248.

108. Singer 1989. I am indebted to Oreskes and Conway (2010) for alerting me to this paper by Singer. Two versions are in circulation. One is archived by the Heartland Institute and cited in Oreskes and Conway (2010). I located a slightly different one, originally carried on Singer's SEPP website and now archived by Greenpeace.

109. Cited in Greider 1992, 24.

110. Cited in Bass 2004.

111. Oreskes 2004.

112. Cited in Gelbspan 2005.

113. Lippmann 1920, 4, 11, 39.

114. Cullen 2006, 2.

115. Speaking to Phillip Adams on March 23, 2011 (Adams 2011).

116. Boykoff and Boykoff 2004, 127, 129–130.

117. Boykoff and Boykoff 2004, 132.

118. Griffiths 2011.

119. Newman 2010. The Howard government had already appointed the global warming "skeptics" Janet Albrechtsen, Ron Brunton, and Keith Windshuttle to the ABC board, so Newman was in like-minded company.

120. Union of Concerned Scientists 2013.

121. Beder 2001.

13 International Brakes on Environmental Priorities

1. Mirowski 2009, 438–439.

2. IMF 2013a; Bretton Woods Project 2010.

3. Beder 2006b, 42. Big finance has recently supplied European heads of state as well. In the course of the European sovereign debt crisis, the man appointed to be Italy's prime minister in 2011 had been a senior finance professional with Goldman Sachs, and the man appointed to lead Greece was also connected to Goldman, having served as Greek Reserve Bank governor at the time when Goldman helped to obscure the Greek debt position, allowing it to join the Eurozone (Roche 2011).

4. Morley 1986.

5. Morley 1986.

6. Sachs 2008.

7. See Fairfax economist Ross Gittins (2002) for a trenchant critique of the IMF's instructions to Australia at that time.

8. Williamson 1989. Williamson did not approve of the term being used as a shorthand for the neoliberal agenda by its critics. His description of the Washington Consensus did, however, set out the major elements of the neoliberal doctrine rather succinctly.

9. Beder 2006b, 45–46.

10. Buckley 2002/2003, 60–61.

11. Economic standstill combined with inflation; see chapter 6.

12. Organisation for Economic Co-operation and Development 2010a, 2010b.

13. World Economic Forum (WEF) 2010a, 2010b.

14. Schwab 2009, 48; Naidoo 2014; Burrow personal communication, 2014. After a distinguished career in the Australian union movement, Sharan Burrow has served in international labor organizations since the early 2000s and was elected General Secretary of the International Trade Union Confederation in 2010. The Grameen Bank provides microcredit to the rural poor of Bangladesh. Yunus and the Grameen Bank were jointly awarded the Nobel Peace Prize in 2006; he has received several invitations to Davos since that time.

15. Vitali, Glattfelder, and Battison 2011.

16. WEF 2010a, 2010b.

17. Machan 1999.

18. WEF 2010c.

19. CNN 2009; G20 2013.

20. Spain is indirectly represented by the EU delegate, Iran not at all.

21. Støre 2010.

22. Duffy 2004.

23. Crozier, Huntington, and Watanuki 1975, 113, 98, 92.

24. Rockefeller 1999.

25. Golub and Townsend 1977.

26. Any web search including the terms "Club of Rome" and either "conspiracy," "Bilderberg," or "Trilateral" will provide access to such claims. All of them fear a conspiracy to establish world government, and many suggest that environmental activists are in league with the business elites.

27. Barker and Mander 1999, 2.

28. Dryden 1995, 12–13.

29. World Trade Organization (WTO) 2010a.

30. Dryden 1995, 13, 25–26, 30, 31–32.

31. Beder 2006b, 111. In his history of the WEF, Schwab (2009, 50) claims that the idea of the Uruguay Round was launched at a Lausanne meeting set up by the WEF between leading trade ministers and the head of GATT.

32. Barker and Mander 1999, 5, 8.

33. Beder 2006b, 113.

34. Arrighi 1994.

35. Beder 2006b, 118.

36. Beder 2006b, 110; Li 2009, 74.

37. Based on Tridico 2011.

38. Montague 1999.

39. Barker and Mander 1999, 25, 27.

40. Montague 1999.

41. WTO 2010b.

42. Barker and Mander 1999, 25, 27.

43. Barker and Mander 1999, 25.

44. European Commission 2009.

45. WTO 2010c.

46. Das 1993, xix.

47. Woolley 2007; Phillips 2006, 266.

48. Sassen 2009.

49. Tett 2009; Bernanke 2004.

50. Patnaik 1999.

51. Omidi 2010; Dennis and Mufson 2010; Taibbi 2010a, 2010b.

52. Carnegy and Peel 2012; Barker and Polity 2013.

53. US Congress 1993, 45. I am indebted to Wallach and Tucker (2010, 7) for alerting me to these comments.

54. Barker and Mander 1999, 37.

55. WTO 2010d.

56. Barker and Mander 1999, 39; WTO 2010e.
57. UN 2009, 39, 104.
58. Reuters 2010; G20 2012.
59. Pell and Eaton 2010.
60. Goodman 2010a.

14 *The Limits to Growth* after Forty Years

1. Simon 1982, 207.
2. Cato Institute 2013; Competitive Enterprise Institute 2013.
3. Bailey 1989a, 1989b.
4. Meadows, Meadows, and Randers 1992.
5. Bailey 1993, 67.
6. UN Environment Programme 2002, chap. 1, 2–3.
7. Bailey 1993, 67.
8. Meadows et al. 1972, 64–68.
9. Meadows et al. 1972, 75.
10. Meadows, Meadows, and Randers 1992, 2.
11. Meadows et al. 1972, 24.
12. Passell, Roberts, and Ross 1972.
13. Meadows 2007, 408–413.
14. *Economist* 1997.
15. Meadows et al. 1972, 64–68.
16. Jaffe and Manning 2000, 16.
17. Ridley 2001.
18. Fullerton 2002.
19. Quiggin 2002. This is not what table 4 claims, and the figures in it are not part of the World3 model anyway. Quiggin's quoted blog is no longer available; a search for "Club of Rome" will locate similar views.
20. Barnett and Morse 1963.
21. Cleveland 1991, 294–295; Zencey 2013; McNeill 2001.
22. Krauss 2008.
23. Campbell 1996; 1997; Campbell and Laherrère 1998.
24. Williams 2011a.
25. Campbell and Laherrère 1998, 82.
26. Or "tight oil," see chapter 2, note 27.
27. Kohler 2012; Krauss and Lipton 2012; Luce 2011.
28. Hughes 2013, 22; Berman and Pittinger 2011.

29. Osborn et al. (2011) linked methane contamination of groundwater to the drilling and fracking process, as did a leaked US Environmental Protection Agency (2013) presentation dealing with wells near Dimock, Pennsylvania. A draft EPA (2011) report on water near Pavillion, Wyoming, found fracking fluids in water, as well as methane; the final release of this study has been delayed as the EPA negotiates with Wyoming state authorities.

30. Howarth, Santoro, and Ingraffea 2011.

31. Meadows et al. 1972, 102.

32. Meadows 2007, 404.

33. Turner 2008, 397–402.

34. See Meadows et al. 1972, 129 (standard run), 147 (comprehensive technology run), 168–172 (stabilized world runs); Turner 2008, 402–410.

35. This involves compensating discrepancies in the birth and death rates.

36. Turner 2008, 410.

37. Huesemann 2003, 30.

38. Hall and Day 2009, 233; 235.

39. Hall and Day 2009, 236–7; McNeill 2001; Cleveland 1991.

40. Jackson 2009b, 48–57.

41. Ma 2011.

42. UN Department of Economic and Social Affairs 2013.

43. Jackson 2009b, 8.

44. Huesemann 2003, 31–32.

15 Conclusion

1. Chakrabarty 2009.

2. "Human processes—primarily the manufacture of fertilizer for food production and the cultivation of leguminous crops—convert around 120 million tonnes of N_2 from the atmosphere per year into reactive forms—which is more than the combined effects from all Earth's terrestrial processes" (Rockström et al. 2009, 474).

3. Vitousek et al. 1986.

4. Rockström et al. 2009.

5. Beckerman 1972; Socolow cited in Kaysen 1972; Simon 1982.

6. O'Connor 1994, 159.

7. *Sydney Morning Herald* 2008b; Graham Bradley, then President of the Business Council of Australia, speaking to Fran Kelly on December 1, 2010 (Kelly 2010).

8. Arndt 1978.

9. Krueger 2004.

10. Escobar 1995, 80.

11. Mshana 2004.

12. Milios 2005, 212.

13. UN Development Programme 2005, 18; Fuentes-Nieva and Galasso 2014.

14. Citigroup 2006, 1.

15. Edward 2006, 1677.

16. The UNICEF report (Ortiz and Cummins 2011) is calculated using statistics from the World Bank, UNU-WIDER, and Eurostat; this may explain the variation from Edward's work with density curves. Such calculations are always affected by uncertainties in the surveys of third world countries and by revisions in the methods of the data collectors.

17. World Commission on Environment and Development 1987, 5–6.

18. World Bank 2007; 2012.

19. Mshana 2007, 7.

20. Li 2009.

21. Sachs 1999, 39.

22. Meadows, Meadows, and Randers 2004, xx.

23. Weaver 1977, 188.

24. The vote on the first Wilderness Act of 1964, which designated nine million acres, was passed 73 to 12 in the US Senate and 373 to 1 in the House (Oreskes 2011).

25. Oreskes 2011.

26. Passell, Roberts, and Ross 1972.

27. Rockström et al. 2009.

28. Daly 2008.

29. See the appendix (Richard Sanders) for a brief explanation of this practice.

References

AAP. 2011. Abbott consistent on climate science: Lib. *Sydney Morning Herald*, March 15. http://news.smh.com.au/breaking-news-national/abbott-consistent-on-climate-science-lib-20110315-1buz4.html.

Abdallah, Saamah, Juliet Michaelson, Laura Stoll et al. 2012. *The Happy Planet Index: 2012 Report*. new economics foundation. http://www.happyplanetindex.org/assets/happy-planet-index-report.pdf.

Ackerman, Ruthie. 2008. Firestone's Superbowl fumble. *Nation*, February 18. http://www.thenation.com/article/firestones-super-bowl-fumble.

Adam, David. 2006. Royal Society tells Exxon: Stop funding climate change denial. *Guardian*, September 20. http://www.guardian.co.uk/environment/2006/sep/20/oilandpetrol.business.

Adams, Jonathan, and Kathleen McLaughlin. 2009. Silicon sweatshops. *Global Post*, November 17. http://www.globalpost.com/dispatch/china-taiwan/091103/silicon-sweatshops-globalpost-investigation.

Adams, Phillip. 2008. Loss of faith in the markets. *Late Night Live*, ABC Radio National, March 25, audio. http://www.abc.net.au/rn/latenightlive/stories/2008/2198972.htm.

Adams, Phillip. 2011. The BBC's changing role. *Late Night Live*, ABC Radio National, March 23. Audio. http://www.abc.net.au/radionational/programs/latenightlive/the-bbcs-changing-role/2992188.

Agarwala, A., and S. Singh. 1958. *The Economics of Underdevelopment*. Bombay: Oxford University Press.

AIMS. 1972. *30 Years of Aims of Industry*. London: Aims of Industry. http://www.powerbase.info/images/d/da/Aims-of-Industry.pdf.

Allen, Frederick. 1931. *Only Yesterday: An Informal History of the Nineteen Twenties*. New York: Harper & Row.

Allen, Geoff. 1976. The capitalist counter-offensive. *Age*, March 31.

American Council on Science and Health. 2013. About. http://www.acsh.org/about.

American Enterprise Institute. 2010. History. http://www.aei.org/about/history.

Amin, Samir. 2003. World poverty, pauperization and capital accumulation. *Monthly Review* 55 (5): 1–9.

Amnesty International. 2008. Rights razed: Forced evictions in Cambodia. http://www.amnesty.org/en/library/asset/ASA23/002/2008/en/337d65d1-6a08-431b-9a58-21131d6b617f/asa230022008en.html.

Anderegg, William, James Prall, Jacob Harold, and Stephen Schneider. 2010. Expert credibility in climate change. *Proceedings of the National Academy of Sciences of the United States of America*, April 9. Abstract. www.pnas.org/cgi/doi/10.1073/pnas.1003187107.

Anderson, Peter. 2011. Going it alone will not solve global problems. *West Australian*, July 12. http://www.acci.asn.au/Research-and-Publications/Media-Centre/Opinion-Pieces-Letters/Going-it-alone-will-not-solve-global-problems---Pu.

Andrews, Edmund. 2001. Bush angers Europe by eroding pact on warming. *New York Times*, April 1. http://www.nytimes.com/2001/04/01/world/01GERM.html.

Angus, Ian, and Simon Butler. 2011. *Too Many People? Population, Immigration, and the Environmental Crisis*. Chicago: Haymarket Books.

Apte, Prakash. 2008. Dharavi: India's model slum. *Planetizen*, September 28. http://www.planetizen.com/node/35269.

Arnalds, Ólafur. 2009. Soils and the living Earth. In *Soils, Society and Global Change: Proceedings of the International Forum Celebrating the Centenary of Conservation and Restoration of Soil and Vegetation in Iceland*, ed. Harriet Bigas, Gudmundur Gudbrandsson, Luca Montanarella, and Andrés Arnalds, 40–45. Luxembourg: European Communities. http://eusoils.jrc.ec.europa.eu/esdb_archive/eusoils_docs/other/EUR23784.pdf.

Arndt, H. W. 1978. *The Rise and Fall of Economic Growth*. Melbourne: Longman Cheshire.

Arndt, H. W. 1981. Economic development: A semantic history. *Economic Development and Cultural Change* 29 (3): 457–466.

Arnold, Ron. 1981. *At the Eye of the Storm: James Watt and the Environmentalists*. Washington, DC: Regnery Gateway.

Arrighi, Giovanni. 1994. *The Long Twentieth Century: Money, Power and the Origins of Our Times*. London: Verso.

Arrighi, Giovanni, Beverly Silver, and Benjamin Brewer. 2003. Industrial convergence, globalization, and the persistence of the north-south divide. *Studies in Comparative International Development* 38 (1): 3–31.

Athanasiou, Tom. 1996. *Divided Planet: The Ecology of Rich and Poor*. Boston: Little, Brown.

Athanasiou, Tom, and Paul Baer. 2002. *Dead Heat*. New York: Seven Stories Press.

Auffhammer, Maximilian, and Richard Carson. 2008. Forecasting the path of China's CO_2 emissions using province-level information. *Journal of Environmental Economics and Management* 55 (3): 229–247.

Austin, Andrew. 2002. Advancing accumulation and managing its discontents: The US antienvironmental countermovement. *Sociological Spectrum* 22 (1): 71–105.

Australian Treasury. 2008. Australia's low pollution future: The economics of climate change mitigation. Canberra: Australian Government. www.erambiente .net/economia/ALPF_the_economics_of_CC_mitigation_AUgov_ott08.pdf.

Badkar, Mamta. 2011. 500 Indian guest workers sue American company for human trafficking after Hurricane Katrina. *Business Insider*, February 22. http:// www.businessinsider.com/indian-workers-sue-american-company-feburary -22-2011-2.

Bagla, Pallava. 2002. Pachauri to head IPCC. *ScienceNow*, April 19. http://news .sciencemag.org/2002/04/pachauri-head-ipcc.

Bagnall, Diana. 2004. Profile of Greg Lindsay. *Bulletin*, September 22, 22–25.

Bailey, Ronald. 1989a. Dr. Doom. *Forbes* 144 (October 16): 45

Bailey, Ronald. 1989b. Hi there, Bambi. *Forbes* 144 (October 16): 46.

Bailey, Ronald. 1993. *Eco-scam: The False Prophets of Ecological Apocalypse*. New York: St. Martin's Press.

Bailey, Ronald. 2002. *Global Warming and Other Eco Myths: How the Environmental Movement Uses False Science to Scare Us to Death*. Roseville, CA: Prima Publishing.

Balazovic, Todd. 2011. China's social unrest: The story behind the stories. *China Daily*, June 27. http://www.chinadaily.com.cn/2011-06/27/content_12819896 .htm.

Barker, Alex, and James Polity. 2013. Brussels proposes €30bn "Tobin tax." *Financial Times*, February 14. http://www.ft.com/intl/cms/s/0/d1b16f7a-7601 -11e2-8eb6-00144feabdc0.html#axzz2dyEVlHgO.

Barker, Debi, and Jerry Mander. 1999. *Invisible Government—The World Trade Organization: Global Government for the New Millennium*. Sausalito, CA: International Forum on Globalization.

Barnett, Harold, and Chandler Morse. 1963. *Scarcity and Growth: The Economics of Natural Resource Availability*. Baltimore, MD: Johns Hopkins University Press.

Barney, Gerald, ed. 1982. *The Global 2000 Report to the President: Entering the Twenty-First Century*. Vols. 1–3. Harmondsworth, UK: Pelican.

Barney, Gerald, P. Freeman, and C. Ulinsky, eds. 1981. *Global 2000: Implications for Canada*. Toronto: Pergamon Press.

Bass, Gary. 2004. Word problem. *New Yorker*, May 3. http://www.newyorker .com/archive/2004/05/03/040503ta_talk_bass.

Beck, Roy, and Leon Kolankiewicz. 2000. The environmental movement's retreat from advocating US population stabilization (1970–1998): A first draft of history. *Journal of Policy History* 12 (1): 123–156.

Beckerman, Wilfred. 1972. Economists, scientists, and environmental catastrophe. *Oxford Economic Papers* 24 (3): 327–344.

Beckerman, Wilfred. 1974. *Two Cheers for the Affluent Society: A Spirited Defense of Economic Growth*. New York: St. Martin's Press.

Beckerman, Wilfred. 1995. *Small Is Stupid: Blowing the Whistle on the Greens*. London: Duckworth.

Beder, Sharon. 2001. Neoliberal think tanks and free market environmentalism. *Environmental Politics* 10 (2): 128–133. http://www.uow.com.au/arts/sts/sbeder/thinktanks.html.

Beder, Sharon. 2002a. *Global Spin: The Corporate Assault on Environmentalism*. White River Junction, VT: Chelsea Green.

Beder, Sharon. 2002b. bp: Beyond Petroleum? In *Battling Big Business: Countering Greenwash, Infiltration and Other Forms of Corporate Bullying*, ed. Eveline Lubbers, 26–32. North Carlton, VIC: Scribe. http://www.herinst.org/sbeder/PR/bp.html.

Beder, Sharon. 2004a. Consumerism: An historical perspective. *Pacific Ecologist* 9 (Spring): 42–48.

Beder, Sharon. 2004b. A SLAPP in the face of democracy. Democratic Audit of Australia. http://arts.anu.edu.au/democraticaudit/papers/20041220_beder_slapps.pdf.

Beder, Sharon. 2006a. *Free Market Missionaries: The Corporate Manipulation of Community Values*. London: Earthscan.

Beder, Sharon. 2006b. *Suiting Themselves: How Corporations Drive the Global Agenda*. London: Earthscan.

Begley, Sharon. 2007. The truth about denial. *Newsweek* 150 (7): 36–43, August 12. http://www.newsweek.com/2007/08/13/the-truth-about-denial.html.

Bello, Walden. 2011. A tale of two countries: Family planning in the Philippines and Thailand. *Inquirer* (Philippines). http://opinion.inquirer.net/8329/a-tale-of-two-countries-family-planning-in-the-philippines-and-thailand-2.

Ben-Ari, Nirit. 2002. Poverty is worsening in African LDCs. *Africa Recovery* 16 (2–3): 9.

Bengali, Shashank. 2006. Firestone takes heat over plantation labor. *Seattle Times,* December 28. http://seattletimes.nwsource.com/html/nationworld/2003498268_firestone28.html.

Berman, Arthur, and Lynn Pittinger. 2011. US shale gas: Less abundance, higher cost. *Oil Drum,* August 5. http://www.theoildrum.com/node/8212.

Berman, Morris. 1981. *The Reenchantment of the World*. Ithaca, NY: Cornell University Press.

Bernanke, Ben. 2004. The great moderation. New York: Federal Reserve Board. http://www.federalreserve.gov/BOARDDOCS/SPEECHES/2004/20040220/default.htm.

Bernays, Edward. 1947. The engineering of consent. *Annals of the American Academy of Political and Social Science* 250:113–120.

Bernays, Edward. 1961 [1923]. *Crystallizing Public Opinion*. New York: Liveright.

Bernays, Edward. 2005 [1928]. *Propaganda*. Brooklyn, NY: IG Publishing.

Biggins, D. R. 1978. Scientific knowledge and values: Imperatives in ecology. *Ethics in Science & Medicine* 6:49–57.

Birdsall, Nancy. 2007. Inequality matters: Why globalization doesn't lift all boats. *Boston Review*, March/April. http://bostonreview.net/nancy-birdsall-inequality-matters.

Bleizeffer, Dustin. 2006. Trading coal for oil may prove profitable. *Jackson Hole Star-Tribune*, December 10.

Blom, Philipp. 2008. *The Vertigo Years: Change and Culture in the West, 1900–1914*. London: Weidenfeld and Nicolson.

Blumenthal, Sidney. 1986. *The Rise of the Counter-Establishment: The Conservative Ascent to Political Power*. New York: Union Square Press.

Boserup, Mogens. 1978. Fear of Doomsday: Past and present. *Population and Development Review* 4 (1): 133–143.

Boulding, Kenneth. 1966. The economics of the coming Spaceship Earth. In Daly and Townsend 1993, 297–309.

Bowden, Witt. 1965. *Industrial Society in England towards the End of the Eighteenth Century*. New York: Macmillan.

Boykoff, Maxwell, and Jules Boykoff. 2004. Balance as bias: Global warming and the US prestige press. *Global Environmental Change* 14:125–136.

BP. 2012. *BP Statistical Review of World Energy June 2012*. http://www.bp.com.

Brandt, Karl. 1950. Review of *Road to Survival* [by W. Vogt]. *Land Economics* 26 (1): 88–90.

Braudel, Fernand. 1981. *The Structures of Everyday Life: The Limits of the Possible*. New York: Harper & Row.

Bretton Woods Project. 2010. Analysis of World Bank voting reforms: Governance remains illegitimate and outdated. London: Bretton Woods Project. http://www.brettonwoodsproject.org/art-566281.

Brewster, Kerry. 2008. Treasury releases modelling on emissions trading scheme. *7.30 Report*, ABC TV, October 30. http://www.abc.net.au/7.30/content/2008/s2406107.htm.

Brinkley, Alan. 1989. The New Deal and the idea of the state. In *The Rise and Fall of the New Deal Order (1930–1980)*, ed. Steve Fraser and Gary Gerstle, 85–121. Princeton, NJ: Princeton University Press.

Bristol, Nellie. 2008. Mechai Viravaidya: Thailand's "Condom King." *Lancet* 371 (9607): 109.

Broad, Robin. 2006. Research, knowledge, and the art of "paradigm maintenance": The World Bank's Development Economics vice-presidency (DEC). *Review of International Political Economy* 13 (3): 387–419.

Broder, David. 2002. Thanks to two think tanks—Heritage and Cato. *Sun Sentinel*, May 8, A21. http://articles.sun-sentinel.com/2002-05-08/news/0205070485_1_cato-institute-heritage-foundation-white-house.

Brown, George. 1996. *Environmental Science under Siege: Fringe Science and the 104th Congress*. http://democrats.science.house.gov/Media/File/Reports/environment_science_report_23oct96.pdf.

Brown, Lester. 2008. *Plan B 3.0: Mobilizing to Save Civilization*. New York: W. W. Norton.

Brugmann, Jeb. 2009. *Welcome to the Urban Revolution: How Cities Are Changing the World*. St. Lucia, QLD: University of Queensland Press.

Brulle, Robert. 2013. Institutionalizing delay: Foundation funding and the creation of US climate change counter-movement organizations. *Climatic Change* (December). DOI 10.1007/s10584-013-1018-7.

Brzoska, Michael. 1983. The military-related external debt of Third World countries. *Journal of Peace Research* 20 (3): 271–277.

Buckley, Ross. 2002/2003. The rich borrow and the poor repay: The fatal flaw in international finance. *World Policy Journal* 19 (4): 59–64.

Buckley, Ross. 2009. When cashed-up IMF offers you help, steer well clear. *Sydney Morning Herald*, May 2. http://www.smh.com.au/opinion/when-cashedup-imf-offers-to-help-steer-well-clear-20090501-aq4y.html?page=-1.

Buell, Frederick. 2003. *From Apocalypse to Way of Life: Environmental Crisis in the American Century*. New York: Routledge.

Burke, William. 1993. The Wise Use Movement: Right-Wing Anti-Environmentalism. *Public Eye* 7 (2). http://www.publiceye.org/magazine/v07n2/wiseuse.html.

Burton, Bob. 1999. BHP admits Ok Tedi mine is environmental disaster. *Asia Times*, August 13. http://www.atimes.com/oceania/AH13Ah01.html.

Burton, Bob. 2004. Atlas Economic Research Foundation: The think-tank breeders. *PR Watch Newsletter* 11 (3): 14. http://www.prwatch.org/node/323.

Bush, George W. 2003. President discusses the future of Iraq. Press release, White House, February 26. http://georgewbush-whitehouse.archives.gov/news/releases/2003/02/20030226-11.html.

Business Standard. 2006. Orissa red carpet for industry. *Business Standard*, Bhubaneswar Bureau, February 6. http://www.business-standard.com/article/economy-policy/orissa-red-carpet-for-industry-106020601040_1.html.

Cahill, Damien. 2004. The radical neoliberal movement as a hegemonic force in Australia, 1976–1996. PhD diss., University of Wollongong. http://ro.uow.edu.au/theses/193.

Cahill, Damien. 2005. From the fringes: The emergence from obscurity of the radical neoliberal movement in Australia. *Hummer* 4 (3). http://asslh.org.au/hidden/hummer/vol-4-no-3/fringes.

Cahill, Damien. 2010. Business mobilisation, the New Right and Australian Labor governments in the 1980s. *Labour History* 98:7–24.

Campbell, Al. 2005. The birth of neoliberalism in the United States: A reorganisation of capitalism. In Saad-Filho and Johnston 2005, 187–198.

Campbell, Colin. 1996. The twenty-first century: The world's endowment of conventional oil and its depletion. http://www.oilcrisis.com/campbell/cen21.htm.

Campbell, Colin. 1997. *The Coming Oil Crisis*. Brentwood, UK: Multi-Science Publishing and Petroconsultants S.A.

Campbell, Colin. 1999. The imminent peak of world oil production. Presentation to a House of Commons All-Party Committee, July 7. http://www.hubbertpeak.com/campbell/commons.htm.

Campbell, Colin, and Jean Laherrère. 1998. The end of cheap oil. *Scientific American* 278 (3): 78–83.

Canadian Parliament. 1999. Sub-Committee on International Trade, Trade Disputes and Investment of the Standing Committee on Foreign Affairs and International Trade. http://www2.parl.gc.ca/HousePublications/Publication.aspx?DocId=1039554&Language=E&Mode=1&Parl=36&Ses=1.

Canan, Penelope, and George Pring. 1988. Strategic lawsuits against public participation. *Social Problems* 35 (5): 506–519.

Carey, Alex. 1987a. Conspiracy or groundswell. In *The New Right's Australian Fantasy*, ed. Ken Coghill, 3–19. Fitzroy, VIC: McPhee Gribble/Penguin Books.

Carey, Alex. 1987b. The ideological management industry. In *Communications and the Media in Australia*, ed. E. Wheelwright and K. Buckley, 156–179. Sydney: Allen and Unwin.

Carey, Alex. 1997. *Taking the Risk out of Democracy: Corporate Propaganda versus Freedom and Liberty*. Urbana: University of Illinois Press.

Carey, John, and Adrienne Carter (with Assif Shameen). 2007. Food vs. fuel. *Business Week*, February 5. http://www.businessweek.com/magazine/content/07_06/b4020093.htm.

Carmichael, Robert, and Lon Nara. 2001. Capital's worst slum fires make thousands homeless. *Phnom Penh Post*, December 7. http://www.phnompenhpost.com/national/capitals-worst-slum-fires-make-thousands-homeless.

Carnegy, Hugh, and Quentin Peel. 2012. Financiers attack Sarkozy "Tobin tax" plan. *Financial Times*, January 8. http://www.ft.com/intl/cms/s/0/ca7ab274-3a03-11e1-a8dc-00144feabdc0.html.

Carr, C. C. 1949. Translating the American economic system. *Public Relations Journal* 5 (6): 1–4.

Carson, Rachel. 1965 [1962]. *Silent Spring*. Harmondsworth, UK: Penguin.

Cato Institute. 2013. Experts: Fellows. http://www.cato.org/people/fellows.

Chakrabarty, Dipesh. 2009. The climate of history: Four theses. *Critical Inquiry* 35 (Winter): 197–222.

Chandler, Alfred. 1977. *The Visible Hand: The Managerial Revolution in American Business*. Cambridge, MA: Belknap Press of Harvard University Press.

Chang, Ha-Joon. 2002. *Kicking Away the Ladder: Development Strategy in Historical Perspective*. London: Anthem Press.

Chang, Pin-tsun. 2009. The rise of Chinese mercantile power in VOC Dutch East Indies. *Chinese Southern Diaspora Studies*, vol. 3. http://csds.anu.edu.au/volume_3_2009/01_CSDS_2009_Chang.pdf.

Chenery, H. 1964. Objectives and criteria of foreign assistance. In *The United States and the Developing Economies*, ed. G. Ranis, 80–91. New York: Norton.

Cherrington, David. 1980. *The Work Ethic: Working Values and Values that Work*. New York: Amacom.

Chew, Sing. 2001. *World Ecological Degradation: Accumulation, Urbanization, and Deforestation 3000 B.C. to A.D. 2000*. Walnut Creek, CA: Altamira.

China Daily. 2006. Pollution costs US$200b each year. *China Daily*, June 5. http://www.chinadaily.com.cn.

Chomsky, Noam. 2001. The fifth freedom: Noam Chomsky interviewed by Stephen Marshall. *Guerrilla News Network*, November. http://www.chomsky.info/interviews/200111--04.htm.

Chu, Henry. 2008. Where every inch counts. *Los Angeles Times*, September 8. http://articles.latimes.com/2008/sep/08/world/fg-dharavi8.

Ciriacy-Wantrup, S., and R. Bishop. 1975. Common property as a concept in natural resources policy. *Natural Resources Journal* 15 (4): 713–727.

Citigroup. 2006. Equity Strategy—Revisiting plutonomy: The rich getting richer. March 5. http://rebel-alliance.org/2011/08/19/leaked-from-citigroup-plutonomy-part-2-2006.

Clark, Greg. 2008. Survival of the richest. *Perspective*, ABC Radio National, August 11. http://www.abc.net.au/radionational/programs/perspective/survival-of-the-richest/3199368.

Clausen, Rebecca. 2007. Healing the rift: Metabolic restoration in Cuban agriculture. *Monthly Review* 59 (1): 40–52.

Cleveland, Cutler. 1991. Natural resource scarcity and economic growth revisited: Economic and biophysical perspectives. In Costanza 1991, 289–317.

CNN. 2009. Officials: G-20 to supplant G-8 as international economic council. September 25. http://edition.cnn.com/2009/US/09/24/us.g.twenty.summit/index.html.

CNN *Money*. 2012. Global 500, 2011. http://money.cnn.com/magazines/fortune/global500/2011/full_list/index.html; and http://money.cnn.com/magazines/fortune/global500/2011/maps/index.html.

Cochran, Thomas, and William Miller. 1961 [1942]. *The Age of Enterprise: A Social History of Industrial America*. New York: Harper.

Cockett, Richard. 1995. *Thinking the Unthinkable: Think-Tanks and the Economic Counter-Revolution, 1931–1983*. London: Fontana.

Cohen, Donald, and Peter Dreier. 2009. Crying wolf: The same old song. *Talking Points Memo*, April 29. http://crywolfproject.org/commentary/crying-wolf-same-old-song-health-care-and-unions.

Cole, H. S. D., Marie Jahoda, K. L. R. Pavitt, et al., eds. 1973. *Models of Doom: A Critique of The Limits to Growth*. New York: Universe.

Committee on Recent Economic Changes. 1929. *Report of the Committee on Recent Economic Changes*. New York: McGraw-Hill. http://www.nber.org/chapters/c4950.pdf.

Competitive Enterprise Institute. 2013. Ronald Bailey: Adjunct analyst. http://cei.org/contributor/ronald-bailey.

Connor, Tim, and Kelly Dent. 2006. *Offside! Labour rights and sportswear production in Asia*. Oxfam Australia. http://resources.oxfam.org.au/pages/view.php?ref=177&search=&offset=0&order_by=popularity&sort=ASC&archive=0.

Consumer Reports. 1994. The ACSH: Forefront of science or just a front? *Consumer Reports*, May. http://research.greenpeaceusa.org/?a=view&d=4698.

Cook, John, Dana Nuccitelli, Sarah Green, et al. 2013. Quantifying the consensus on anthropogenic global warming in the scientific literature. *Environmental Research Letters* 8:02424.

Cordell, Dana, and Stuart White. 2008. *The Australian Story of Phosphorus: Sustainability Implications of Global Fertilizer Scarcity for Australia*. Sydney: Institute for Sustainable Futures, University of Technology Sydney. http://www.susana.org/docs_ccbk/susana_download/2-1032-en-the-australian-story-of-phosphorus-2008.pdf.

Costanza, Robert. 1991. *Ecological Economics: The Science and Management of Sustainability*. New York: Columbia University Press.

Cowdrick, E. 1927. The new economic gospel of consumption. *Industrial Management* 74:209–211.

CPI Strategic. 2009. Community attitudes towards climate change. http://www.abc.net.au/4corners/special_eds/20091109/ETS/docs/cpi.pdf.

Creel, George. 1918. Public opinion in war time. *Annals of the American Academy of Political and Social Science* 78:185–194.

Creel, George. 1920. *How We Advertised America: The First Telling of the Amazing Story of the Committee on Public Information That Carried the Gospel of Americanism to Every Corner of the Globe*. London and New York: Harper & Bros. http://archive.org/details/howweadvertameri00creerich.

Crick, Bernard. 2007. Florida farm workers face near-slavery conditions. *Alternet*, December 5. http://www.alternet.org/rights/69748.

Crosby, Alfred. 1986. *Ecological Imperialism: The Biological Expansion of Europe, 900–1900*. Cambridge: Cambridge University Press.

Crozier, Michel, Samuel Huntington, and Joji Watanuki. 1975. *The Crisis of Democracy*. New York: New York University Press.

Cubby, Ben. 2012. Scientist denies he is mouthpiece of US climate-sceptic think tank. *Sydney Morning Herald*, February 16. http://www.smh.com.au/environment/climate-change/scientist-denies-he-is-mouthpiece-of-us-climatesceptic-think-tank-20120215-1t6yi.html.

Cullen, Peter. 2006. Science and politics: Speaking truth to power. Wentworth Group, Australia. Paper presented at the North American Benthological Society Annual Conference, Anchorage, Alaska, June. www.wentworthgroup.org/docs/Speaking_Truth_To_Power1.pdf.

Cushman, John. 1998. Industrial group plans to battle climate treaty. *New York Times*, April 26. http://www.nytimes.com/1998/04/26/us/industrial-group-plans-to-battle-climate-treaty.html.

Cushman, John. 2001. After "Silent Spring," industry put spin on all it brewed. *New York Times*, March 25. http://www.nytimes.com/2001/03/26/us/after-silent-spring-industry-put-spin-on-all-it-brewed.html.

Daly, Herman. 1978. *Steady-State Economics: The Economics of Biophysical Equilibrium and Moral Growth*. San Francisco: W. H. Freeman.

Daly, Herman. 1991a. *Steady-State Economics: Second Edition with New Essays*. Washington, DC: Island Press.

Daly, Herman. 1991b. Elements of environmental macroeconomics. In Costanza 1991, 32–46.

Daly, Herman. 1993. Sustainable growth: An impossibility theorem. In Daly and Townsend 1993, 267–273.

Daly, Herman. 1996. *Beyond Growth: The Economics of Sustainable Development*. Boston: Beacon Press.

Daly, Herman. 1999. *Ecological Economics and the Ecology of Economics: Essays in Criticism*. Northampton, MA: Edward Elgar.

Daly, Herman. 2008. A Steady-State Economy. London: Sustainable Development Commission, UK. http://www.theoildrum.com/node/3941.

Daly, Herman, and John Cobb. 1989. *For the Common Good*. Boston: Beacon Press.

Daly, Herman, and Kenneth Townsend, eds. 1993. *Valuing the Earth: Economics, Ecology, Ethics*. Cambridge, MA: MIT Press.

Darling, Frank Fraser. 1970. *Wilderness and Plenty*. London: British Broadcasting Corporation.

Das, Dilip. 1993. *International Finance: Contemporary Issues*. London: Routledge.

Davies, James, Susanna Sandström, Anthony Shorrocks, and Edward Wolff. 2008. *The World Distribution of Household Wealth*. Helsinki: UNU-WIDER. http://www.wider.unu.edu/publications/working-papers/discussion-papers/2008/en_GB/dp2008-03.

Davis, Bob, and Douglas Belkin. 2008. Food inflation, riots spark worries for world leaders. *Wall Street Journal*, April 14, A1.

Davis, Mike. 2004. Planet of slums. *New Left Review* 26 (March–April): 5–34.

De Launey, Guy. 2006. Cambodia's "out of control" evictions. *BBC News*, September 29. http://news.bbc.co.uk/2/hi/asia-pacific/5392474.stm.

DeLong, J. Bradford. 1997. *Slouching Towards Utopia? The Economic History of the Twentieth Century*. Draft. Berkeley: University of California. http://www.j-bradford-delong.net/TCEH/Slouch_wealth2.html.

DeLong, J. Bradford. 1998. Estimating world GDP, one million B.C.–present. Faculty paper, Department of Economics, University of California at Berkeley. http://www.j-bradford-delong.net/TCEH/1998_Draft/World_GDP/Estimating_World_GDP.html.

Den Biggelaar, Christoffel, Rattan Lal, Keith Wiebe, and Vince Breneman. 2003. The global impact of soil erosion on productivity. Part I. Absolute and relative erosion-induced yield losses. *Advances in Agronomy* 81:1–48.

Dennis, Brady, and Steven Mufson. 2010. Bankers lobby against financial regulatory overhaul. *Washington Post*, March 19, A18. http://www.washingtonpost.com/wp-dyn/content/article/2010/03/18/AR2010031805370.html.

Department of Climate Change and Energy Efficiency. 2011. *Securing a Clean Energy Future*. Canberra: Australian Government. http://www.acci.asn.au/Files/Government-Carbon-Tax-Plan.

Department of Foreign Affairs and Trade. 2007. *APEC and the Rise of the Global Middle Class*. Canberra: Australian Government. http://www.scribd.com/doc/50807056/null.

Department of Foreign Affairs and Trade. 2012. Agriculture and the WTO. http://www.dfat.gov.au/trade/negotiations/trade_in_agriculture.html#pagenav1.

Diamond, Jared. 2005. *Collapse: How Societies Choose to Fail or Survive*. Camberwell, VIC: Allen Lane.

Dobbs, Lou. 2008. *Lou Dobbs Tonight*. CNN, December 16. http://transcripts.cnn.com/TRANSCRIPTS/0812/16/ldt.01.html.

Dolny, Michael. 1996. The think tank spectrum: For the media, some thinkers are more equal than others. *Extra!*, May–June. http://www.fair.org/index.php?page=1357.

Douthwaite, Richard. 1999. *The Growth Illusion*. Gabriola Island, BC: New Society Publishers.

Dowie, Mark. 1977. Pinto madness. *Mother Jones*, September–October. http://www.motherjones.com/politics/1977/09/pinto-madness.

Dryden, Steve. 1995. *Trade Warriors: USTR and the American Crusade for Free Trade*. New York: Oxford University Press.

Dryzek, John. 1997. *Politics of the Earth: Environmental Discourses*. New York: Oxford University Press.

Duffy, Jonathan. 2004. Bilderberg: The ultimate conspiracy theory. *BBC News Online Magazine*, June 3. http://news.bbc.co.uk/2/hi/uk_news/magazine/3773019.stm.

Duménil, Gérard, and Dominique Lévy. 2005. The Neoliberal (Counter-)Revolution. In Saad-Filho and Johnston 2005, 9–19.

Duncan, Tim. 1985. New Right: Where it stands and what it means. *Bulletin*, December 10, 38–41.

Dunlap, Riley, and Aaron McCright. 2010. Climate change denial: Sources, actors and strategies. In *Routledge Handbook of Climate Change and Society*, ed. Constance Lever-Tracy, 240–259. London: Routledge.

Durning, Alan. 1994. Redesigning the forest economy. In *State of the World 1994: A Worldwatch Institute Report on Progress toward a Sustainable Society*, ed. Lester Brown, 22–40. New York: W. W. Norton.

Easterly, William. 2001. The lost decades: Developing countries' stagnation in spite of policy reform 1980–1998. *Journal of Economic Growth* 6:135–157.

Ebenstein, Alan. 2003. *Hayek's Journey: The Mind of Friedrich Hayek*. New York: Palgrave Macmillan.

Ecologist. 1993. *Whose Common Future? Reclaiming the Commons*. London: Earthscan.

Economist. 1972. Limits to misconception. *Economist,* March 11, 21–22.

Economist. 1997. Plenty of gloom. *Economist*, December 18. http://www.economist.com/node/455855?story_id=E1_QVVRVV.

Economist. 2003. A greener Bush. *Economist,* February 13. http://www.economist.com/node/1576767.

Edney, Julian. 2005. Greed (Part I). *post-autistic economics review* 31. http://www.paecon.net/PAEReview/issue31/Edney31.htm.

Edward, Peter. 2006. Examining inequality: Who really benefits from global growth? *World Development* 34 (10): 1667–1695.

Edwards, David. 2004. Climate catastrophe. *Znet*, January 10. http://www.countercurrents.org/en-edwards100104.htm.

Edwards, Lee. 1997. *The Power of Ideas: The Heritage Foundation at 25*. Ottawa, IL: Jameson Books.

Egan, Carmel. 2007. Bitter life of chocolate's child slaves. *Age*, November 4. http://www.theage.com.au/articles/2007/11/03/1193619205911.html?page=fullpage#contentSwap1.

Egan, Timothy. 1991. Fund-raisers tap anti-environmentalism. *New York Times*, December 19. http://www.nytimes.com/1991/12/19/us/fund-raisers-tap-anti-environmentalism.html.

Ehrlich, Paul. 1968. *The Population Bomb*. London: Ballantine.

Ehrlich, Paul, and John Holdren. 1971. Impact of population growth. *Science* 171 (3977): 1212–1217.

Ehrlich, Paul, Anne Ehrlich, and John Holdren. 1993 [1977]. Availability, entropy, and the laws of thermodynamics. In Daly and Townsend 1993, 69–73.

Eilperin, Juliet. 2005. Climate official's work is questioned. *Washington Post*, December 5. http://www.washingtonpost.com/wp-dyn/content/article/2005/12/04/AR2005120400891.html.

Ekins, Paul. 1991. The sustainable consumer society: A contradiction in terms? *International Environmental Affairs* 3 (4): 243–258.

Elkington, John, and Mark Lee. 2005. Will hard-won environmental and social gains survive China's economic rise? *Grist*, August 24. http://grist.org/article/china1.

Elmer-DeWitt, Philip. 1992. Summit to save the Earth: Rich vs. poor. *Time*, June 1. http://www.time.com/time/magazine/article/0,9171,975656-9,00.html.

Environment News Service. 2008. Growth in China's CO_2 emissions double previous estimates. March 11. http://www.ens-newswire.com/ens/mar2008/2008-03-11-01.asp.

Enzensberger, H. 1974. A critique of political ecology. *New Left Review* 84:3–32.

Escobar, Arturo. 1995. *Encountering Development: The Making and Unmaking of the Third World*. Princeton, NJ: Princeton University Press.

European Commission. 2009. European Communities—Measures concerning meat and meat products (hormones). http://trade.ec.europa.eu/doclib/docs/2009/january/tradoc_142145.pdf.

Evans, Suzannah. 2009. The business of extinction. *Seed Magazine*, July 16. http://seedmagazine.com/content/article/finding_fish/P2.

Ewen, Stuart. 1996. *PR! A Social History of Spin*. New York: Basic Books.

Farrelly, Elizabeth. 2006. In the battle to be green, the human factor can work wonders. *Sydney Morning Herald*, October 3. http://www.smh.com.au/news/opinion/in-the-battle-to-be-green-the-human-factor-can-work-wonders/2006/10/03/1159641321963.html?page=fullpage#contentSwap1.

Fernandes, Walter. 2006. Liberalisation and development-induced displacement. *Social Change* 36 (1): 104–123.

Fisher, Joseph. 1949. Untitled review of *Road to Survival* [by W. Vogt]. *American Economic Review* 39 (3): 822–825.

Fones-Wolf, Elizabeth. 1994. *Selling Free Enterprise: The Business Assault on Labor and Liberalism, 1945–60*. Urbana: University of Illinois Press.

Food and Agriculture Organization of the United Nations. 2008. Food Outlook: The FAO Price Index. Rome: FAO, June. http://www.fao.org/docrep/010/ai466e/ai466e16.htm.

Food and Agriculture Organization of the United Nations. 2009. More people than ever are victims of hunger. Rome: FAO, June 20. http://www.fao.org/fileadmin/user_upload/newsroom/docs/Press%20release%20june-en.pdf.

Food and Agriculture Organization of the United Nations. 2010. *The State of Food Insecurity in the World*. Rome: FAO. http://www.fao.org/docrep/013/i1683e/i1683e.pdf.

Food and Agriculture Organization of the United Nations. 2012. *The State of Food Insecurity in the World*. Rome: FAO. http://www.fao.org/docrep/016/i3027e/i3027e00.htm.

Food and Agriculture Organization of the United Nations. 2013. World food situation: FAO Food Price Index. Rome: FAO. http://www.fao.org/worldfoodsituation/wfs-home/foodpricesindex/en.

Forero, Juan. 2009. Rain forest residents, Texaco face off in Ecuador. National Public Radio (US), April 30. http://www.npr.org/templates/story/story.php?storyId=103233560.

Forrester, Jay. 1971. *World Dynamics*. Cambridge, MA: Wright-Allen Press.

Forrester, Jay, Gilbert Low, and Nathaniel Mass. 1974. The debate on "World Dynamics": A response to Nordhaus. *Policy Sciences* 5 (2): 169–190.

Fortune. 1938. Business-and-Government. October, 50–51.

Fortune. 1949. Business is still in trouble. May, 67–71, 196, 198, 200.

Foster, John Bellamy. 1998. Malthus's essay on population at age 200: A Marxian view. *Monthly Review* 50 (7): 1–18.

Frank, André. 1978. The post-war boom: Boom for the West, bust for the South. *Millennium* 7 (2): 153–161.

Fraser, Steve. 2008. The specter of Wall Street: Wall Street's comeback as the place Americans love to hate. *Baltimore Chronicle*, October 2. http://www.baltimorechronicle.com/2008/100208Fraser.html.

Fraser, W. Hamish. 1981. *The Coming of the Mass Market*. Hamden, CT: Archon Books.

Freeman, Christopher. 1973. Malthus with a computer. In Cole et al. 1973, 5–13.

Frieden, Jeffry. 2006. *Global Capitalism: Its Fall and Rise in the Twentieth Century*. New York: Norton.

Friedman, Milton. 1962. *Capitalism and Freedom*. Chicago: University of Chicago Press.

Friedman, Milton. 1980. *Free to Choose*, vol. 1. PBS. http://www.freetochoosemedia.org/freetochoose/detail_ftc1980_transcript.php?page=1.

Friedman, Milton, and Rose Friedman. 1998. *Two Lucky People*. Chicago: University of Chicago Press.

Friedman, Thomas. 1999. *The Lexus and the Olive Tree*. London: HarperCollins.

Fuchs, Sandor. 1970. Ecology movement exposed. *Progressive Labor* 7:50–63.

Fuentes-Nieva, Ricardo, and Nick Galasso. 2014. Working for the few: Political capture and economic inequality. Oxfam Briefing Paper 178, January. Oxford: Oxfam International. http://www.oxfam.org/sites/www.oxfam.org/files/bp-working-for-few-political-capture-economic-inequality-200114-en.pdf.

Fullerton, Ticky. 2002. Interview with Chris Murphy, September 19. http://www.abc.net.au/4corners/stories/s718263.htm.

G20. 2012. Financial regulation. http://www.g20mexico.org/en/financial-regulation.

G20. 2013. What is the G20. http://www.g20.org/docs/about/about_G20.html.

Gadrey, Jean. 2005. What's wrong with GDP and growth? The need for alternative indicators. In *A Guide to What's Wrong with Economics*, ed. Edward Fullbrook, 262–278, 318–319. London: Anthem Press.

Galbraith, John Kenneth. 1958. How much should a country consume? In *Perspectives on Conservation*, ed. Henry Jarrett, 89–99. Baltimore, MD: Johns Hopkins University Press.

Galeano, Eduardo. 2008. *Open Veins of Latin America: Five Centuries of the Pillage of a Continent*. Gurgaon, India: Three Essays Collective.

Garnaut, Ross. 2008. *National Press Club Address Draft Report Launch, 4 July 2008*. Garnaut Climate Change Review. http://www.garnautreview.org.au/ca25734e0016a131/pages/all-reports--resources.html.

Gelbspan, Ross. 2005. Snowed: How the media ignores global warming. *Mother Jones*, May–June. http://motherjones.com/politics/2005/05/snowed.

George, Susan. 1976. *How the Other Half Dies: The Real Reasons for World Hunger*. Harmondsworth, UK: Penguin.

George, Susan. 1992. *The Debt Boomerang: How Third World Debt Harms Us All*. Boulder, CO: Westview.

Georgescu-Røegen, Nicholas. 1971. *The Entropy Law and the Economic Process*. Cambridge, MA: Harvard University Press.

Georgescu-Røegen, Nicholas. 1975. Energy and economic myths. *Southern Economic Journal* 41 (3): 347–381.

Gershwin, Lisa-Ann. 2013. *Stung! On Jellyfish Blooms and the Future of the Ocean*. Chicago: University of Chicago Press.

Ghandi, Mahatma. 1928. Discussion with a Capitalist. *Young India*, December 12. http://www.gandhiashramsevagram.org/mkgandhi/cwmg/VOL043.PDF, p. 412.

Gilding, Paul. 2009. The great disruption. *Background Briefing*, ABC Radio National, June 14. http://www.abc.net.au/rn/backgroundbriefing/stories/2009/2592909.htm.

Gilding, Paul. 2011. *The Great Disruption: How the Climate Crisis will Transform the Global Economy*. London: Bloomsbury.

Gillette, Robert. 1972. The limits to growth: Hard sell for a computer view of Doomsday. *Science* 175 (4026): 1088–1092.

Gittins, Ross. 2002. Beware international experts bearing gratuitous advice. *Age*, September 28, Business, 1. http://www.theage.com.au/articles/2002/09/27/1032734326302.html.

Gittins, Ross. 2008. Carbon trading: Big business vote of no confidence in itself. *Sydney Morning Herald*, August 25. http://www.smh.com.au/business/carbon-trading-big-business-vote-of-no-confidence-in-itself-20080824-41eb.html.

Golden, Frederic. 1982. Storm over a deadly downpour. *Time*, December 6. http://www.time.com/time/magazine/article/0,9171,923118,00.html.

Goldenberg, Suzanne. 2009. Copenhagen: The key players and how they rated. *Observer*, December 20. http://www.guardian.co.uk/environment/2009/dec/20/copenhagen-obama-brown-climate.

Goldman, Michael. 2005. *Imperial Nature: The World Bank and Struggles for Social Justice in the Age of Globalization*. New Haven, CT: Yale University Press.

Goldsmith, Edward, and Robert Allen (with contributions by John Davoll, Sam Lawrence, and Michael Allaby). 1972. *A Blueprint for Survival*. Harmondsworth, UK: Penguin.

Golub, Robert, and Joe Townsend. 1977. Malthus, multinationals and the Club of Rome. *Social Studies of Science* 7 (2): 210–222.

Goodman, Amy. 2009a. "We are not begging for aid": Chief Bolivian negotiator says developed countries owe climate debt. *Democracy Now!*, December 9. http://www.democracynow.org/2009/12/9/we_are_not_begging_for_aid.

Goodman, Amy. 2009b. Bolivian President Evo Morales: "Shameful" for West to spend trillions on war and only $10 billion for climate change. *Democracy Now!*, December 16. http://www.democracynow.org/2009/12/16/bolivian_president_evo_morales_shameful_for.

Goodman, Amy. 2010a. Senate passes sweeping financial reform bill: Lobbying frenzy expected as measure moves to committee. *Democracy Now!*, May 21. http://www.democracynow.org/2010/5/21/senate_passes_sweeping_financial_reform_bill.

Goodman, Amy. 2010b. Wendell Potter on "Deadly spin: An insurance company insider speaks out on how corporate PR is killing health care and deceiving Americans." *Democracy Now!*, November 16. http://www.democracynow.org/2010/11/16/wendell_potter_on_deadly_spin_an.

Goodman, Amy. 2011. On climate change, the message is simple: Get it done. *Guardian*, December 15. http://www.theguardian.com/commentisfree/cifamerica/2011/dec/14/durban-climate-change-conference-2011.

Gordon, Scott. 1973. Today's apocalypses and yesterday's. Papers and Proceedings of the Eighty-fifth Annual Meeting of the American Economics Association. *American Economic Review* 63 (2): 106–110.

Grattan, Michelle. 2010. Joyce questions foreign aid cost in time of debt. *Age*, February 4. http://www.theage.com.au/national/joyce-questions-foreign-aid-cost-in-time-of-debt-20100203-ndl1.html.

Green, Mark, and Andrew Buchsbaum. 1980. *The Corporate Lobbies: Political Profiles of The Business Roundtable and The Chamber of Commerce*. Washington, DC: Public Citizen.

Greenpeace. 2004a. Exxon Secrets factsheet: The Advancement of Sound Science Coalition. http://www.exxonsecrets.org/html/orgfactsheet.php?id=6.

Greenpeace. 2004b. Exxon Secrets factsheet: S. Fred Singer. http://www.exxonsecrets.org/html/personfactsheet.php?id=1.

Greenpeace. 2007. Exxon Secrets factsheet: Competitive Enterprise Institute, CEI. http://www.exxonsecrets.org/html/orgfactsheet.php?id=2.

Greenpeace. 2008. Exxon Secrets factsheet: American Council on Science and Health, ACSH. http://www.exxonsecrets.org/html/orgfactsheet.php?id=8.

Greenpeace. 2013. Organizations in Exxon Secrets database. http://www.exxonsecrets.org/html/listorganizations.php.

Greer, Jed, and Kavaljit Singh. 2000. A brief history of transnational corporations. *Global Policy Forum*. http://www.globalpolicy.org/component/content/article/221/47068.html.

Greider, William. 1992. *Who Will Tell the People: The Betrayal of American Democracy*. New York: Simon and Schuster.

Griffin, Larry, Michael Wallace, and Beth Rubin. 1986. Capitalist resistance to the organization of labor before the New Deal: Why? How? Success? *American Sociological Review* 51 (2): 147–167.

Griffith, Robert. 1983. The selling of America: The Advertising Council and American politics, 1942–1960. *Business History Review* 57 (3): 388–412.

Griffiths, Matthew. 2002. Case study of the impact of CAP on a developing country: Importation of milk solids into Jamaica from the EU. *Trócaire Development Review*, Dublin, 99–106. http://www.trocaire.org/resources/tdr-article/case-study-impact-cap-developing-country-importation-milk-solids-jamaica-eu.

Griffiths, Meredith. 2011. 2010 one of the hottest on record. *AM*, ABC Radio, January 21. http://www.abc.net.au/am/content/2011/s3117917.htm.

Grubb, Michael, Matthias Koch, Koy Thomson, et al. 1993. *The "Earth Summit" Agreements: A Guide and Assessment*. London: Earthscan.

Grumbine, R., and Maharaj Pandit. 2013. Threats from India's Himalaya dams. *Science* 339 (6115): 36–37.

Guereña, Arantxa (with contributions by Luca Chinotti, Sonia Goicoechea, Jean-Denis Crola, and Eric Hazard). 2010. Halving hunger: Still possible? Oxfam Briefing Paper 139. September. http://www.oxfam.org/sites/www.oxfam.org/files/oxfam-halving-hunger-sept-2010.pdf.

Guha, Ramachandra. 2005. Prime ministers and big dams. *Hindu*, December 18. http://ramachandraguha.in/archives/prime-ministers-and-big-dams.html.

Gumbel, Andrew. 2007. US oil company accused of dumping waste in Amazon. *Independent*, May 4. https://www.commondreams.org/archive/2007/05/04/954.

Hall, Charles, and John Day. 2009. Revisiting the Limits to Growth after peak oil. *American Scientist* 97 (May–June): 230–237.

Hamer, Ed. 2007. Bananas: From plantation to plate. *Ecologist*, September 1. http://theecologist-test.net-genie.co.uk/Investigations/FoodandFarming/465261/bananas_from_plantation_to_plate.html.

Hamilton, Clive. 2007. *Scorcher: The Dirty Politics of Climate Change*. Melbourne: Black Inc.

Hammes, David, and Douglas Wills. 2005. Black gold: The end of Bretton Woods and the oil-price shocks of the 1970s. *Independent Review* 9 (4): 501–511.

Hanlon, Joseph. 2002. Defining illegitimate debt and linking its cancellation to economic justice. Oslo: Norwegian Church Aid. http://journal.probeinternational.org/2002/06/01/defining-illegitimate-debt-understanding-issues.

Hanlon, Joseph. 2006. "Illegitimate" loans: Lenders, not borrowers, are responsible. *Third World Quarterly* 27 (2): 211–226.

Hardin, Garrett. 1968. "The Tragedy of the Commons." In Daly and Townsend 1993, 127–143.

Hardin, Garrett. 1974a. Living on a lifeboat. *Bioscience* 24 (10): 561–568.

Hardin, Garrett. 1974b. Lifeboat ethics: The case against helping the poor. *Psychology Today*, September. http://www.garretthardinsociety.org/articles/art_lifeboat_ethics_case_against_helping_poor.html.

Harmer, Wendy. 2008. *Stuff*, ABC TV. http://www.abc.net.au/tv/stuff.

Harris, Nigel. 1986. *The End of the Third World: Newly Industrializing Countries and the End of an Ideology*. Harmondsworth, UK: Penguin.

Harris, Paul. 1999. Common but differentiated responsibility: The Kyoto Protocol and United States policy. *New York University Environmental Law Journal* 27:27–48.

Harris, Ralph. 1978. The conversion of the intellectual. In Ivens 1978, 10–12.

Harris, Ralph. 1997. The plan to end planning: The founding of the Mont Pèlerin Society. *National Review*, June 16.

Hartung, William. 1992. Curbing the arms trade: From rhetoric to restraint. *World Policy Journal* 9 (2): 219–247.

Hay, Peter. 2002. *Main Currents in Western Environmental Thought*. Sydney: UNSW Press.

Hayek, Friedrich von. 1976 [1944]. *The Road to Serfdom*. London: Routledge and Kegan Paul.

Hayek, Friedrich von. 1999 [1945]. *The Road to Serfdom*. Reader's Digest. London: IEA. http://www.iea.org.uk/sites/default/files/publications/files/upldbook43pdf.pdf.

Heartland Institute. 2013. *2009 International Conference on Climate Change*. Chicago: Heartland Institute. http://climateconferences.heartland.org/iccc2.

Helleiner, Eric. 2008. The evolution of the international monetary and financial system. In *Global Political Economy*, ed. John Ravenhill, 213–240. Oxford: Oxford University Press.

Helvarg, David. 1994. *The War against the Greens: The "Wise-Use" Movement, the New Right, and Anti-environmental Violence*. San Francisco, CA: Sierra Club Books.

Henry, James S. 2012. *The Price of Offshore* revisited: New estimates for "missing" global private wealth, income, inequality and lost taxes. Tax Justice Network. http://www.taxjustice.net/cms/front_content.php?idcat=148.

Heritage Foundation. 2010. The Heritage Foundation's 35th Anniversary: A history of achievements. http://www.heritage.org/About/Our-History/35th-Anniversary.

Herrick, Charles, and Dale Jamieson. 2001. Junk science and environmental policy: Obscuring public debate with misleading discourse. *Philosophy & Public Policy Quarterly* 21 (2–3): 11–16.

Hirsch, G. 1975. Only you can prevent ideological hegemony. *Insurgent Sociologist* 5 (3): 64–82.

Hirsch, Robert. 2005. The inevitable peaking of world oil production. *Atlantic Council Bulletin* 16 (3): 1–9.

Hobsbawm, Eric. 1989. *The Age of Empire: 1975–1914*. New York: Vintage.

Howarth, Robert, Renee Santoro, and Anthony Ingraffea. 2011. Methane and the greenhouse-gas footprint of natural gas from shale formations: A letter. *Climatic Change* 106 (4): 679–690.

Huesemann, Michael. 2003. The limits of technological solutions to sustainable development. *Clean Technologies and Environmental Policy* 5 (1): 21–34.

Hughes, J. David. 2013. *Drill, Baby, Drill: Can unconventional fuels usher in a new era of energy abundance?* Post Carbon Institute. http://shalebubble.org/drill-baby-drill.

Humphrys, John. 2004. Interview with Myron Ebell. *Today*, November 4. BBC Radio 4.

Hunnicutt, Benjamin. 1988. *Work without End: Abandoning Shorter Hours for the Right to Work*. Philadelphia, PA: Temple University Press.

Hunnicutt, Benjamin. 1996. *Kellogg's 6-hour Day*. Philadelphia, PA: Temple University Press.

ILO. 2006. The decent work deficit: A new ILO report outlines the latest global employment trends. International Labour Organization, *World of Work*. http://www.ilo.org/wcmsp5/groups/public/---dgreports/---dcomm/documents/publication/dwcms_080599.pdf.

IMF. 2000. *World Economic Outlook*, chap. 5. http://www.imf.org/external/pubs/ft/weo/2000/01/pdf/chapter5.pdf.

IMF. 2009. *World Economic Outlook: Crisis and Recovery*. http://www.imf.org/external/pubs/ft/weo/2009/01/pdf/text.pdf.

IMF. 2013a. IMF quotas. http://www.imf.org/external/np/exr/facts/quotas.htm.

IMF. 2013b. *World Economic Outlook* (April). http://www.imf.org/external/pubs/ft/weo/2013/01/pdf/text.pdf.

International Air Transport Association. 2013. The value of air cargo. http://www.iata.org/whatwedo/cargo/sustainability/Pages/benefits.aspx.

International Energy Agency. 2010. *World Energy Outlook*. Press release, November 9. http://www.iea.org/media/weowebsite/2010/weo2010_london_nov9.pdf.

International Institute for Population Sciences (IIPS) and Macro International. 2007. *National Family Health Survey (NFHS-3), 2005–06: India*. Vol. I. Mumbai: IIPS.

IPA Review. 1955. Economics for the People. *IPA Review* 9 (1): 11–17.

IPCC. 2001. *Climate Change 2001: Synthesis Report*. Cambridge: Cambridge University Press. http://www.grida.no/publications/other/ipcc_tar.

IPCC. 2007. *Climate Change 2007: Synthesis Report*. http://www.ipcc.ch/publications_and_data/ar4/syr/en/main.html.

IPCC. 2013. *Climate Change 2013: Synthesis Report*. http://www.ipcc.ch/report/ar5.

Ivens, Michael, ed. 1978. *International Papers on the Revival of Freedom and Enterprise*. London: AIMS.

Jackson, Liz. 2010. The authentic Mr Abbott. *Four Corners*, ABC TV, March 15. http://www.abc.net.au/4corners/content/2010/s2846485.htm.

Jackson, Tim. 2009a. *Prosperity without Growth: Economics for a Finite Planet*. Abingdon, UK: Earthscan.

Jackson, Tim. 2009b. *Prosperity without Growth: The Transition to a Sustainable Economy*. Sustainable Development Commission (UK). http://www.sd-commission.org.uk/data/files/publications/prosperity_without_growth_report.pdf.

Jacobs, Michael. 2011. Taking stock on climate. *Inside Story*, March 2. http://inside.org.au/taking-stock-on-climate.

Jacques, Peter, Riley Dunlap, and Mark Freeman. 2008. The organisation of denial: Conservative think tanks and environmental scepticism. *Environmental Politics* 17 (3): 349–385.

Jaffe, Amy, and Robert Manning. 2000. The shocks of a world of cheap oil. *Foreign Affairs*, January–February. http://www.foreignaffairs.com/print/55627.

Jishnu, Latha. 2008. Sustainable development's an oxymoron. *Business Standard* (New Delhi), October 19. http://www.business-standard.com/article/opinion/-39-sustainable-development-39-s-an-oxymoron-39-108101901048_1.html.

Johnson, Ian. 2013a. Pitfalls abound in China's push from farm to city. *New York Times*, June 13. http://www.nytimes.com/2013/07/14/world/asia/pitfalls-abound-in-chinas-push-from-farm-to-city.html.

Johnson, Ian. 2013b. China's great uprooting: Moving 250 million into cities. *New York Times*, June 15. http://www.nytimes.com/2013/06/16/world/asia/chinas-great-uprooting-moving-250-million-into-cities.html.

Johnston, David. 2005. Richest are leaving even the rich far behind. *New York Times*, June 5. http://www.nytimes.com/2005/06/05/national/class/HYPER-FINAL.html.

Jones, Alan. 2006. Connecting the dots. *Inside Higher Education*, June 6. http://insidehighered.com/views/2006/06/16/jones.

Jones, Meredith, Shelley Marshall, Richard Mitchell, and Ian Ramsay. 2007. Company directors' views regarding stakeholders. Melbourne: Corporate Governance and Workplace Partnerships Project, Faculty of Law, University of Melbourne. www.law.unimelb.edu.au/82BD8650-4B28-11E2-95000050568D0140.

Jopson, Debra. 2007. Phantom aid never leaves our shores. *Sydney Morning Herald*, May 28. http://www.smh.com.au/news/national/phantom-aid-never-leaves-our-shores/2007/05/27/1180205079584.html?page=fullpage#contentSwap1.

Jubilee Research. 1998. How it all began: Causes of the debt crisis. http://www.globalissues.org/article/224/how-it-all-began-causes-of-the-debt-crisis.

JunkScience.com. 2014. What is junk science? http://junkscience.com/what-is-junk-science.

Kahn, Joseph, and Mark Landler. 2007. China grabs West's smoke-spewing factories. *New York Times*, December 21. http://www.nytimes.com/2007/12/21/world/asia/21transfer.html.

Kahn, Joseph, and Jim Yardley. 2007. As China roars, pollution reaches deadly extremes. *New York Times*, August 26. http://www.nytimes.com/2007/08/26/world/asia/26china.html.

Kaiser, Robert, and Ira Chinoy. 1999. Scaife: Funding father of the Right. *Washington Post*, May 2, A1. http://www.washingtonpost.com/wp-srv/politics/special/clinton/stories/scaifemain050299.htm.

Karl, Jonathan. 2007. Cheney on global warming: Vice President's views at odds with majority of climate scientists. *ABC News* (US), February 23. http://abcnews.go.com/Technology/story?id=2898539&page=1.

Karliner, Joshua. 1997. The World Bank and corporations fact sheet. *CorpWatch*. http://www.corpwatch.org/article.php?id=445.

Kaysen, Carl. 1972. The computer that printed out W*O*L*F*. *Foreign Affairs* 50 (4): 660–668.

Keavney, Jack. 1978. Enterprise Australia: A case study in mobilisation. In Ivens 1978, 66–69.

Keay, Douglas. 1987. Interview for *Woman's Own*, September 23. Margaret Thatcher Foundation. http://www.margaretthatcher.org/document/106689.

Keen, Steve. 2001. *De-bunking Economics: The Naked Emperor of the Social Sciences*. Annandale, NSW: Pluto Press.

Kelly, Fran. 2010. Business Council of Australia president on immigration policy. *Breakfast*, ABC Radio National, December 1. http://www.abc.net.au/radionational/programs/breakfast/business-council-of-australia-president-on/2966918#transcript.

Kempf, Hervé. 2008. *How the Rich Are Destroying the Earth*. White River Junction, VT: Chelsea Green.

Kennedy, Robert F., Jr. 2003. Crimes against nature. *Rolling Stone*, December 11. http://www.commondreams.org/views03/1120-01.htm.

Kettering, Charles. 1929. Keep the consumer dissatisfied. *Nation's Business*, January 16.

Kettles, Nick. 2008. Designing for destruction. *Ecologist* 38 (6): 47.

Khor, Martin. 1997. Effects of globalisation on sustainable development after UNCED. *Third World Resurgence* 81/82 (May–June): 5–11. http://www.twnside.org.sg/title/rio-cn.htm.

Kiernan, Mike. 1972. The coming Apocalypse. *Mother Earth News* 15 (May–June). http://www.motherearthnews.com/print.aspx?id={EEA31E0D-7F69-4903-9259-736D99683D3C}#axzz2efYNz9ux.

Klare, Michael. 2005. The energy crunch to come. *Mother Jones*, March 22. http://www.motherjones.com/politics/2005/03/energy-crunch-come.

Klein, Naomi. 2002. *Fences and Windows, Dispatches from the Front Lines of the Globalization Debate*. London: Flamingo.

Klein, Naomi. 2007. *The Shock Doctrine*. New York: Henry Holt.

Kneese, Allen, and Ronald Ridker. 1972. Predicament of mankind. *Washington Post,* March 2.

Koehler, John. 1973. Untitled review of *The Limits to Growth: A Report for the Club of Rome's Project on the Predicament of Mankind* [by D. H. Meadows, D. L. Meadows, J. Randers, and W. W. Behrens]. *Journal of Politics* 35 (2): 513–514.

Kohler, Alan. 2009. CPRS could add billions to federal budget: AIGN. *Inside Business*. ABC TV, November 29. http://www.abc.net.au/insidebusiness/content/2009/s2756514.htm.

Kohler, Alan. 2012. The death of peak oil. *Business Spectator*, February 29. http://www.businessspectator.com.au/article/2012/2/29/commodities/death-peak-oil.

Kohn, Rachael. 2010. Women hold up half the sky. *Spirit of Things*, ABC Radio National, January 31. http://www.abc.net.au/rn/spiritofthings/stories/2010/2803410.htm.

Kraft, Michael. 1994. Population policy. In *Encyclopaedia of Policy Studies*, ed. Stuart Nagel, 617–644. New York: Marcel Dekker.

Kraft, Michael. 2000. US environmental policy and politics: From the 1960s to the 1990s. *Journal of Policy History* 12 (1): 17–42.

Krakauer, John. 1991. Brownfellas. *Outside*, December. http://byliner.com/jon-krakauer/stories/brownfellas.

Krauss, Clifford. 2008. Commodities' relentless surge: Chinese and US demand push food and minerals steeply higher. *New York Times*, January 15, C1. http://www.nytimes.com/2008/01/15/business/worldbusiness/15commodities.html.

Krauss, Clifford, and Eric Lipton. 2012. US inches toward goal of energy independence. *New York Times*, March 22. http://www.nytimes.com/2012/03/23/business/energy-environment/inching-toward-energy-independence-in-america.html.

Kristof, Nicholas. 2006. Aid: Can it work? *New York Review of Books* 53 (15). http://www.nybooks.com/articles/19374.

Kristol, Irving. 1977. On corporate philanthropy. *Wall Street Journal,* March 21.

Krueger, Anne. 2004. Letting the future in: India's continuing reform agenda. Keynote speech at the Stanford India Conference, June 4. http://www.imf.org/external/np/speeches/2004/060404.htm.

Krupp, Fred. 2002. Cars can get much cleaner. *New York Times*, July 20. http://www.nytimes.com/2002/07/20/opinion/cars-can-get-much-cleaner.html.

Lahsen, Myanna. 2008. Experiences of modernity in the greenhouse: A cultural analysis of a physicist "trio" supporting the backlash against global warming. *Global Environmental Change* 18 (1): 204–219.

Lamoreaux, Naomi. 1985. *The Great Merger Movement in American Business, 1895–1904*. Cambridge: Cambridge University Press.

Landers, Kim. 2007. Pacific garbage patch grows bigger. *World Today*, ABC, November 1. http://www.abc.net.au/worldtoday/content/2007/s2078528.htm.

Landesa Rural Development Institute. 2011. Landesa 6th 17-province China survey, summary. http://www.landesa.org/china-survey-6.

Lane, Edgar. 1950. Some lessons from past congressional investigations of lobbying. *Public Opinion Quarterly* 14 (1): 14–32.

Lane, Sabra. 2011. New lobby launches anti-carbon tax ads. *AM*, ABC Radio, July 21. http://www.abc.net.au/am/content/2011/s3274203.htm.

Lapham, Lewis. 2004. Tentacles of rage: The Republican propaganda mill, a brief history. *Harper's Magazine* 309:1852.

Lappé, Frances Moore, Jennifer Clapp, Molly Anderson, et al. 2013a. How we count hunger matters. *Ethics & International Affairs* 27 (3): 251–259.

Lappé, Frances Moore, Jennifer Clapp, Molly Anderson, et al. 2013b. *Framing Hunger: A Response to The State of Food Insecurity in the World 2012*. Cambridge, MA: Small Planet Institute. http://smallplanet.org/content/jkfls.

Leach, William. 1993. *Land of Desire: Merchants, Power, and the Rise of a New American Culture*. New York: Vintage Books.

Le Bon, Gustave. 1960 (1895). *The Crowd*. Harmondsworth, UK: Penguin.

Lee, Ivy. 1929. *The Press Today: How the News Reaches the Public*. New York: Ivy Lee.

Legacy Tobacco Documents Library. 1969. *Smoking and Health Proposal*. BN 690010951–690010959. http://legacy.library.ucsf.edu/tid/zqy56b00.

Leggett, Jeremy. 2001. *The Carbon War: Global Warming and the End of the Oil Era*. New York: Routledge.

Lever, Chauncey. 1978. Fighting for free enterprise in America: The USIC approach. In Ivens 1978, 71–72.

Lewis, Aaron. 2006. The cost of development. *Dateline*, SBS Television, May 31. http://www.sbs.com.au/dateline.

Lewis, W. Arthur. 1954. Economic development with unlimited supply of labour. In Agarwala and Singh 1958, 400–435.

Li Keqiang. 2012. Li Keqiang expounds on urbanization. Trans. He Shan and Chen Xia. http://www.china.org.cn/china/2013-05/26/content_28934485.htm.

Li, Minqi. 2009. *The Rise of China and the Demise of the Capitalist World-Economy*. New York: Monthly Review Press.

Liberal Party of Australia—Queensland Branch. 2007. Formation of the Liberal Party of Australia. http://web.archive.org/web/20070426002837/http://www.qld.liberal.org.au/history/formation.aspx.

Lindsay, Greg. 1996. The CIS at twenty: Greg Lindsay talks to Andrew Norton. *Policy*, Winter, 16–21.

Link, Henry. 1948. Is freedom the issue? *Public Relations Journal* 4 (2): 1–2, 14.

Lippmann, Walter. 1920. *Liberty and the News*. New York: Harcourt and Brace. http://archive.org/details/libertyandnews01lippgoog.

Lippmann, Walter. 1993 [1927]. *The Phantom Public*. New Brunswick, NJ: Transaction Publishers.

Long, Katherine. 1991. His goal: Destroy environmentalism. Man and group prefer that people exploit the Earth. *Seattle Times*, December 2. http://community .seattletimes.nwsource.com/archive/?date=19911202&slug=1320610.

Long, Stephen. 2009. Rich banks to benefit from loans to poor nations. *PM*, ABC Radio, September 22. http://www.abc.net.au/pm/content/2009/s2693454.htm.

Lorenz, Andreas. 2005. The Chinese miracle will end soon: *Spiegel* interview with China's deputy minister of the environment. *Der Spiegel,* March 7. http://www .spiegel.de/international/spiegel/0,1518,345694,00.html.

Luce, Edward. 2011. America is entering a new age of plenty. *Financial Times*, November 20. http://www.ft.com/cms/s/0/a307107c-1364-11e1-9562 -00144feabdc0.html#ixzz2SenaO2Al.

Luntz, Frank. 2002. Straight Talk. The environment: A cleaner, safer, healthier America. https://www2.bc.edu/~plater/Newpublicsite06/suppmats/02.6.pdf.

Lynch, David. 2004. Discontent in China boils into public protest. *USA Today*, September 14. http://www.usatoday.com/news/world/2004-09-14-china-protest _x.htm.

Lynch, David. 2008. US bends the rules of free markets. *USA Today,* September 19. http://usatoday30.usatoday.com/money/economy/2008-09-18-free-market -bailout_N.htm.

Ma, Haibing. 2011. Energy intensity is rising slightly. *Vital Signs* (September 20). http://vitalsigns.worldwatch.org/vs-trend/energy-intensity-rising-slightly.

MacDougall, A. K. 1980. Advocacy: Business increasingly uses (in both senses) media to push views. *Los Angeles Times*, November 16, 14.

MacFarlane, Ian. 2006. *Boyer Lectures 2006: The Search for Stability*. Sydney: ABC Books.

Machan, Dyan. 1999. Power broker. *Forbes,* November 15. http://www.forbes .com/global/1999/1115/0223108a.html.

Macinko, George. 1973. Man and the environment: A sampling of the literature. *Geographical Review* 63 (3): 378–391.

Maddox, John. 1972. *The Doomsday Syndrome*. New York: McGraw-Hill.

Magnier, Mark. 2013. India passes massive program to feed 800 million in poverty. *Los Angeles Times*, September 2. http://www.latimes.com/world/ worldnow/la-fg-wn-india-food-program-20130902,0,2565188.story.

Maier, Charles. 1987. *In Search of Stability: Explorations in Historical Political Economy*. Cambridge: Cambridge University Press.

Malthus, Thomas. 1976 [1798]. *An Essay on the Principle of Population*. Edited by Philip Appleman. New York: W. W. Norton.

Malthus, Thomas. 1989 [1803]. *An Essay on the Principle of Population, the version published in 1803, with the variora of 1806, 1807, 1817 and 1826*. Edited by Patricia James. Vol. 2. Cambridge: Cambridge University Press.

Mandel, Stephen. 2006. Odious lending: Debt relief as if morals mattered. London: new economics foundation. http://www.i-r-e.org/bdf/docs/a006_odious-lending-and-debt-relief.pdf.

Marchand, Roland. 1998. *Creating the Corporate Soul: The Rise of Public Relations and Corporate Imagery in American Big Business*. Berkeley: University of California Press.

Marcuse, Herbert. 1968. *One Dimensional Man: Studies in the Ideology of Advanced Industrial Society*. London: Routledge and Kegan Paul.

Marcuse, Herbert. 1969. *An Essay on Liberation*. London: Allen Lane.

Marcuse, Herbert. 1970. *Five Lectures: Psychoanalysis, Politics and Utopia*. London: Allen Lane.

Mares, Peter. 2011. Psst! Want a unit of digital data? *National Interest*, ABC Radio National, October 21. http://www.abc.net.au/radionational/programs/nationalinterest/psst-want-a-unit-of-digital-data/3594458#transcript.

Markowitz, Gerald, and David Rosner. 2002a. *Deceit and Denial: The Deadly Politics of Industrial Pollution*. Berkeley and Los Angeles: University of California Press.

Markowitz, Gerald, and David Rosner. 2002b. Industry challenges to the principle of prevention in public health: The precautionary principle in historical perspective. *Public Health Reports* 117 (November–December): 510–512. http://www.ncbi.nlm.nih.gov/pmc/articles/PMC1497488.

Marshall Institute. 2009. Science for better public policy. http://search.atomz.com/search/?sp-q=sound+science&sp-a=sp100240f4&sp-f=ISO-8859-1.

Marshall Institute. 2013. About the Marshall Institute. http://www.marshall.org/category.php?id=6.

Martin, Frederick. 1911. *The Passing of the Idle Rich*. Garden City, NY: Doubleday.

Maruyama, Sadami. 1996. Responses to Minamata disease. In *The Long Road to Recovery: Community Responses to Industrial Disaster*, ed. James Mitchell, 41–59. Tokyo: United Nations University Press.

Marx, Karl. 1954a [1848]. *The Communist Manifesto*. Chicago: Henry Regnery Co.

Marx, Karl. 1954b [1887]. *Capital*. Vol. 1. Moscow: Progress Publishers.

Marx, Karl. 1959 [1894]. *Capital*. Vol. 3. Moscow: Progress Publishers.

Matthews, Jonathan. 2008. India's Maoist revolution. *SBS Dateline*, July 23. http://www.sbs.com.au/dateline/story/transcript/id/552344/n/India-s-Maoist-Revolution.

May, Robert. 2007. Relations among nations on a finite planet. Lowy Institute Lecture, November 15, Sydney. http://www.lowyinstitute.org/publications/lowy-lecture-series-relations-among-nations-finite-planet-lord-robert-may.

Mayer, Jane. 2010. The billionaire brothers who are waging a war against Obama. *New Yorker,* August 30, http://www.newyorker.com/reporting/2010/08/30/100830fa_fact_mayer?currentPage=all.

McCright, Aaron, and Riley Dunlap. 2000. Challenging global warming as a social problem: An analysis of the conservative movement's counter-claims. *Social Problems* 47 (4): 499–522.

McGinnis, Robert. 1973. Untitled review of *The Limits to Growth: A Report for the Club of Rome's Project on the Predicament of Mankind* [by D. H. Meadows, D. L. Meadows, J. Randers, and W. W. Behrens]. *Demography* 10 (2): 295–299.

McKendrick, Neil, John Brewer, and J. H. Plum. 1982. *The Birth of a Consumer Society: The Commercialization of Eighteenth Century England*. London: Europa.

McLaughlin, Kathleen. 2012. Chinese land rights again proven major source of unrest. *Global Post*, February 7. http://www.globalpost.com/dispatches/globalpost-blogs/china/chinese-land-rights-protests-wukan.

McNeill, John. 2001. *Something New Under the Sun: An Environmental History of the Twentieth-Century World*. London: Penguin.

McNeill, John. 2003. Historical perspectives on global ecology. *World Futures* 59:263–274.

Meadows, Dennis. 2007. Evaluating past forecasts: Reflections on one critique of "The Limits to Growth." In *Sustainability or Collapse?* ed. Robert Costanza, Lisa Graumlich, and Will Steffen, 399–415. Cambridge, MA: MIT Press.

Meadows, Donella, Dennis Meadows, Jørgen Randers, and William Behrens III. 1972. *The Limits to Growth: A Report for the Club of Rome's Project on the Predicament of Mankind*. New York: Signet.

Meadows, Donella, Dennis Meadows, Jørgen Randers, and William Behrens III. 1973. A response to Sussex. In Cole et al. 1973, 217–240.

Meadows, Donella, Dennis Meadows, and Jørgen Randers. 1992. *Beyond the Limits: Confronting Global Collapse; Envisioning a Sustainable Future*. Post Mills, VT: Chelsea Green.

Meadows, Donella, Dennis Meadows, and Jørgen Randers. 2004. *The Limits to Growth: The 30-Year Update*. Post Mills, VT: Chelsea Green.

Michaux, Simon. 2013. Peak mining & implications for natural resource management. http://www.youtube.com/watch?v=TFyTSiCXWEE.

Middleton, Neil, Phil O'Keefe, and Sam Moyo. 1993. *The Tears of the Crocodile: From Rio to Reality in the Developing World*. London: Pluto Press.

Milios, John. 2005. European integration as a vehicle of neoliberal Hegemony. In Saad-Filho and Johnston 2005, 208–214.

Mill, John Stuart. 1848. *Principles of Political Economy*. Vol. 2. London: John W. Parker.

Millennium Ecosystem Assessment. 2005. *Ecosystems and Human Well-being: Synthesis*. Washington DC: Island Press. http://www.maweb.org/documents/document.356.aspx.pdf.

Mirowski, Philip. 1984. Physics and the "marginalist revolution." *Cambridge Journal of Economics* 8:361–379.

Mirowski, Philip. 2009. Postface: Defining neoliberalism. In Mirowski and Plehwe 2009, 417–455.

Mirowski, Philip, and Dieter Plehwe, eds. 2009. *The Road from Mont Pèlerin*. Cambridge, MA: Harvard University Press.

Mitchell, Neil. 1989. *The Generous Corporation: A Political Analysis of Corporate Power*. New Haven, CT: Yale University Press.

Monbiot, George. 2006. The denial industry. *Guardian*, September 19. http://www.theguardian.com/environment/2006/sep/19/ethicalliving.g2.

Monbiot, George. 2009. The population myth. *Guardian*, September 29. http://www.monbiot.com/2009/09/29/the-population-myth.

Montague, Peter. 1999. The WTO and free trade. *Rachel's Environment & Health Weekly* 673, October 21. http://www.ratical.org/co-globalize/REHW673.txt.

Montgomery, David. 2007. *Dirt: The Erosion of Civilizations*. Berkeley and Los Angeles: University of California Press.

Mooney, Chris. 2004. The fraud of "sound science." *Gadflyer*, May 12. http://www.alternet.org/story/18696.

Mooney, Chris. 2005. Some like it hot. *Mother Jones*, May–June. http://motherjones.com/environment/2005/05/some-it-hot.

Mooney, Chris. 2010. Address to the National Press Club (Australia), March 10. http://www.npc.org.au.

Moore, Bette, and Gary Carpenter. 1987. Main players. In *The New Right's Australian Fantasy*, ed. Ken Coghill, 3–19. Fitzroy, VIC: McPhee Gribble/Penguin Books.

More, Thomas. 1999 [1516]. *Utopia*. Edited by David Sacks. Boston: Bedford/St. Martins Press.

Morford, Mark. 2005. Global warmin' is fer idjuts. *San Francisco Chronicle*, June 10. http://www.sfgate.com/entertainment/morford/article/Global-Warmin-Is-Fer-Idjuts-Exxon-writes-2663534.php.

Morley, Morris. 1986. Behind the World Bank loans. *Sydney Morning Herald*, December 9, 17.

Mother Jones. 2005. Put a tiger in your think tank. *Mother Jones*, May/June. http://motherjones.com/politics/2005/05/put-tiger-your-think-tank#.

Mshana, Rogate. 2004. The current financial system is one of the main causes of inequality in the global economy. Geneva: World Council of Churches. http://archived.oikoumene.org/po/resources/documents/wcc-programmes/public-witness-addressing-power-affirming-peace/poverty-wealth-and-ecology/finance-speculation-debt/r-mshana-on-current-financial-system.html.

Mshana, Rogate. 2007. Poverty, wealth and ecology: The impact of economic globalization: A background to the study process. Geneva: World Council of Churches. http://archived.oikoumene.org/po/resources/documents/wcc-programmes/public-witness-addressing-power-affirming-peace/poverty-wealth-and-ecology/finance-speculation-debt/r-mshana-on-current-financial-system.html.

Mugliston, Michael. 1998. Climate Change Kyoto outcome. Department of Foreign Affairs and Trade, February 12. http://www.apec.org.au/docs/muglistn.pdf.

Mulama, Joyce. 2006. Flying toilets still airborne. *IPS News*, October 24. http://ipsnews.net/africa/nota.asp?idnews=35222.

Murdoch, Rupert. 2008. Lecture 5: The global middle class roars. *Boyer Lectures*, ABC Australian Broadcasting Corporation. http://abc.com.au/rn/boyerlectures/stories/2008/2397948.htm#transcript.

Murray, Georgina. 2006. *Capitalist Networks and Social Power in Australia and New Zealand*. Hampshire, UK: Ashgate.

Musgrove, Mike. 2006. Sweatshop conditions at iPod factory reported. *Washington Post*, June 16. http://www.washingtonpost.com/wp-dyn/content/article/2006/06/15/AR2006061501898.html.

Mydans, Seth. 2008. A corner of Indonesia, sinking in a sea of mud. *New York Times*, December 18. http://www.nytimes.com/2008/12/19/world/asia/19mud.html.

Nadeau, Robert. 2008. Brother, can you spare me a planet? Mainstream economics and the environmental crisis. *Scientific American*, March 19. http://www.scientificamerican.com/article.cfm?id=brother-can-you-spare-me-a-planet.

Naidoo, Kumi. 2014. Davos: The shifting nature of power and the shifting power of nature. Greenpeace, January 22. http://www.greenpeace.org/international/en/news/Blogs/makingwaves/world-economic-forum/blog/47959.

National Research Council. 1983. *Changing Climate: Report of the Carbon Dioxide Assessment Committee*. Washington, DC: National Academy Press.

Nef, John. 1969. Another challenge: Two Industrial Revolutions. In *The Industrial Revolution*, ed. C. Stewart Doty, 22–28. Hinsdale, IL: Dryden Press.

Newman, Maurice. 2010. Address to ABC staff. *Australian*, March 11. http://www.theaustralian.com.au/media/maurice-newman-speech/story-e6frg996-1225839427099.

Newport, Frank. 2010. Americans' global warming concerns continue to drop. Gallup Politics, March 11. http://www.gallup.com/poll/126560/americans-global-warming-concerns-continue-drop.aspx.

Nierenberg, William. 1984. *Report of the Acid Rain Peer Review Panel*, Chicago: EPA. http://nepis.epa.gov/Exe/ZyPURL.cgi?Dockey=20013U0E.txt.

Nordhaus, William. 1973. World dynamics: Measurement without data. *Economic Journal* 83 (332): 1156–1183.

Nordhaus, William, and James Tobin. 1973. Is growth obsolete? In *The Measurement of Economic and Social Performance*, ed. Milton Moss, 509–564. Cambridge, MA: National Bureau of Economic Research. http://www.nber.org/chapters/c3621.pdf.

Norgaard, Richard. 1994. *Development Betrayed: The End of Progress and a Co-evolutionary Revisioning of the Future*. New York: Routledge.

O'Connor, James. 1994. Is sustainable capitalism possible? In *Is Capitalism Sustainable? Political Economy and the Politics of Ecology*, ed. James O'Connor, 152–175. New York: Guilford Press.

Ogle, Greg. 2010. Gunns backs down. *New Matilda*, February 5. http://newmatilda.com/2010/02/05/gunns-backs-down.

Ohlin, Goran. 1974. The new breed of Malthusians. *Family Planning Perspectives* 6 (3): 158–159.

Omidi, Maryam. 2010. G20 faces uphill battle on global reform. *Financial News*, November 10. http://www.efinancialnews.com/story/2010-11-10/industry-sceptical-g20-financial-reform.

Ong, Elisa, and Stanton Glantz. 2001. Constructing "sound science" and "good epidemiology": Tobacco, lawyers, and public relations firms. *American Journal of Public Health* 91 (11): 1749–1757.

Oreskes, Naomi. 2004. Beyond the ivory tower: The scientific consensus on climate change. *Science* 306 (5702).

Oreskes, Naomi. 2011. Merchants of doubt. *Science Show*, ABC Radio National, January 8. http://www.abc.net.au/rn/scienceshow/stories/2011/3101369.htm#transcript.

Oreskes, Naomi, and Erik Conway. 2008. Challenging knowledge: How climate science became a victim of the Cold War. In *Agnotology: The Making and Unmaking of Ignorance*, ed. Robert Proctor and Londa Schiebinger, 55–89. Stanford, CA: Stanford University Press.

Oreskes, Naomi, and Erik Conway. 2010. *Merchants of Doubt: How a Handful of Scientists Obscured the Truth on Issues from Tobacco Smoke to Global Warming*. New York: Bloomsbury Press.

Organisation for Economic Co-operation and Development. 2010a. History. http://www.oecd.org/pages/0,3417,en_36734052_36761863_1_1_1_1_1,00.html.

Organisation for Economic Co-operation and Development. 2010b. What we do and how. http://www.oecd.org/pages/0,3417,en_36734052_36761681_1_1_1_1_1,00.html.

Ormerod, Paul. 1994. *The Death of Economics*. London: Faber.

O'Rorke, Fergus. 2013. China's revived Green GDP program still faces challenges. *CleanBiz Asia*, March 28. http://www.cleanbiz.asia/news/chinas-revived-green-gdp-program-still-faces-challenges?page=show.

Ortiz, Isabel, and Matthew Cummins. 2011. Global inequality: Beyond the bottom billion. New York: UNICEF. http://www.unicef.org/socialpolicy/index_58230.html.

Osborn, Fairfield. 1948. *Our Plundered Planet*. Boston: Little, Brown.

Osborn, Stephen, Avner Vengosh, Nathaniel Warner, and Robert Jackson. 2011. Methane contamination of drinking water accompanying gas-well drilling and hydraulic fracturing. *Proceedings of the National Academy of Sciences of the United States of America* 108 (20): 8172–8176.

Osnos, Evan. 2006. That low-priced cashmere sweater has a hidden cost. *Chicago Tribune*, December 28. http://seattletimes.nwsource.com/html/nationworld/2003498352_cashmere282.html.

Ostrom, Elinor. 1990. *Governing the Commons: The Evolution of Institutions for Collective Action*. New York: Cambridge University Press.

Packard, Vance. 1959. *The Status Seekers: An Exploration of Class Behavior in America*. New York: David McKay.

Packard, Vance. 1963 (1960). *The Waste Makers*. New York: Cardinal.

Padel, Felix, and Samarendra Das. 2007. Agya, what do you mean by development? In *Caterpillar and the Mahua Flower: Tremors in India's Mining Fields*, ed. Rakesh Kalshian, 24–46. New Delhi: Panos South Asia. http://www.panossouthasia.org/pdf/Caterpillar%20and%20the%20Mahua%20Flower.pdf.

Page, Jeremy. 2007. Indian children suffer more malnutrition than in Ethiopia. *Times Online*, February 22. http://lists.goanet.org/pipermail/goanet-goanet.org/2007-February/140715.html.

Parenti, Michael. 1986. *Inventing Reality: The Politics of the Mass Media*. New York: St. Martin's Press.

Passell, Peter, Marc Roberts, and Leonard Ross. 1972. Review of Limits to Growth. *New York Times Book Review*, April 2, 1, 10, 12–13.

Patnaik, Prabhat. 1999. The real face of financial liberalisation. *Frontline Magazine* 16 (4): 13–26. http://www.frontline.in/static/html/fl1604/16041010.htm.

Patnaik, Utsa. 2007. Neoliberalism and rural poverty in India. *Economic and Political Weekly* 42 (30): 3132–3150.

Patnaik, Utsa. 2009. Origins of the food crisis in India and developing countries. *Monthly Review* 61 (3): 63–77.

Pauly, Daniel. 2010. *5 Easy Pieces: How Fishing Impacts Marine Ecosystems*. Washington, DC: Island Press.

Pauly, Daniel. 2011. Beyond duplicity and ignorance in global fisheries. *Pacific Ecologist* 20:32–36.

Pavitt, K. 1973. Malthus and other economists: Some doomsdays revisited. In Cole et al. 1973, 137–158.

Paxman, Jeremy. 2006. Debate between a climate scientist and oil industry funded lobbyist. *Newsnight*, BBC, September 20. http://www.youtube.com/watch?v=YyKUblhXJw8.

Payer, Cheryl. 1991. *Lent and Lost: Foreign Credit and Third World Development*. London: Zed Books.

PBS. 1980. Who protects the consumer? *Free to Choose*, vol. 7. http://www.freetochoosemedia.org/freetochoose/detail_ftc1980_transcript.php?page=7

PBS. 2002. *Commanding Heights: The Battle for the World Economy*. US: Heights Productions. http://www.pbs.org/wgbh/commandingheights/lo/story/tr_menu.html.

PBS. 2006. Hot Politics interviews Frank Luntz. PBS *Frontline*, November 13. http://www.pbs.org/wgbh/pages/frontline/hotpolitics/interviews/luntz.html.

Pearce, Fred. 2010. On World Population Day, take note: Population isn't the problem. *Grist*, July 11. http://www.grist.org/article/2010-07-11-on-world-population-day-take-note-population-isnt-the-problem.

Pearse, Guy. 2007. *High and Dry: John Howard, Climate Change and the Selling of Australia's Future*. Camberwell, VIC: Penguin.

Pearse, Guy. 2009. *Quarry Vision: Coal, Climate Change and the End of the Resources Boom. Quarterly Essay 33*. Collingwood, VIC: Black Inc.

Pell, M. B., and Joe Eaton. 2010. Five lobbyists for each member of Congress on financial reforms. Centre for Public Integrity, May 21. http://www.publicintegrity.org/articles/entry/2096.

Pellow, David. 2007. *Resisting Global Toxics: Transnational Movements for Environmental Justice*. Boston: MIT Press.

Perlez, Jane, and Raymond Bonner. Below a mountain of wealth, a river of waste. *New York Times*, December 27. http://www.nytimes.com/2005/12/27/international/asia/27gold.html.

Perlez, Jane, and Kirk Johnson. 2005. Behind gold's glitter: Torn lands and pointed questions. *New York Times*, October 24. http://www.nytimes.com/2005/10/24/international/24GOLD.html.

Perlo-Freeman, Sam, and Catalina Perdomo. 2008. The developmental impact of military budgeting and procurement: Implications for an arms trade treaty. Prepared for Oxfam GB. http://www.sipri.org/research/armaments/milex/publications/unpubl_milex/mili_budget.

Phelps, Glenn, and Steve Crabtree. 2013. Worldwide, median household income about $10,000. Gallup, December 16. http://www.gallup.com/poll/166211/worldwide-median-household-income-000.aspx.

Phillips, Kevin. 2006. *American Theocracy: The Peril and Politics of Radical Religion, Oil, and Borrowed Money in the 21st Century*. New York: Viking.

Plehwe, Dieter. 2009. Introduction. In Mirowski and Plehwe 2009, 1–42.

Polanyi, Karl. 1964. *1944. The Great Transformation: The Political and Economic Origins of Our Time*. Boston: Beacon Press.

Ponting, Clive. 1993. *A Green History of the World: The Environment and the Collapse of Great Civilizations*. New York: Penguin.

Ponting, Clive. 2002. The burden of the past. *Global Dialogue* 4 (1): 1–10.

Powell, Lewis. 1971. Confidential memorandum: Attack on American free enterprise system. August 23. http://reclaimdemocracy.org/corporate_accountability/powell_memo_lewis.html.

Pimentel, David. 2003. Ethanol fuels: Energy balance, economics, and environmental impacts are negative. *National Resources Research* 12 (2): 127–134. http://www.energyjustice.net/ethanol/pimentel2003.pdf.

PR Watch. 2004. Atlas Economic Research Foundation: The think-tank breeders. *PR Watch Newsletter* 11 (3). http://www.prwatch.org/node/323.

Prakash, Om. 1991. Restrictive trading regimes: VOC and the Asian spice trade in the seventeenth century. In *Emporia, Commodities and Entrepreneurs in Asian Maritime Trade, c. 1400–1750*, ed. Roderich Ptak and Dietmar Rothermund, 107–126. Stuttgart: Steiner Verlag.

Proctor, Robert. 2012. *Golden Holocaust: Origins of the Cigarette Catastrophe and the Case for Abolition*. Berkeley: University of California Press.

Productivity Commission (Australia). 2011. *Carbon Emission Policies in Key Economies*. Research Report. Canberra: Australian Government. www.pc.gov.au/__data/assets/pdf_file/0003/109830/carbon-prices.pdf.

Public Citizen. 2003. The US threats against Europe's GMO policy and the WTO SPS agreement. http://www.citizen.org/documents/GMObackgrndr.pdf.

Quiggin, John. 2002. Ecologists vs. economists. November 5. http://johnquiggin.com.

Rampton, Sheldon, and John Stauber. 2001. *Trust Us, We're Experts*. New York: Penguin Putnam.

Ranelagh, John. 1991. *Thatcher's People: An Insider's Account of the Politics, the Power and the Personalities*. London: HarperCollins.

Ray, Shantana, and Shoma Chaudhury. 2008. My vision is to get 85 percent of India into cities. *Tehelka Magazine* 5 (21). http://www.tehelka.com/story_main39.asp?filename=Ne310508cover_story.asp.

Redclift, Michael. 2005. Sustainable development (1987–2005): An oxymoron comes of age. *Sustainable Development* 13 (4): 212–227.

Reddy, Sanjay. 2008. The World Bank's new poverty estimates: Digging deeper into a hole. Faculty paper. New York: Columbia University. http://www.columbia.edu/~sr793/response.pdf.

Reddy, Sanjay, and Thomas Pogge. 2005. How *not* to count the poor. Faculty paper. New York: Columbia University. www.columbia.edu/~sr793/count.pdf.

Reich, Charles. 1970. *The Greening of America*. New York: Bantam.

Renner, Michael. 2008. Vehicle production rises, but few cars are "green." *Vital Signs Online*, Worldwatch Institute, May 21. http://www.worldwatch.org/node/5461?emc=el&m=218549&l=4&v=6e60d326d8.

Reuters. 2010. Factbox: Outcome of the Seoul G20 summit. November 12. http://www.reuters.com/article/2010/11/12/us-g20-outcomes-idUSTRE6AB1G920101112.

Revkin, Andrew. 2005. Ex-Bush aide who edited climate reports to join ExxonMobil. *New York Times*, June 15. http://www.nytimes.com/2005/06/15/science/14cnd-climate.html.

Revkin, Andrew. 2009. Industry ignored its scientists on climate. *New York Times*, April 24. http://www.nytimes.com/2009/04/24/science/earth/24deny.html.

Ricardo, David. 1952 [1817]. *The Works and Correspondence of David Ricardo*. Vol. 7. Cambridge: Cambridge University Press.

Riddell, Roger. 2009. Is aid working? Is this the right question to be asking? *Open Democracy*, November 20. http://www.opendemocracy.net/roger-c-riddell/is-aid-working-is-this-right-question-to-be-asking.

Ridley, Matt. 2001. Technology and the environment: The case for optimism. Prince Philip Lecture, May 8. http://www.agbioworld.org/newsletter_wm/index.php?caseid=archive&newsid=1073.

Rippa, S. Alexander. 1988. *Education in a Free Society: An American History*. New York: Longman.

Roche, Marc. 2011. Our friends from Goldman Sachs. *Le Monde*, November 16. http://www.presseurop.eu/en/content/article/1177241-our-friends-goldman-sachs.

Rockefeller, David. 1999. Looking for new leadership. *Newsweek* (Atlantic Edition) 133 (5): 63.

Rockström, Johan, Will Steffen, Kevin Noone, et al. 2009. Planetary boundaries: Exploring the safe operating space for humanity. *Ecology and Society* 14 (2): 1–32.

Roosevelt, Franklin D. 1941. Annual address to Congress: The "Four Freedoms." http://www.fdrlibrary.marist.edu/pdfs/fftext.pdf.

Roosevelt, Franklin D. 1944. *Rendezvous with Destiny: Addresses and Opinions of Franklin Delano Roosevelt*. Edited by J. B. Hardman. Whitefish, MT: Kessinger.

Ross, Eric. 1998. Malthusianism, counterrevolution, and the Green Revolution. *Organization & Environment* 11 (4): 446–450.

Ross, Eric. 2000. *The Malthus Factor: Poverty, Politics and Population in Capitalist Development*. Sturminster Newton, UK: The Corner House.

Rostow, Walt. 1960. *The Stages of Economic Growth: A Non-Communist Manifesto*. Cambridge: Cambridge University Press.

Rowell, Andrew. 1996. *Green Backlash: Global Subversion of the Environment Movement*. New York: Routledge.

Roy, Arundhati. 2002. The ladies have feelings so ... Shall we leave it to the experts? *Power Politics*, 1–34. Cambridge, MA: South End Press.

Ruddiman, William. 2005. *Plows, Plagues & Petroleum: How Humans Took Control of Climate*. Princeton, NJ: Princeton University Press.

Ruggiero, Renato. 1997. Charting the trade routes of the future: Towards a borderless economy. Geneva: World Trade Organization. http://www.wto.org/english/news_e/sprr_e/sanfra_e.htm.

Sachs, Jeffrey. 2008. A new deal for Poor farmers. *Project Syndicate*, March 20. http://www.project-syndicate.org/print_commentary/sachs141/English.

Sachs, Wolfgang. 1999. Sustainable development and the crisis of nature: On the political anatomy of an oxymoron. In *Living with Nature: Environmental Politics as Cultural Discourse*, ed. Maarten Hajer and Frank Fischer, 23–41. Oxford: Oxford University Press.

Saad-Filho, Alfredo, and Deborah Johnston, eds. 2005. *Neoliberalism: A Critical Reader*. London: Pluto Press.

Sample, Ian. 2007. Scientists offered cash to dispute climate study. *Guardian*, February 2. http://www.guardian.co.uk/environment/2007/feb/02/frontpagenews.climatechange.

Sandbach, Francis. 1978. A further look at the environment as a political issue. *International Journal of Environmental Studies* 12:99–110.

Sassen, Saskia. 2009. Too big to save: The end of financial capitalism. *Open Democracy*, April 1. http://www.opendemocracy.net/article/too-big-to-save-the-end-of-financial-capitalism-0.

Saul, John Ralston. 1997. *The Unconscious Civilization*. Ringwood, VIC: Penguin.

Sayle, Murray. 2007. Overloading Emoh Ruo: The rise and rise of hydrocarbon civilization. *Griffith Review* 12:11–51.

Schor, Juliet. 1991. *The Overworked American: The Unexpected Decline of Leisure*. New York: Basic Books.

Schumpeter, Joseph. 1975 [1942]. *Capitalism, Socialism and Democracy*. New York: Harper.

Schwab, Klaus. 2009. *The World Economic Forum: A Partner in Shaping History*. Geneva: World Economic Forum. http://www3.weforum.org/docs/WEF_First40Years_Book_2010.pdf.

Science and Environmental Policy Project. 2004. Board of directors. http://www.sepp.org.

Scott, Vernon. 1946. The conflict of two faiths. *Public Relations Journal* 2 (11): 10–14, 33.

Scott, Walter. 1950. *Greater Production: Its Problems and Possibilities*. Sydney: Law Book Co. of Australasia.

Seabrook, Jeremy. 2002. Sustainable development is a hoax: We cannot have it all. *Guardian*, August 5. http://www.guardian.co.uk/Archive/Article/0,4273,4475689,00.html.

Sengupta, Somini. 2006. India digs deeper but wells are drying up. *New York Times*, September 30. http://www.nytimes.com/2006/09/30/world/asia/30water2.html.

Shabecoff, Philip. 1983. Haste of global warming trend opposed. *New York Times*, October 21. http://www.nytimes.com/1983/10/21/us/haste-of-global-warming-trend-opposed.html.

Shaxson, Nicholas, John Christensen, and Nick Mathiason. 2012. Inequality: You don't know the half of it. Tax Justice Network. http://www.taxjustice.net/cms/upload/pdf/Inequality_120722_You_dont_know_the_half_of_it.pdf.

Sheeran, Josette. 2008. The world food crisis. Speech to the Peterson Institute, Washington, May 6. Audio. http://archive.org/details/PetersonInstituteTheWorldFoodCrisis.

Shiva, Vandana. 2002. *Water Wars: Privatization, Pollution, and Profit*. Cambridge, MA: South End Press.

Shiva, Vandana. 2008. *Soil Not Oil: Environmental Justice in an Age of Climate Crisis*. Cambridge, MA: South End Press.

Shnayerson, Michael. 2007. A convenient untruth. *Vanity Fair*, May. http://www.vanityfair.com/politics/features/2007/05/skeptic200705?printable=true¤tPage=all.

Silk, Leonard. 1972. On the imminence of disaster. *New York Times*, March 14.

Simmons, Matthew. 2000. Revisiting "The Limits to Growth": Could the Club of Rome have been correct, after all? GreatChange.org. http://greatchange.org/ov-simmons,club_of_rome_revisted.pdf.

Simmons, Matthew. 2005. *Twilight in the Desert: The Coming Saudi Oil Shock and the World Economy*. Hoboken, NJ: Wiley.

Simon, Julian. 1981. *The Ultimate Resource*. Princeton, NJ: Princeton University Press.

Simon, Julian. 1982. Answer to Malthus? Julian Simon interviewed by William Buckley. *Population and Development Review* 8 (1): 205–208.

Simon, Julian, and Herman Kahn, eds. 1984. *The Resourceful Earth: A Response to Global 2000*. Oxford: Blackwell.

Sinding, Steven. 2007. Overview and perspective. In *The Global Family Planning Revolution: Three Decades of Population Policies and Programs*, ed. Warren Robinson and John Ross, 1–12. Washington, DC: World Bank.

Singer, S. Fred. 1989. My adventures in the ozone layer. *National Review*, June. http://research.greenpeaceusa.org/?a=download&d=3291.

Singer, S. Fred 2000. *National Assessment of the Potential Impact of Climate Change: Climate Change Impacts on the United States*, Hearing before the Senate Committee on Commerce, Science and Transportation, Testimony of Prof. S. Fred Singer, July 18. http://stephenschneider.stanford.edu/Publications/PDF_Papers/Singer.pdf.

Smil, Vaclav. 2006. Energy at the crossroads. OECD Global Science Forum, May 17–18. http://home.cc.umanitoba.ca/~vsmil/pdf_pubs/oecd.pdf.

Smith, Adam. 1852 [1776]. *An Inquiry into the Nature and Causes of the Wealth of Nations*. London: Nelson and Sons.

Smith, Adam. 1853 [1759]. *The Theory of Moral Sentiments*. London: Henry G. Bohn.

Smith, Barry. 2002. Nitrogenase reveals its inner secrets. *Science* 297 (5587): 1654–1655.

Smith, Bruce. 1995. *The Emergence of Agriculture*. New York: Scientific American Library.

Smith, James. 1991. *The Idea Brokers: Think Tanks and the Rise of the New Policy Élite*. New York: Free Press.

Snowe, Olympia, and John D. Rockefeller IV. 2006. Rockefeller and Snowe demand that Exxon Mobil end funding of campaign that denies global climate change. ClimateScienceWatch.org, October 30. http://www.climatesciencewatch

.org/2006/10/31/senators-snowe-and-rockefeller-to-exxonmobil-stop-funding-denialists.

Soares, Claire. 2005. How EU, US "dumping" hurts West African farmers. *Christian Science Monitor*, December 15. http://www.csmonitor.com/2005/1215/p04s01-wogi.html.

Soemarwoto, O., and G. Conway. 1992. The Javanese Homegarden. *Journal for Farming Systems Research-Extension* 2 (3): 95–118. http://ciesin.org/docs/004-194/004-194.html.

Sokal, Michael. 1981. The origins of the Psychological Corporation. *Journal of the History of the Behavioral Sciences* 17:54–67.

Solo, Robert. 1974. Arithmomorphism and entropy. *Economic Development and Cultural Change* 22 (3): 510–517.

Solomon, Norman. 2003. "Globalization" and its malcontents. *Media Monitors*, February 21. http://www.mediamonitors.net/solomon114.html.

Solow, Robert. 1973. Is the end of the world at hand? *Challenge* (March–April): 39–50.

Solow, Robert. 1974. The economics of resources or the resources of economics. *American Economic Review* 64 (2): 1–14.

Soule, George. 1947. *Prosperity Decade, From War to Depression: 1917–1929*. New York: Rinehart.

State Environment Protection Administration of China. 2006. Green GDP Accounting Study Report 2004 issued, September 11. http://www.gov.cn/english/2006-09/11/content_384596.htm.

Steinhart, John. 1973. *Energy Reorganization Act of 1973: Hearings, Ninety-third Congress, Committee on Government Operations, United States Congress, House*. Washington, DC: US Government Printing Office.

Stern, Nicholas. 2007. *The Economics of Climate Change: The Stern Review*. Cambridge: Cambridge University Press.

Stone, Diane. 1991. Old Guard versus new partisans: Think tanks in transition. *Australian Journal of Political Science* 26:197–215.

Støre, Jonas. 2010. Norway takes aim at G-20: "One of the greatest setbacks since World War II." *Der Spiegel*, June 22. http://www.spiegel.de/international/europe/0,1518,702104,00.html.

Stretton, Hugh. 2000. *Economics: A New Introduction*. London: Pluto Press.

Strong, Maurice. 1972. The Stockholm Conference. *Geographical Journal* 138 (4): 411–417.

Swan, Norman. 2010. Famine and civil rights in India. *Health Report*, ABC Radio National, January 25. http://www.abc.net.au/rn/healthreport/stories/2010/2798686.htm#transcript.

Sydney Morning Herald. 2008a. GE Venture's workers face toxins. *Sydney Morning Herald*, March 26. http://news.smh.com.au/technology/report-ge-ventures-workers-face-toxins-20080326-21kh.html.

Sydney Morning Herald. 2008b. Australia must lift game. *Sydney Morning Herald*, April 16. http://www.smh.com.au/small-business/australia-must-lift-game-20090619-cq9w.html.

Symons-Brown, Bonny. 2009. Unions deny ETS will cost mining jobs. *Brisbane Times*, May 22. http://news.brisbanetimes.com.au/breaking-news-national/unions-denies-ets-will-cost-mining-jobs-20090522-bhhi.html.

Taibbi, Matt. 2009. The Great American bubble machine. *Rolling Stone* 1082–1083 (July 9–23). http://www.rollingstone.com/politics/news/the-great-american-bubble-machine-20100405?print=true.

Taibbi, Matt. 2010a. Wall Street's war. *Rolling Stone* 1106 (June 10). http://www.rollingstone.com/politics/news/wall-streets-war-20100526?print=true.

Taibbi, Matt. 2010b. Wall Street's big win. *Rolling Stone* 1111 (August 6). http://www.rollingstone.com/politics/news/wall-streets-big-win-20100804?print=true.

Tainter, Joseph. 1988. *The Collapse of Complex Societies*. Cambridge: Cambridge University Press.

Talberth, John, Clifford Cobb, and Noah Slattery. 2007. *The Genuine Progress Indicator 2006*. Oakland, CA: Redefining Progress. http://www.rprogress.org/publications/2007/GPI%202006.pdf.

Taylor, John. 2005a. China orders media silence on village clash. *ABC News*, June 14. http://www.abc.net.au/news/2005-06-14/china-orders-media-silence-on-village-clash/1592832.

Taylor, John. 2005b. Footage emerges of Chinese battle. *PM*, ABC Radio, June 16. http://www.abc.net.au/pm/content/2005/s1394020.htm.

Taylor, John. 2005c. Chinese social unrest a growing problem. *PM*, ABC Radio, July 7. http://www.abc.net.au/pm/content/2005/s1409542.htm.

Tett, Gillian. 2009. Lost through destructive creation. *Financial Times* (London), March 9. http://www.ft.com/cms/s/0/0d55351a-0ce4-11de-a555-0000779fd2ac.html#axzz1DcMR9bAK.

Thorsen, Dag. 2009. The neoliberal challenge: What is neoliberalism? Manuscript, Department of Political Science, University of Oslo. http://folk.uio.no/daget/neoliberalism2.pdf.

Time. 1948. Eat hearty. *Time,* November 8. http://www.time.com/time/printout/0,8816,853337,00.html.

Trembath, B. 2007. APEC fights poverty, says Howard. *PM*, ABC Radio, September 4. http://www.abc.net.au/pm/content/2007/s2024057.htm.

Tribe, Keith. 2009. Liberalism and Neoliberalism in Britain, 1930–1980. In Mirowski and Plehwe 2009, 68–97.

Tridico, Pasquale. 2011. Varieties of capitalism and responses to the financial crisis: The European social model versus the US model. Working Paper 129. University Roma Tre and New York University. http://www.academia.edu/2626178/Varieties_of_Capitalism_and_responses_to_the_Financial_Crisis_the_European_Social_Model_versus.

Truman, Harry. 1949. Inaugural address, January 20, 1949. Harry S. Truman Library and Museum. http://www.trumanlibrary.org/whistlestop/50yr_archive/inagural20jan1949.htm.

Turner, Graham. 2008. A comparison of *The Limits to Growth* with 30 years of reality. *Global Environmental Change* 18:397–411.

UN. 1948. *Universal Declaration of Human Rights*. http://un.org/Overview/rights.html.

UN. 2009. *Report of the Commission of Experts of the President of the United Nations General Assembly on Reforms of the International Monetary and Financial System*. http://www.un.org/ga/econcrisissummit/docs/FinalReport_CoE.pdf.

UN Centre on Transnational Corporations. 2013. UNCTC evolution. http://unctc.unctad.org/aspx/UNCTCEvolution.aspx.

UN Conference on Environment and Development. 1992a. *Agenda 21*. New York: UNCED. http://www.unep.org/Documents.Multilingual/Default.Print.asp?documentid=52.

UN Conference on Environment and Development. 1992b. *The Rio Declaration*. New York: UNCED. http://www.unep.org/Documents.Multilingual/Default.Print.asp?documentid=78&articleid=1163.

UN Conference on Trade and Development. 2004. *Debt Sustainability: Oasis or Mirage*. New York/Geneva: UN. http://unctad.org/en/Docs/gdsafrica20041_en.pdf.

UN Department of Economic and Social Affairs. 1951. *Measures for the Economic Development of Underdeveloped Countries*. New York: UN.

UN Department of Economic and Social Affairs. Population Division. 2006. Highlight: *World Urbanization Prospects: The 2005 Revision*. New York: UN. http://www.un.org/esa/population/publications/WUP2005/2005WUPHighlights_Final_Report.pdf.

UN Department of Economic and Social Affairs. 2013. Highlights: *World Population Prospects: The 2012 Revision*. New York: UN. http://esa.un.org/unpd/wpp/Documentation/pdf/WPP2012_Highlights.pdf.

UN Development Programme. 2005. Summary: *Human Development Report 2005*. New York: UNDP. http://hdr.undp.org/en/media/hdr05_summary.pdf.

UN Development Programme. 2010. Summary: *Human Development Report 2010*. New York: UNDP. http://hdr.undp.org/en/media/HDR10%20EN%20summary_without%20table.pdf.

UN Environment Programme. 2002. *Global Environment Outlook 3. London/Sterling*. VA: Earthscan.

UN Environment Programme. 2007. *Global Environment Outlook: Environment for Development (GEO-4)*. http://www.unep.org/geo/geo4/report/GEO-4_Report_Full_en.pdf.

UN Environment Programme. 2012. *Global Environment Outlook (GEO-5)*. http://www.unep.org/geo/geo5.asp.

UN Human Settlements Programme. 2003a. *The Challenge of Slums: Global Report on Human Settlements*. London and Sterling, VA: Earthscan Publications. http://www.unhabitat.org/pmss/listItemDetails.aspx?publicationID=1156.

UN Human Settlements Programme. 2003b. *Slums of the World: The Face of Urban Poverty in the New Millennium?* New York: UN. http://www.unhabitat.org/pmss/listItemDetails.aspx?publicationID=1124.

UN Millennium Development Goals. 2009. *Millennium Development Goals Report 2009*. New York: UN. http://www.un.org/millenniumgoals/pdf/MDG_Report_2009_ENG.pdf.

UN Millennium Development Goals. 2013. *Millennium Development Goals Report 2013*. New York: UN. http://www.un.org/millenniumgoals/pdf/report-2013/mdg-report-2013-english.pdf.

UNICEF. 2009. A matter of magnitude. United Nations Children's Fund. http://www.unicef.org/infobycountry/files/amatterofmagnitude.pdf.

Union of Concerned Scientists. 2013. Debunking misinformation about stolen climate emails in the "Climategate" manufactured controversy. http://www.ucsusa.org/global_warming/science_and_impacts/global_warming_contrarians/debunking-misinformation-stolen-emails-climategate.html.

US Congress. 1939. *Violations of Free Speech and Rights of Labor: Report of the Committee on Education and Labor, Part III. The National Association of Manufacturers*. Washington, DC: US Government Printing Office.

US Congress. 1993. *Financial Services Chapter of NAFTA: Hearing before the Committee on Banking, Finance and Urban Affairs, House of Representatives*, September. Washington, DC: US Government Printing Office. http://archive.org/stream/financialservice00unit/financialservice00unit_djvu.txt.

US Energy Information Administration. 2013. *International Energy Outlook 2013*. Washington, DC: US Department of Energy. http://www.eia.gov/forecasts/ieo/pdf/0484(2013).pdf

US Environmental Protection Agency. 2011. Draft investigation of ground water contamination near Pavillion, Wyoming. EPA. http://www2.epa.gov/sites/production/files/documents/EPA_ReportOnPavillion_Dec-8-2011.pdf.

US Environmental Protection Agency. 2013. Isotech-stable isotope analysis: Determining the origin of methane and its effects on the aquifer. EPA. http://www.desmogblog.com/2013/08/05/censored-epa-pennsylvania-fracking-water-contamination-presentation-published-first-time.

US Senate. 2007. Examining the case for the California waiver: Statement of Edmund G. Brown Jr., California Attorney General, May 22. http://epw.senate.gov/public/index.cfm?FuseAction=Hearings.Testimony&Hearing_ID=964ce44f-802a-23ad-4b22-acc2fa62e4d7&Witness_ID=cac5d320-f60e-4d24-8836-2250906875bf.

Verney, R. 1972. The management of natural resources. *Geographical Journal* 138 (4): 417–420.

Vidal, John. 2007. CO_2 output from shipping twice as much as airlines. *Guardian*, March 3. http://www.guardian.co.uk/frontpage/story/0,2025725,00.html.

Vidal, John. 2008. True scale of CO_2 emissions from shipping revealed. *Guardian*, February 13. http://www.guardian.co.uk/environment/2008/feb/13/climatechange.pollution.

Vidal, John. 2013. China and India "water grab" dams put ecology of Himalayas in danger. *Observer*, August 10. http://www.theguardian.com/global-development/2013/aug/10/china-india-water-grab-dams-himalayas-danger.

Vitali, Stephania, James Glattfelder, and Stefano Battison. 2011. The network of global corporate control. *PLoS ONE* 6 (10). http://www.plosone.org/article/info%3Adoi%2F10.1371%2Fjournal.pone.0025995.

Vitousek, Peter, Paul Ehrlich, Anne Ehrlich, and Pamela Matson. 1986. Human appropriation of the products of photosynthesis. *Bioscience* 36 (6): 368–373.

Vogel, David. 1990. *Fluctuating Fortunes: The Political Power of Business in America*. New York: Basic Books.

Vogt, William. 1948. *Road to Survival*. New York: William Sloane.

Voyer, Roger, and Mark Murphy. 1984. *Global 2000: Canada, a View of Canadian Economic Development Prospects, Resources and the Environment*. Toronto: Pergamon.

Wackernagel, Mathis, and William Rees. 1996. *Our Ecological Footprint: Reducing Human Impact on the Earth*. Gabriola Island, BC: New Society Publishers.

Wackernagel, Mathis, Larry Onisto, Alejandro Linares, et al. 1997. *Ecological Footprints of Nations: How Much Nature Do They Use? How Much Nature Do They Have?* Toronto: International Council for Local Environmental Initiatives.

Wade, Robert. 2004. Is globalization reducing poverty and inequality? *World Development* 32 (4): 567–589.

Wade, Robert. 2007. Globalisation: Emancipating or reinforcing? *Open Democracy*, January 29. http://www.stwr.org/globalization/globalisation-emancipating-or-reinforcing.html.

Waldman, Amy. 2005. Mile by mile, India paves a smoother road to its future. *New York Times*, December 4. http://www.nytimes.com/2005/12/04/international/asia/04highway.html?pagewanted=all.

Walker, Brian, and David Salt. 2006. *Resilience Thinking: Sustaining Ecosystems and People in a Changing World*. Washington, DC: Island Press.

Walker, Joe. 1998. Memo: Global Climate Science Communications Action Plan. http://www.euronet.nl/users/e_wesker/ew@shell/API-prop.html.

Walker, Martin, Paul Brown, Jan Rocha, and John Vidal. 1992. Rio Summit Crumbling. *Guardian*, June 1, 1.

Walker, S. and Paul Sklar. 1938. Business finds its voice: Parts I–III. *Harper's Magazine* 176:113–123, 317–329, 428–440.

Wall, Wendy. 2008. *Inventing the "American Way": The Politics of Consensus from the New Deal to the Civil Rights Movement*. Oxford: Oxford University Press.

Wallach, Lori, and Todd Tucker. 2010. Memo: Answering critical questions about conflicts between financial re-regulation and WTO rules. *Public Citizen*, May 21. http://www.citizen.org/documents/Memo_to_Senate_Finance_on_GATS_Regulation.pdf.

Wallerstein, Immanuel. 1974. *The Modern World-System I: Capitalist Agriculture and the Origins of the European World-Economy in the Sixteenth Century*. New York: Academic Press.

Walls, James. 1973. Book review: Ecodoom. *Family Planning Perspectives* 5 (1): 64.

Walt, Vivienne. 2008. The world's growing food-price crisis. *Time*, February 27. http://www.time.com/time/world/article/0,8599,1717572,00.html.

Walters, Brian. 2003. *Slapping on the Writs: Defamation, Developers and Community Activism*. Sydney: UNSW Press.

Ward, Bob. 2009. Why ExxonMobil must be taken to task over climate denial funding. *Guardian*, July 1. http://www.guardian.co.uk/environment/cif-green/2009/jul/01/bob-ward-exxon-mobil-climate.

Warren, Matthew. 2008. Power producers warn on emission targets. *Australian*, May 24. http://www.theaustralian.com.au/news/power-producers-warn-on-targets/story-e6frg6no-1111116429504.

Wasley, Andrew. 2009. UK companies linked to devastating Indian mine. *Ecologist*, June 19. http://www.theecologist.org/trial_investigations/272286/uk_companies_linked_to_devastating_indian_mine.html.

Wasley, Andrew. 2011. Scandal of the "tomato slaves" harvesting crop exported to UK. *Ecologist*, September 1. http://www.theecologist.org/News/news_analysis/1033179/scandal_of_the_tomato_slaves_harvesting_crop_exported_to_uk.html.

Weaver, Paul. 1977. Corporations are defending themselves with the wrong weapon. *Fortune*, June, 186–196.

Whitwell, Greg. 1989. *Making the Market: The Rise of Consumer Society*. Melbourne: McPhee Gribble.

Williams, Raymond. 1995 [1982]. Socialism and ecology. *Capitalism, Nature, Socialism* 6 (1): 41–57.

Williams, Robyn. 2010. Climate change skepticism: Its sources and strategies. *Science Show*, ABC Radio National, April 3. http://www.abc.net.au/rn/scienceshow/stories/2010/2859986.htm#transcript.

Williams, Robyn. 2011a. Peak oil: Just around the corner. *Science Show*, ABC Radio National, April 23. http://www.abc.net.au/rn/scienceshow/stories/2011/3198227.htm#transcript.

Williams, Robyn. 2011b. We are 7 billion. *Science Show*. ABC Radio National, October 29. http://www.abc.net.au/radionational/programs/scienceshow/we-are-7-billion/3599632#transcript.

Williamson, John. 1989. What Washington means by policy reform. Washington, DC: Peterson Institute for International Economics. http://www.piie.com/publications/papers/paper.cfm?ResearchID=486.

Wolf, Eric. 1997. *Europe and the People without History*. Berkeley: University of California Press.

Woodley, Naomi. 2011. New study flags $60 carbon price. *World Today*, ABC Radio, March 15. http://www.abc.net.au/worldtoday/content/2011/s3164251.htm.

Woodward, David, and Andrew Simms. 2006. *Growth Isn't Working*. London: new economics foundation. www.neweconomics.org/page/-/files/Growth_Isnt_Working.pdf.

Woolley, Paul. 2008. Financial sector dysfunctionality: Is society well served by its financial institutions? *Big Ideas*, ABC Radio National, February 3. http://www.abc.net.au/rn/bigideas/stories/2008/2149972.htm#transcript.

World Bank. 2007. *Global Economic Prospects: Managing the Next Wave of Globalization*. Washington, DC: World Bank. https://openknowledge.worldbank.org/handle/10986/7157.

World Bank. 2008. *Framework Document for a Global Food Crisis Response Program*. Washington, DC: World Bank. http://www-wds.worldbank.org/external/default/WDSContentServer/WDSP/IB/2008/06/30/000333038_20080630001046/Rendered/PDF/438410BR0REVIS10and0IDAR20081016212.pdf.

World Bank. 2012. *World Development Indicators 2012*. Washington, DC: World Bank. http://data.worldbank.org/sites/default/files/wdi-2012-ebook.pdf.

World Coal Association. 2013. Coal Facts 2013. http://www.worldcoal.org/resources/coal-statistics.

World Commission on Dams. 2000. *Dams and Development: A New Framework for Decision-Making*. London and Sterling, VA: Earthscan. http://www.internationalrivers.org/resources/dams-and-development-a-new-framework-for-decision-making-3939.

World Commission on Environment and Development. 1987. *Our Common Future*. Oxford: Oxford University Press.

World Economic Forum. 2010a. History. http://www.weforum.org/history-0.

World Economic Forum. 2010b. History Index. http://www3.weforum.org/tools/history/index.html.

World Economic Forum. 2010c. The Leadership Team. http://www.weforum.org/content/leadership-team-klaus-schwab.

World Resources Institute. 2005. *EarthTrends*. Energy and Resources. http://earthtrends.wri.org.

World Trade Organization. 2010a. Fiftieth anniversary of the multilateral trading system. http://www.wto.org/english/thewto_e/minist_e/min96_e/chrono.htm.

World Trade Organization. 2010b. Venezuela, Brazil versus US: Gasoline. http://www.wto.org/english/tratop_e/envir_e/edis07_e.htm.

World Trade Organization. 2010c. Mexico etc. versus US: "Tuna-dolphin." http://www.wto.org/english/tratop_e/envir_e/edis04_e.htm.

World Trade Organization. 2010d. Understanding on commitments in financial services. http://www.wto.org/english/tratop_e/serv_e/21-fin_e.htm.

World Trade Organization. 2010e. Services: Rules for growth and investment. http://www.wto.org/english/thewto_e/whatis_e/tif_e/agrm6_e.htm.

Worldwatch Institute. 2006. *Vital Signs 2006–2007*. New York: W. W. Norton.

Worldwatch Institute. 2011. *Vital Signs 2011*. Washington, DC: Worldwatch Institute.

Yeung Yue-man, and Fu-chen Lo. 1996. Global restructuring and emerging urban corridors in Pacific Asia. In *Emerging World Cities in Pacific Asia*, ed. Fu-Chen Lo and Yue-Man Yeung. Tokyo: UN University Press.

Zencey, Eric. 2013. Energy as master resource. In *State of the World 2013: Is Sustainability Still Possible*? Worldwatch, 73–83. Washington, DC: Island Press.

Index